endorsed by edexcel

Edexcel Physics for A2

Tim Akrill
Graham George

HODDER EDUCATION
AN HACHETTE UK COMPANY

This book is dedicated by Graham George to his surgeon Jimmy Adshead MA, MD, FRCS, and his team at the Lister Hospital, Stevenage, together with the many physicists whose discoveries, research and development made his diagnosis and recent life-saving operation possible.

Hachette UK's policy is to use papers that are natural, renewable and recyclable products and made from wood grown in sustainable forests. The logging and manufacturing processes are expected to conform to the environmental regulations of the country of origin.

Orders: please contact Bookpoint Ltd, 130 Milton Park, Abingdon, Oxon OX14 4SB. Telephone: (44) 01235 827720. Fax: (44) 01235 400454. Lines are open 9.00–5.00, Monday to Saturday, with a 24-hour message answering service. Visit our website at www.hoddereducation.co.uk

First published in 2009 by
Hodder Education,
An Hachette UK company
338 Euston Road
London NW1 3BH

Impression number 5 4 3
Year 2013 2012 2011 2010

Cover photo Edward Kinsman/Science Photo Library
Illustrations by Barking Dog Art
Typeset in 10.5/12pt Goudy by Fakenham Photosetting Limited, Fakenham, Norfolk
Printed in Italy

A catalogue record for this title is available from the British Library

ISBN: 978 0340 88807 0

Introduction

Welcome to *Edexcel Physics for A2*. The Edexcel specification was developed from the best of the Edexcel concept-led and the Salters Horners context-led specifications. Although this book has been specifically written to cover the concept approach to the new specification, it also makes a very valuable resource for the context approach as it is illustrated throughout by many contextual worked examples. The authors both have vast experience of teaching, examining and writing about physics. Both have examined for Edexcel for over 30 years and both have been Chief Examiners for A-level Physics.

As in the AS book, a key aspect of the text is the emphasis on practical work. Virtually all the experiments suggested in the specification are described here in such a way that students can carry out the experiments in a laboratory environment. However, most experiments are illustrated with typical data for the reader to work through, sometimes in Exercises provided, without requiring access to a laboratory. In addition, Chapter 18 is *A Guide to Practical Work*. This is an extension of Chapter 2 in the AS book, with the emphasis on the further requirements for A2 practical assessment. Before carrying out a practical activity, teachers should identify the hazards and assess the risks. This can be done by consulting a model (generic) risk assessment provided by CLEAPSS to subscribing authorities.

The numerous Worked examples allow you to test yourself as you go along, and at the end of each chapter you will find exam-style Review questions to further help you check your progress. Numerical answers to these Review questions are given at the end of the book and full written answers are provided on the student website (see below for how to access this website). This makes the Review questions ideal for teachers to set for homework! The website also contains more exam-style practice questions for you to try, together with the answers in an interactive format with audio.

Throughout the book there are Tips in the margin. These may be reminders, for example to use SI units, or they may be warnings to avoid common errors, or they may be hints about short cuts in performing calculations.

At the end of each of Unit 4 and Unit 5 there is a Unit Test, which has been written to be as close as possible to the style of actual Edexcel Test papers. Detailed answers to these tests, together with mark allocation guides, are also given on the website. These will give you a useful benchmark of the standard you have reached.

The combination of the book and the dedicated website should provide you with all the guidance and information needed to prepare you to face your exams with confidence.

Student website

The website mentioned above can be found at **www.hodderplus.co.uk/edexcelphysicsa2**.

User name: edexcelphysicsa2

Password: gravity1

Please note that the user name and password are both case sensitive.

Contents

Unit 4 1

Topic 1 Review

1 Review and revision **2**

Topic 2 Further mechanics

2 Linear momentum **4**

3 Momentum and energy **13**

4 Motion in a circle **21**

Topic 3 Electric and magnetic fields

5 Electric fields **30**

6 Capacitance **39**

7 Magnetic fields **53**

Topic 4 Particle physics

8 Electrons and nuclei **68**

9 Particle physics **81**

Unit 4 Test **92**

Unit 5 95

Topic 5 Thermal energy

10 Specific heat capacity **96**

11 Internal energy and absolute zero **106**

12 Gas laws and kinetic theory **114**

Topic 6 Nuclear decay

13 Nuclear decay **128**

Topic 7 Oscillations

14 Oscillations **150**

Topic 8 Astrophysics and cosmology

15 Universal gravitation 175

16 Astrophysics 185

17 Cosmology 197

Topic 9 Practical work

18 A guide to practical work 212

Unit 5 Test 221

A2 Data sheet 225

Numerical Answers to Review Questions 227

Index 230

Acknowledgements

The Publishers would like to thank the following for permission to reproduce copyright material:

Photo credits
p.1 © CERN; **p.4** PA Photos/Rui Vieira; **p.7** Science Photo Library/Edward Kinsman; **p.8** Getty Images/Jamie McDonald; **p.11** NASA; **p.13** *t* Alamy/Gnomus, *b* Science Photo Library; **p.19** Science Photo Library; **p.21** Fred Espenak; **p.24** Getty Images/Stu Forster; **p.26** Alamy/Corbis RF; **p.27** PA Photos/John Walton/Empics Sport; **p.30** Alamy/The Print Collector; **p.32** NASA; **p.33** Rex Features/Alix/Phanie; **p.35** Science Photo Library/Hank Morgan; **p.43** Science Photo Library/Andrew Lambert; **p.56** PA Photos/John Birdsall; **p.63** Rex Features/APHM – GARO; **p.70** Science Photo Library/Andrew Lambert; **p.74** Science Photo Library/Fermilab; **p.75** ©CERN; **p.76** *t* Science Photo Library/Lawrence Berkeley Laboratory, *b* James L Cronin; **p.77** Science Photo Library/Powell, Fowler & Perkins; **p.81** SLAC National Accelerator Laboratory; **p.82** Science Photo Library/Carl Anderson; **p.84** Science Photo Library/ Lawrence Berkeley Laboratory; **p.85** Science Photo Library/Wellcome Dept of Cognitive Neurology; **p.88** Science Photo Library/Omikron; **p.89** *t* Science Photo Library/Andrew Lambert, *b* A. P. French/Department of Physics, Massachusetts Institute of Technology, Cambridge, MA, USA; **p.95** Science Photo Library/Allan Morton/Dennis Milon; **p.96** Science Photo Library/Tony McConnell; **p.97** Corbis/Bettmann; **p.110** Science Photo Library/David Parker/IMI/Univ. of Birmingham High TC Consortium; **p.111** *t* Science Photo Library/CERN, *c* Rex Features/China Span Keren Su/Sunset, *b* NASA/ESA/The Hubble Heritage Team(STScI/Aura); **p.115** Rex Features/Zena Holloway; **p.120** USAF/Airman Frank Snider; **p.121** Science Photo Library/Segre Collection/American Institute of Physics; **p.125** *t* Alamy/The London Art Archive, *b* Science Photo Library/Jean-Loup Charmet; **p.128** *l* Alamy/North Wind Picture Archives, *r* Science Photo Library; **p.130** Royal Cornwall Hospitals Trust; **p.134** *l* Royal Society, P.A.Kapitza November 1924, Vol.106, no.739, 602-622 "Alpha Ray Tracks in a Strong Magnetic Field", Proceedings of The Royal Society London, *r* Kurie, Richardson & Paxton, 1936; **p.136** Science Photo Library/US Dept of Energy; **p.138** *l* Science Photo Library/RVI Medical Physics, Newcastle/Simon Fraser, *r* Science Photo Library; **p.145** Topfoto; **p.156** Science Photo Library/Edward Kinsman; **p.171** *l* Science Photo Library/Martin Bond, *c* Science Photo Library/Adrian Bicker, *r* PA Photos/AP; **p.172** *t* PA Photos/Kamran Jebreili/AP, *b* Science Photo Library/Mehau Kulyk; **p.175** NASA; **p.179** NASA/JSC; **p.185** NASA/JPL; **p.188** Harvard College Observatory; **p.191** Courtesy Royal Society; **p.197** NASA/ESA/STScI/J. Hester & P. Scowen (Arizona State University); **p.202** Corbis/Hulton-Deutsch Collection.

t = top; *b* = bottom, *l* = left, *c* = centre, *r* = right

Every effort has been made to trace all copyright holders, but if any have been inadvertently overlooked the Publishers will be pleased to make the necessary arrangements at the first opportunity.

Unit 4

Topic 1 Review

1 Review and revision

Topic 2 Further mechanics

2 Linear momentum

3 Momentum and energy

4 Motion in a circle

Topic 3 Electric and magnetic fields

5 Electric fields

6 Capacitance

7 Magnetic fields

Topic 4 Particle physics

8 Electrons and nuclei

9 Particle physics

1 Review and revision

An elderly physicist was asked how much he had in the bank. 'Fifteen million, three hundred thousand, one hundred and four,' he replied. Well, Turkish Lira may have altered in value since you read these words, perhaps a year ago, at the beginning of the AS book, and the physicist may now be richer or poorer, but the reply is still meaningless without its unit. Units are crucial in science. Now is the moment for you to review the first chapter of the AS book!

1.1 Physical quantities and units

In this review you should recall what physicists mean when they refer to:

- **physical quantities**, e.g. mass
- **base units in SI**, e.g. ampere
- **derived** units, e.g. pascal
- **prefixes**, e.g mega.

You should revise:

- the way in which **vector quantities** can be **resolved** or **added**
- how to check that the units are the same on both sides of a physical equation.

Tip

List the meaning of micro (μ), milli (m), kilo (k), mega (M) and any other prefixes you know.

Tip

Make a list of all the vector quantities you have come across in your AS course.

Figure 1.1 ▲

In the next chapter you will need to use your knowledge of **velocity** and **acceleration** and recall methods for analysing motion using light gates and displacement sensors. As always in physics, graphs containing lots of information will be used from time to time.

You will use forces to apply **Newton's laws of motion** and this will lead to a new concept, **momentum**. This is not the 'momentum' widely and loosely used by sports commentators but a special physical quantity. **Energy conservation** will, of course, be important.

REVIEW QUESTIONS

1 Which of the following

a) is a base unit?

A hertz B joule C metre D volt

b) is a derived unit?

A ampere B kilogram C second D volt

c) is **not** a unit for a physical quantity?

A mass B pascal C ohm D watt

d) is a vector quantity?

A work B power C length D velocity

e) is a prefix for 10^6?

A kilo B mega C giga D pico

f) is a way of writing 'micro'?

A 10^{-3} B 10^{-5} C 10^{-6} D 10^{-9}

g) is a physical quantity that can be resolved?

A angle B density
C displacement D frequency

2 A firework rocket is fired at an angle of 75° to the horizontal. The initial speed of the rocket is $40\,\mathrm{m\,s^{-1}}$. Calculate:

a) its initial vertical velocity

b) its initial horizontal velocity.

3 What are the units of the coefficient of viscosity η in Poiseuille's equation describing the flow of a liquid through a pipe:

$$\frac{V}{t} = \frac{\pi p a^4}{8\eta l}$$

where the volume flowing per unit time is V/t, the pipe is of radius a and length l, and p is the pressure drop along the pipe?

4 When an elephant treads on a mouse, the downward force of the elephant's foot on the mouse is equal to the upward force of the mouse on the elephant's foot. Explain this statement.

2 Linear momentum

When you push on the ground, Newton's third law states that the ground pushes you back. The push on the sprinters in Figure 2.1 gives them forward momentum. An understanding of momentum in physics and its conservation is central to explaining how animals and vehicles move.

In this chapter you will develop your understanding of Newtonian mechanics, especially in the way bodies interact with one another. This will involve studying collisions, recoils and impulsive forces.

Figure 2.1 ▲
On your marks, get set, go!

2.1 Momentum

Definition

Momentum = mass × velocity

$p = mv$

Tip

Be careful to leave a gap between N and s, otherwise it looks as if it might mean newtons (plural).

When you push a loaded supermarket trolley to link with an empty stationary one, what happens depends not only on the speed with which you launch the loaded trolley, but also on just how loaded it is.

The product of a body's mass and velocity is useful in analysing such collisions: this product is called the body's **momentum**, or, more precisely, its linear momentum, p.

The unit of momentum is therefore kg m s^{-1}, but this can also be expressed as N s.

$$\text{N s} \equiv \text{kg m s}^{-1}$$

Because velocity is a vector, momentum is also a vector.

Tip

Be careful to spot the difference between a question that asks for the size or magnitude of a body's momentum and one that asks for its magnitude *and* direction.

Worked example

1 How large is the momentum of a child of mass 25 kg walking at $0.38\,\text{m s}^{-1}$?
2 What is the momentum of a Eurostar train of mass 650 tonnes moving due south at a speed of $60\,\text{m s}^{-1}$?

Answer
1 Child's momentum $= 25\,\text{kg} \times 0.38\,\text{m s}^{-1} = 9.5\,\text{kg m s}^{-1}$
2 Eurostar's momentum $= 650 \times 10^3\,\text{kg} \times 60\,\text{m s}^{-1} = 3.9 \times 10^7\,\text{kg m s}^{-1}$ due south.

2.2 Collisions

The reason momentum is important is that it is conserved in interactions between bodies. In physics quantities like energy or charge or momentum, which are conserved, help enormously in understanding how the world around us behaves. A study of the history of science will support this, and we will see later that it can even lead scientists to suggest the existence of undiscovered elementary particles.

Experiment

A simple collision between trolleys on a friction-compensated slope

Trolley A of mass m_A is given a push so that, after its release, its interrupter card cuts through a light beam with the trolley moving at a constant velocity down the friction-compensated slope. The time for which the light beam is interrupted is recorded electronically and the (constant) velocity u of trolley A is calculated.

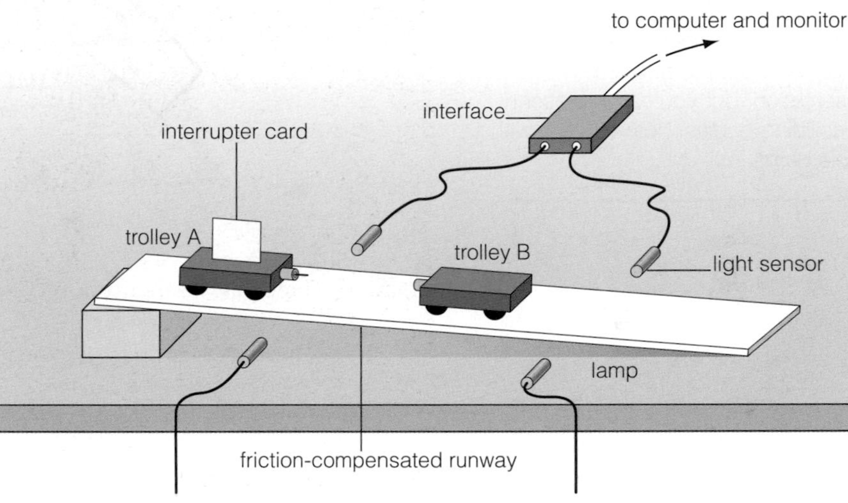

Figure 2.2 ▲

Trolley A has a cork attached with a pin sticking out of it that 'couples' to a cork attached to trolley B of mass m_B when A hits B. After collision the two move off together at a new constant velocity v down the slope. This velocity is calculated as the interrupter card on trolley A passes through the second light beam.

If momentum is conserved in this collision, then $m_A u$ should equal $(m_A + m_B)v$. The easiest case is when $m_A = m_B$, for which v should equal $\frac{1}{2}u$. By varying how fast trolley A is set moving, and by adding masses to either or both trolleys, the relation that conserves the trolleys' momentum

$$m_A u = (m_A + m_B)v$$

can be tested for a variety of different values in the expression. Instead of using light gates, the pin and cork can be got rid of and a transmitter can be attached to the back of the first trolley and a receiver fixed at the top of the slope. Using this technique a displacement–time graph of the collision can be drawn, from which the software can make a speed–time graph as $v = \Delta s/\Delta t$.

The experiment can, in principle, be tried with the supermarket trolleys mentioned at the beginning of this chapter, but measuring their speeds is quite difficult!

Safety note

Care should be taken with positioning runways and blocks, which are heavy masses. A 'catch box' should be used if trolleys are likely to run off the end of the bench.

Worked example

A loaded supermarket trolley of unknown total mass is rolled into a stationary stack of two empty trolleys, each of mass 8.0 kg. The speed of the loaded trolley before they link together is $2.2\,\mathrm{m\,s^{-1}}$ and the speed of the linked trolleys after the collision is $1.0\,\mathrm{m\,s^{-1}}$. Calculate the mass of the shopping in the loaded trolley.

Answer
Suppose the mass of the shopping is m.

Before

After

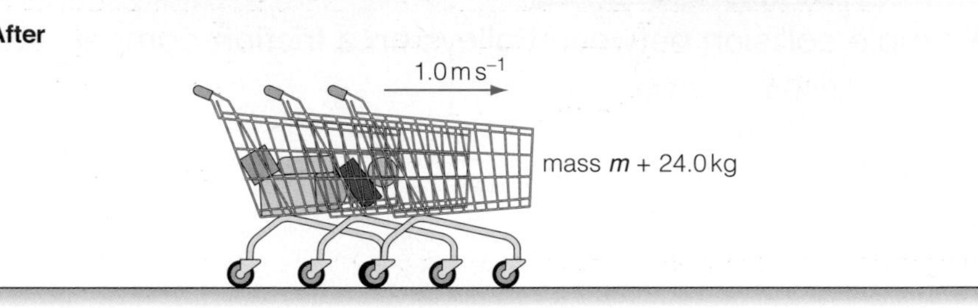

Figure 2.3 ▲

Using the fact that momentum is conserved, i.e. that

$$\begin{aligned} \text{momentum before the collision} &= \text{momentum after the collision} \\ (m + 8.0\,\text{kg}) \times 2.2\,\text{m s}^{-1} &= (m + 24.0\,\text{kg}) \times 1.0\,\text{m s}^{-1} \\ \Rightarrow m \times (2.2\,\text{m s}^{-1} - 1.0\,\text{m s}^{-1}) &= 24.0\,\text{kg m s}^{-1} - 17.6\,\text{kg m s}^{-1} \\ \text{i.e.} \quad m \times 1.2\,\text{m s}^{-1} &= 6.4\,\text{kg m s}^{-1} \\ m &= 5.3\,\text{kg} \end{aligned}$$

Tip

It really pays to sketch the situation before and after when solving problems about momentum conservation, but you can represent the colliding bodies, here trolleys, with simple blobs.

Safety note

If an airgun is going to be used, stringent safety precautions need to be followed. A model risk assessment should be consulted.

The experiment on page 5 can be developed to measure the speed of a small fast-moving object, for example a rifle bullet. The bullet might be fired so as to lodge in a piece of wood attached to a free-running trolley or equivalent. If the mass of the bullet and wood + trolley are known, then measuring the speed of the trolley + wood + bullet after the collision enables the initial speed of the bullet to be found.

Worked example

An arrow of mass 120 g is fired into a block of wood with a mass of 0.60 kg resting on a wall. The arrow and wood fly off the wall at $8.0\,\text{m s}^{-1}$. Calculate the speed of the arrow.

Answer
Momentum is conserved. Let the speed of the arrow be v.

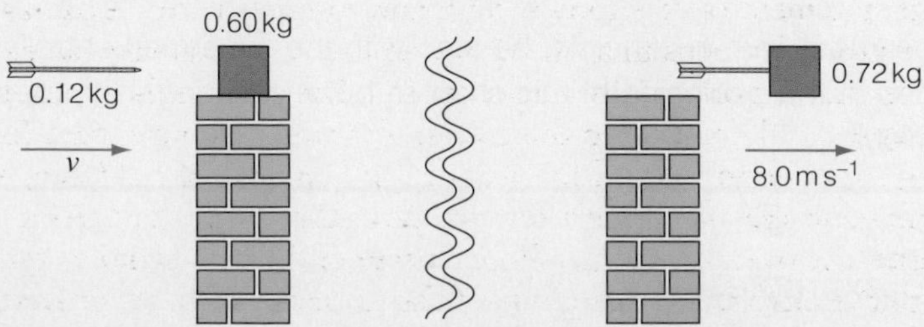

Figure 2.4 ▲

$$\begin{aligned} 0.12\,\text{kg} \times v &= (0.12\,\text{kg} + 0.60\,\text{kg}) \times 8.0\,\text{m s}^{-1} \\ \Rightarrow v &= 48\,\text{m s}^{-1} \end{aligned}$$

Collision experiments can also be simulated using gliders on linear air tracks where the friction is negligible. To get the gliders to stick together, a bit of Blu-Tack can be stuck to the gliders where they strike one another. More complex collisions can also be performed on the air track. Figure 2.5 shows a plan view of a demonstration of gliders that bounce off each other when they collide.

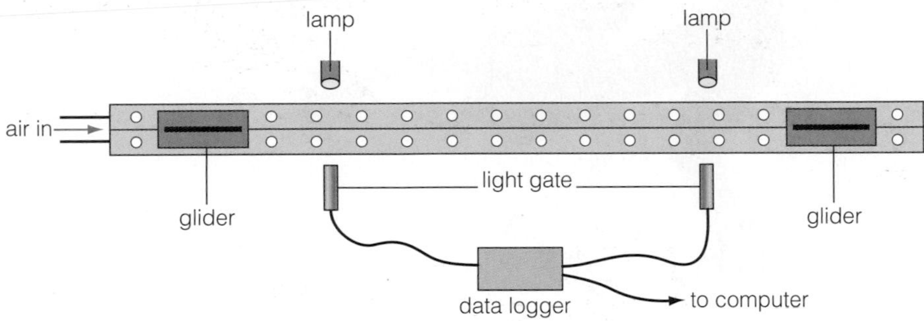

Figure 2.5 ▲
Testing momentum conservation using gliders on an air track

The two gliders can be fitted with small magnetic buffers that are set to repel. The gliders are first set moving towards each other. When they collide they bounce apart and, provided the computer software can store the initial velocities and then record the velocities after the bounce, the **principle of conservation of momentum** can be tested. The experiment works just as well for other buffers such as pieces of cork or small pencil erasers, showing that the principle works even when kinetic energy is lost – see page 15 – provided no external force acts on the gliders.

In practice, other forces such as friction often act, but forces that are perpendicular to the line in which momentum is being measured, for example the weight of the gliders and the upward air force on them, can be ignored.

Definition

The principle of conservation of linear momentum states that in any interaction between bodies, linear momentum is conserved, provided no resultant external force acts on the bodies.

2.3 Momentum and impulse

Let's have a further look at **Newton's second law of motion** for the case of a single force F acting on a body of fixed mass m, namely $F = ma$.

$$\text{acceleration } a = \frac{\Delta v}{\Delta t}$$

So we can write the second law as

$$F = m\frac{\Delta v}{\Delta t}$$

or as $F\,\Delta t = m\,\Delta v$

or even as $F\,\Delta t = m(v_2 - v_1)$

where the velocities before and after the action of the force are v_1 and v_2.

The last statement of Newton's second law, together with **Newton's third law**, tells us what happens in a situation where two ice skaters push each other apart as they move horizontally across the ice. The push of the man on the woman is equal and opposite to the push of the woman on the man (Newton's third law), so the term $F\,\Delta t$, which is called the **impulse** of the force, will be exactly the same size during each interval Δt as they push apart. As a consequence the overall change of momentum of the man will be equal but opposite to the change of momentum of the woman. We have 'proved' that the conservation of momentum in this case is a consequence of Newton's second and third laws.

Figure 2.6 shows an instant during a tennis racket hitting a ball. Here the forces can be very large and the time during which they act very small. This is sometimes called an impulsive force.

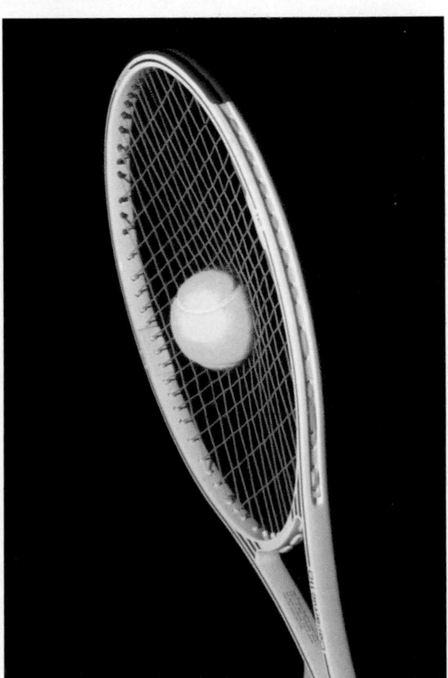

Figure 2.6 ▲
An impulsive force

Figure 2.7 ▲
If the skaters push one another apart, the man's change of momentum will be equal and opposite to the woman's change of momentum.

The case of two skaters pushing one another apart is an example of **recoil**. The word recoil is often applied to a large gun after it fires a shell, but it is equally what happens when a nucleus emits an alpha particle or when a rocket is fired in space.

Worked example

An alpha particle is emitted from a polonium nucleus at a speed of $1.8 \times 10^7\,\text{m s}^{-1}$. The relative masses of the alpha particle and the remaining nucleus are 4.002 and 212.0. Calculate the recoil velocity of the nucleus.

Answer
Let the masses be $4.002m$ and $212.0m$ respectively and the recoil speed be v.

Figure 2.8 ▲

As momentum will be conserved,

$$\text{momentum before} = 0, \text{ so momentum after} = 0$$
$$4.002m \times 1.8 \times 10^7\,\text{m s}^{-1} - 212.0mv = 0$$

The m cancels, giving $v = 3.4 \times 10^5\,\text{m s}^{-1}$, in the reverse direction to the alpha particle.

2.4 Impulsive forces

When a large force acts on a body for a short time, for example the tennis racket and tennis ball on page 7, the average size of the force can be estimated if the time of contact is known. The **impulse–momentum equation**:

$$F\,\Delta t = m\,\Delta v$$

can be used to estimate the size of the force if we can measure the change of momentum $m\,\Delta v$ of the body.

Definition
The impulse of a force $= \sum F\Delta t$

Worked example

A gymnast first touches a trampoline moving downwards at 9.0 m s^{-1} and leaves it moving upwards at the same speed. She has a mass of 55 kg and is in contact with the trampoline surface for only 0.75 s during a bounce.

What is the average resultant force acting on her during the bounce?

Answer

Her change of momentum is 55 kg × 18 m s^{-1} = 990 kg m s^{-1} upwards

Therefore

$$F \times 0.75\,\text{s} = 990\,\text{kg m s}^{-1}$$
$$\Rightarrow F = 1320\,\text{kg m s}^{-2}$$

So the average resultant force on her is 1300 N to 2 sig. fig.

Tip

The unit of $F\Delta t$ is N s and the unit of $m\Delta v$ is kg m s^{-1}. These are equivalent because N ≡ kg m s^{-2}.

Figure 2.9 ▲

Calculations of this kind can reveal just how large the forces involved in violent collisions can be – collisions such as a car crashing into a wall.

The impulse–momentum equation also helps us to analyse force versus time graphs, where $\sum F\,\Delta t$ represents the area under the graph. Such graphs are now routinely produced using modern force platforms in sports science laboratories.

Worked example

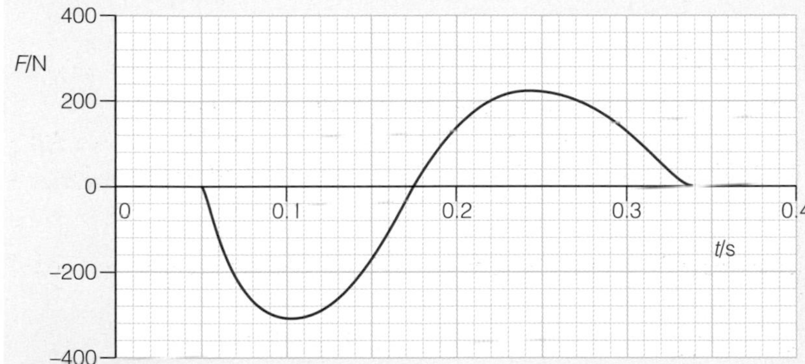

Figure 2.10 ▲

Figure 2.10 shows how the horizontal force F on a jogger's foot varies with time t during a single stride. F is taken to be positive when it is in the direction of the jogger's motion.

a) Describe how F varies during this stride.

b) Estimate the average backward force on the jogger after his foot touches the ground.

c) The area above the axis is the same as the area below the axis. Explain what this tells you about the motion of the jogger.

d) Estimate the area under the graph during the period when the horizontal force of the jogger is forward.

Answer

a) After the jogger's foot first touches the ground at 0.05 s, the horizontal force of the ground is backwards, slowing the runner down. This reaches a maximum at $t \approx 0.1$ s. At $t = 0.175$ s the forward force of the ground on the jogger is zero and after that there is a forward force on him until his foot leaves the ground at $t \approx 0.34$ s.

b) The average backward force $\approx$ 120 N, because a horizontal line drawn at this level splits the area of the curve below the axis into two approximately equal areas.

c) The area under the graph represents the jogger's change of momentum. Since the backward change in momentum is the same as the forward change of momentum, the runner continues forward at the same velocity as a result of this stride.

d) The area is nearly $2\frac{1}{2}$ 'big' squares. One big square is equivalent to a change of momentum of $200\,\text{N} \times 0.05\,\text{s} = 10\,\text{N s} = 10\,\text{kg m s}^{-1}$. So the area under the graph $\approx 25\,\text{kg m s}^{-1}$.

Worked example

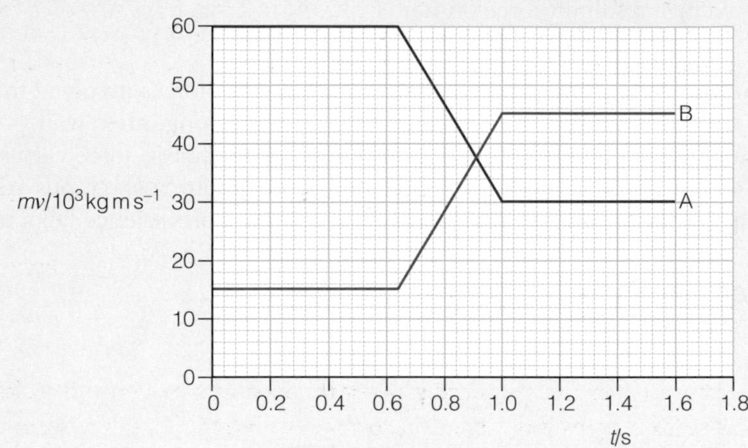

Figure 2.11 ▲

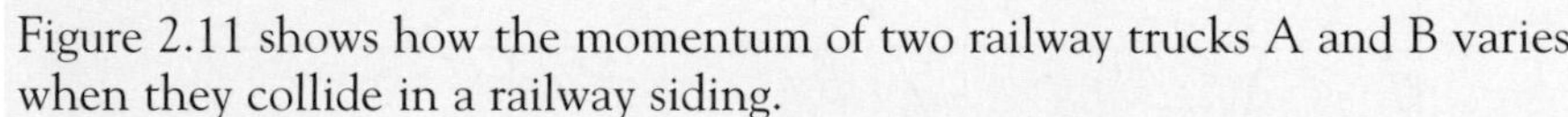

Figure 2.11 shows how the momentum of two railway trucks A and B varies when they collide in a railway siding.

a) Calculate the force that truck A exerts on truck B.

b) Calculate the force that truck B exerts on truck A.

Comment on your answers to a) and b).

Answer

a) The force on truck B is given by the rate of change of momentum of truck B.

$$F_B = \frac{\Delta(mv)_B}{\Delta t} = \frac{(45\,000\,\text{kg m s}^{-1} - 15\,000\,\text{kg m s}^{-1})}{(1.0\,\text{s} - 0.64\,\text{s})}$$

$\Rightarrow F_B = 83\,000\,\text{kg m s}^{-2}$ or $83\,000\,\text{N}$

b) The force on truck A is given by the rate of change of momentum of truck A.

$$F_A = \frac{\Delta(mv)_A}{\Delta t} = \frac{(30\,000\,\text{kg m s}^{-1} - 60\,000\,\text{kg m s}^{-1})}{(1.0\,\text{s} - 0.64\,\text{s})}$$

$\Rightarrow F_A = -83\,000\,\text{kg m s}^{-2}$ or $-83\,000\,\text{N}$

The two forces are equal in size but opposite in direction, as we would expect from Newton's third law.

Definition

One form of Newton's second law of motion states that the resultant force acting on a body is equal to the rate of change of momentum of the body.

$$F = \frac{\Delta(mv)}{\Delta t}$$

For a body of fixed mass this is equivalent to $F = m\,\Delta v/\Delta t$. We usually calculate the acceleration $\Delta v/\Delta t$ before using the second law, so the rate of change of momentum $\Delta(mv)/\Delta t$ is usually used only when dealing with impulsive forces as described in this section (see the Worked example alongside) and with rockets (see the next section).

2.5 Rockets

The equation $F = m\,\Delta v/\Delta t$ can sometimes be thought of 'the other way round', i.e.

$$F = \frac{v\,\Delta m}{\Delta t}$$

Clearly the units on the right of the two equations are the same: kg × m s^{-1} × s^{-1} and m s^{-1} × kg × s^{-1}, giving kg m s^{-2} or N in each case. The second equation is useful when thinking about things like firework rockets, where a stream of matter $\Delta m/\Delta t$ is ejected at a speed v.

Worked example

A rocket is ready for take-off. It ejects waste gases, the result of exploding a mixture of oxygen and hydrogen, at the rate of 1.2 × 10^4 kg s^{-1}.

If the exhaust speed is 4.0 × 10^3 m s^{-1}, calculate the upward thrust of the ejected gases on the rocket.

Answer

The downward push F of the rocket on the ejected gases is, by Newton's second law, equal to $v\,\Delta m/\Delta t$.

$$\therefore F = (4.0 \times 10^3\,\text{m s}^{-1}) \times (1.2 \times 10^4\,\text{kg s}^{-1})$$
$$\Rightarrow F = 4.8 \times 10^7\,\text{N or } 48\,\text{MN}$$

By Newton's third law, the upward push of the ejected gases on the rocket is equal in size but opposite in direction to this. It is 48 MN, a force big enough to lift a mass of 4800 tonnes.

Figure 2.12 ▲

REVIEW QUESTIONS

1 a) How big is the momentum of a rifle bullet of mass 6.0 g moving at 450 m s^{-1}?

b) What is the momentum of a canal barge of mass 12 tonnes moving due east at a speed of 1.5 m s^{-1}?

c) Calculate the size of the momentum of a 8700 kg truck moving at 50 km h^{-1}.

2 The momentum of a male Olympic sprinter is about:

A 1800 kg m s^{-1} B 900 kg m s^{-1}
C 300 kg m s^{-1} D 90 kg m s^{-1}

3 Which of these expressions does **not** have the units of energy?

A $\frac{1}{2}mv$ B $2mv^2$ C $2p^2/m$ D $\frac{1}{2}pv$

4 Which of the following physical quantities is **not** subject to a law of conservation?

A charge B energy C impulse D momentum

5 A force of 250 N acts on a body of mass 80 kg moving at 6.0 m s^{-1} for 12 s. The change of momentum of the body is:

A 480 N s B 1500 N s C 3000 N s D 5760 N s

6 In a nuclear experiment an unknown particle moving at 390 km s^{-1} makes a head-on collision with a carbon nucleus. After the collision a single particle continues at a speed of 30 km s^{-1}.

Express the mass of the unknown particle as a fraction of the mass of a carbon nucleus.

7 The relationship $m_A u = (m_A + m_B)v$ might be applied to a collision where:

A a heavy object A strikes a stationary light object B

B a heavy object A sticks to a stationary light object B

C a heavy object B sticks to a stationary light object A

D a heavy object B strikes a stationary light object A

8 A 'baddie' in a cartoon is shot in the chest. What would happen to him in real life if he were to be shot like this?

A He would slump to the floor.

B He would fall forward.

C He would fall backwards.

D He would not move at first.

9 A man is stranded in the middle of a frozen pond. He is unable to grip the ice with his feet. To move towards the edge of the pond he should:

A hurl himself towards the edge

B fall towards the edge and repeat this movement

C lift one leg and lower it towards the edge

D throw his coat away from the edge

10 What is the change of momentum of a tennis ball of mass 55 g that bounces vertically on the ground if its speed before and after impact is $6\,\text{m s}^{-1}$ and $4\,\text{m s}^{-1}$ respectively?

11 a) Use a graphical method to find the resultant of a momentum of $50\,\text{kg m s}^{-1}$ north and one of $100\,\text{kg m s}^{-1}$ west.

b) Confirm your answer by calculating the result of the same addition.

12 A rowing boat has a mass of 300 kg and is floating at rest in still water. A cox of mass 50 kg starts to walk from one end of the boat to the other at a steady speed of $1.2\,\text{m s}^{-1}$. What happens to the boat?

13 A golf ball of mass 45 g is struck by a golf club with a force that varies with time as follows:

t/ms	0	0.5	1	1.5	2	2.5	3	3.5	4
F/N	0	700	1800	2500	2100	1500	800	300	0

Table 2.1 ▲

Plot a graph of F (up) against t (along) and find the impulse exerted by the club on the ball. Hence find the initial speed of the ball after it is struck.

14 Figure 2.13 shows the horizontal force exerted by a wall on a ball that is hit hard at it horizontally and rebounds. The mass of the ball is 57.5 g.

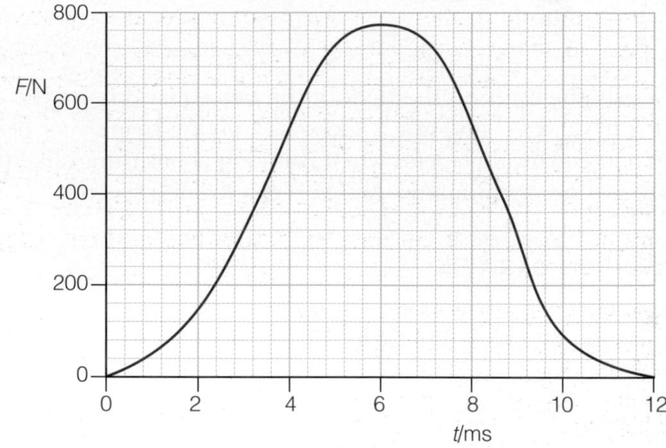

Figure 2.13 ▲

By estimating the area under the graph, deduce the change of velocity of the ball during this impact.

15 Imagine one billion people in India are organised to jump up from the ground at noon local time on a certain day. The average mass of the people is 60 kg and the average speed at which they leave the ground is $2\,\text{m s}^{-1}$. The Earth has a mass of 6×10^{24} kg.

Does the Earth recoil? Explain your answer.

If you think it recoils, calculate the speed of recoil.

16 A stationary helicopter of mass 1200 kg hovers over the scene of a road accident. The whirling helicopter blades force air downwards at a speed of $20\,\text{m s}^{-1}$.

Calculate the rate – in kg s^{-1} – at which air is pushed downwards.

17 A rocket car can accelerate from 0 to 100 m.p.h. in a very short time. Outline the physics principles behind this method of propulsion.

18 A vertical wall has a surface area of $120\,\text{m}^2$. It lies perpendicular to a wind that blows at $25\,\text{m s}^{-1}$. The density of air is $1.3\,\text{kg m}^{-3}$. What force does the air exert on the wall, assuming that the air striking the wall is brought momentarily to rest?

19 By suitable scale drawing, deduce what must be added to a momentum of $64\,\text{kg m s}^{-1}$ to the east to produce a resulting momentum of $64\,\text{kg m s}^{-1}$ to the north-west.

3 Momentum and energy

In the previous chapter you learnt about momentum but we didn't mention energy, so here is an opportunity to revise your understanding of energy and to consider energy in new situations. You will learn that kinetic energy is usually *not* conserved in collisions. These are known as inelastic collisions. However, elastic collisions in which kinetic energy *is* conserved do occur – for example between molecules and atoms, and also sometimes between larger objects such as snooker balls. Elastic collisions between particles of equal mass result in special outcomes that are important to both atomic physicists and snooker players!

Figure 3.1 ▲
An easy 'pot'

3.1 Work and energy

Remember that work and energy are both measured in joules and both are scalar quantities.

$$1\,\text{J} \equiv 1\,\text{N m} \equiv 1\,\text{kg m}^2\,\text{s}^{-2}$$

You can calculate the **work done by a force** as

$$\Delta W = F_{av}\,\Delta x$$

Using this you can calculate how much mechanical energy an object has been given. This energy might be stored as

	gravitational potential energy	$(\Delta GPE = mg\,\Delta h)$
or as	**elastic potential energy**	$(\Delta EPE = \frac{1}{2}k\,\Delta x^2)$
or as	**kinetic energy**	$(KE \text{ or } E_k = \frac{1}{2}mv^2)$

Mechanical energy sometimes appears to be lost to the surroundings, often as a result of work done against air resistance or contact friction. When this happens we say that there has been a transfer of mechanical energy to **internal energy** – thermal energy of the particles of the surroundings (see page 106).

The **principle of conservation of energy** states that energy is never created or destroyed, but it can be transferred from one form into another. Other forms of energy include chemical energy and radiant energy, which you met in your AS studies. Yet more forms, such as nuclear energy and electrostatic potential energy, you will meet later in this book.

Figure 3.2 ▲
James Joule

Note

James Joule wasn't French, but was the son of a Manchester brewer – so we should really pronounce the unit as 'jowl' not 'jool'!

Some useful equations

Look at the equations for kinetic energy E_k and linear momentum p.

$$E_k = \frac{1}{2}mv^2 \qquad p = mv$$

For an object of mass m we can eliminate the velocity v and get equations linking kinetic energy to momentum. Squaring the second equation and rearranging the first gives:

$$p^2 = m^2v^2 \quad \text{and} \quad v^2 = \frac{2E_k}{m}$$

$$\Rightarrow p^2 = m^2 \times \frac{2E_k}{m}$$

So, with no v:

$$p = \sqrt{(2mE_k)} \quad \text{or} \quad E_k = \frac{p^2}{2m}$$

Worked example

1 Calculate the momentum of a 58 g tennis ball with kinetic energy of 75 J.
2 Calculate the kinetic energy of an arrow of mass 0.12 kg with a momentum of 5.2 kg m s^{-1}.

Answer

1 Using $p = \sqrt{(2mE_k)}$ gives the momentum of the ball as

$$p = \sqrt{(2 \times 0.058\,\text{kg} \times 75\,\text{J})} = 2.9\,\text{kg m s}^{-1}$$

2 Using $E_k = p^2/2m$ gives the kinetic energy of the arrow as

$$E_k = \frac{(5.2\,\text{kg m s}^{-1})^2}{2 \times 0.12\,\text{kg}} = 113\,\text{J}$$

Tip

Look to see if the equations are homogeneous with respect to units: the key is that J ≡ N m ≡ kg m^2 s^{-2}, so all is well.

These relationships work in the microscopic world as well as in everyday life. But for fast-moving atomic or nuclear particles the equations only work if the particles are moving at less than about 10% of the speed of light, $c = 3.0 \times 10^8$ m s^{-1}. Such particles are called non-relativistic particles. We will consider particles moving at almost the speed of light in Chapter 9.

The energies of particles such as electrons and protons, and of ions such as alpha particles, are usually quoted in a unit much smaller than the joule, the **electron-volt** (eV).

$$1\,\text{eV} \equiv 1.6 \times 10^{-19}\,\text{J}$$

Definition

The electron-volt is the energy gained by a particle of unit electronic charge when accelerated by a p.d. of 1 volt.

The energy of *both* an electron *and* a proton that have each been accelerated through 150 V will be 150 eV. However, because they have very different masses, they will have very different momentums.

Worked example

What is the momentum of:
a) an electron of mass 9.1×10^{-31} kg,
b) a proton of mass 1.7×10^{-27} kg,
each of which has a kinetic energy of 150 eV?

Answer

For both, 150 eV = $150 \times 1.6 \times 10^{-19}$ J = 2.4×10^{-17} J.
Using $p = \sqrt{(2mE_k)}$ gives:
a) $p_{\text{electron}} = \sqrt{(2 \times 9.1 \times 10^{-31}\,\text{kg} \times 2.4 \times 10^{-17}\,\text{J})} = 6.6 \times 10^{-24}\,\text{N s}$
b) $p_{\text{proton}} = \sqrt{(2 \times 1.7 \times 10^{-27}\,\text{kg} \times 2.4 \times 10^{-17}\,\text{J})} = 2.9 \times 10^{-22}\,\text{N s}$

The same applies to large objects. The young person called Sam in Figure 3.3 has a mass of 32 kg, and the tennis ball has a mass of 0.055 kg. Let's work out their momentums if each has a kinetic energy of 100 J.

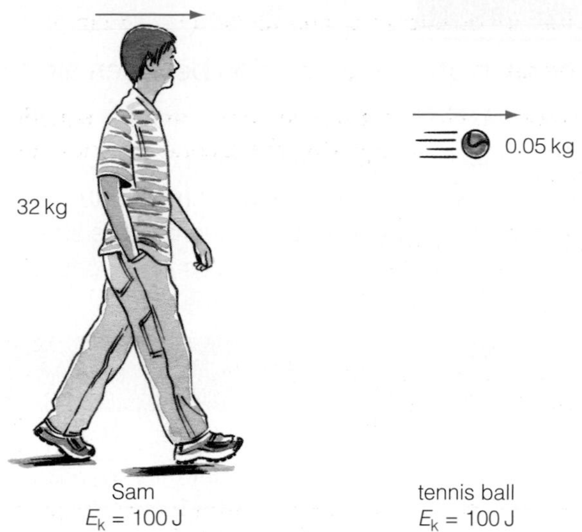

Figure 3.3 ▲

Using $p = \sqrt{(2mE_k)}$

for Sam $\quad p = \sqrt{(2 \times 32\,\text{kg} \times 100\,\text{J})} = 80\,\text{kg m s}^{-1}$
for the ball $\quad p = \sqrt{(2 \times 0.055\,\text{kg} \times 100\,\text{J})} = 3.3\,\text{kg m s}^{-1}$

So their momentums are very different, though their kinetic energies are the same.

Tip

Of course, you could work out the speed of Sam and of the ball from $E_k = \frac{1}{2}mv^2$ and then plug the speeds into $p = mv$. Trying to *remember* formulas like $p = \sqrt{(2mE_k)}$ is not essential. Make sure you understand the basic principles.

3.2 Elastic and inelastic collisions

Look back at the simple experiment of colliding trolleys fitted with corks and a pin, shown in Figure 2.2 on page 5. Let's have a look at what happens to the kinetic energy of the trolleys in such a collision.

First suppose the mass of each trolley is m, i.e. $m_A = m_B = m$. The equation expressing the conservation of momentum then becomes

$$mu = 2mv \qquad \text{that is} \qquad v = \frac{1}{2}u$$

Total KE before collision $= \frac{1}{2}mu^2 + 0 = \frac{1}{2}mu^2$

Total KE after collision $= \frac{1}{2}(2m)v^2 = mv^2 = m(\frac{1}{2}u)^2 = \frac{1}{4}mu^2$

so that the total KE has fallen from $\frac{1}{2}mu^2$ to $\frac{1}{4}mu^2$, i.e. $KE_{after} = \frac{1}{2}KE_{before}$

Momentum has been conserved but some kinetic energy has been lost, half the initial kinetic energy in this case. This result is the same for all collisions where objects of equal mass form linked collisions, as in Figure 2.2. Such a collision is an example of a **non-elastic** or an **inelastic** collision.

But remember, **total energy must be conserved**. The loss of kinetic energy is equal to the gain in internal energy – in this case as the pin enters the cork. The Sankey diagram in Figure 3.4 emphasises this energy conservation.

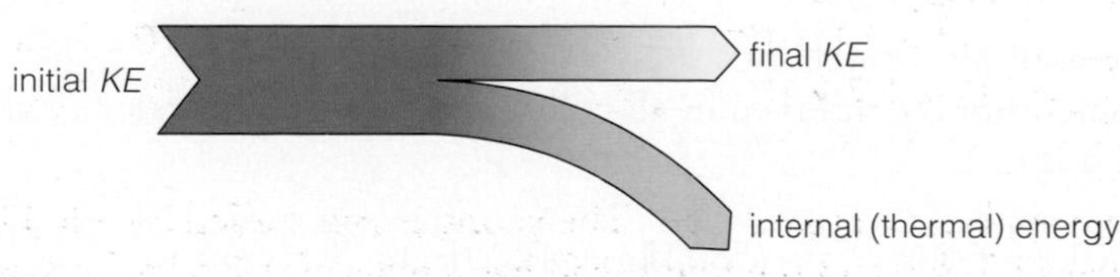

Figure 3.4 ▲
Total energy is conserved

Experiment

Analysing energy conservation in a collision between air gliders

The set-up is as in Figure 2.5 on page 7.

The table shows data for gliders P and Q that move towards each other on the air track, collide and bounce back. The change of momentum Δp and the change of kinetic energy ΔE_k have been calculated and are shown in the last two columns. Left-to-right has been chosen as positive for u, v and Δp. Glider P initially moves to the right; Q moves to the left.

	Mass/kg	Initial velocity u/m s^{-1}	Final velocity v/m s^{-1}	Δp/kg m s^{-1}	ΔE_k/J
P	0.20	0.14	−0.26	−0.080	0.0048 more
Q	0.25	−0.22	0.10	0.080	0.0048 less

Table 3.1 ▲

You can see that the change of momentum of glider P is 0.080 kg m s^{-1} to the left, and that of glider Q is 0.080 kg m s^{-1} to the right, i.e. momentum is conserved. Is kinetic energy conserved?

In the experiment above there was no loss of kinetic energy, as all the 0.0048 J lost by glider Q is gained by glider P. This type of collision is called an **elastic** collision. Each glider had a tiny magnet with like poles as a buffer, creating an elastic bounce.

In the everyday world almost all collisions are inelastic, but in the sub-atomic world elastic collisions are not uncommon.

You could set up a spreadsheet to analyse any two-body collision, if provided with the data for the masses and velocities of the two bodies.

Worked example

A 1200 kg car is stationary on an icy road. The car is hit from behind by a skidding lorry of mass 5600 kg that is moving at 18 km h^{-1} (5.0 m s^{-1}). The two vehicles remain locked together after the crash.

a) What type of collision is this crash?

b) What fraction of the lorry's kinetic energy becomes internal energy as a result of the collision?

Answer

Figure 3.5 ▲

a) The collision is inelastic.

b) Momentum is conserved in all collisions. Suppose the velocity after the crash is v.

$$5600\,\text{kg} \times 5.0\,\text{m s}^{-1} = (5600\,\text{kg} + 1200\,\text{kg})v$$
$$\Rightarrow v = 4.12\,\text{m s}^{-1}$$

$$KE_{before} - KE_{after} = \frac{1}{2} \times 5600\,kg \times (5.0\,m\,s^{-1})^2 - \frac{1}{2} \times 6800\,kg \times (4.12\,m\,s^{-1})^2$$
$$= 70\,000\,J - 57\,700\,J = 12\,300\,J$$

which is 12 300 J/70 000 J = 0.176 or, to 2 sig. fig., 18% of the lorry's original *KE*.

Worked example

A proton of mass m moving at a speed $10u$ makes a head-on collision with a stationary helium nucleus of mass $4m$. After the collision the helium nucleus moves forward at $4u$.

Discuss whether this collision is elastic or inelastic.

Answer

Suppose the speed of the proton after the collision is v. As momentum is conserved,

$$m \times 10u = mv + 4m \times 4u$$
$$\Rightarrow mv = -6mu$$

So the proton bounces backwards at a speed of $6u$.

Kinetic energy of the proton before collision $KE_{before} = \frac{1}{2}m \times (10u)^2 = 50mu^2$

Kinetic energy of the proton and the helium nucleus after collision

$$KE_{after} = \frac{1}{2}m \times (-6u)^2 + \frac{1}{2} \times 4m \times (4u)^2 = 18mu^2 + 32mu^2 = 50mu^2$$

As both KE_{before} and KE_{after} are equal to $50mu^2$, no kinetic energy is lost and so the collision is elastic.

Tip

When, as in these two worked examples, the problem is not split into separate parts, try to see 'where you are going' before lunging into a solution.

Exercise

Two 'bumper' cars G and B are moving along the same line at a fairground. Car B driven by a boy bumps into the back of car G in which are two girls. Table 3.2 gives the masses of the cars and their speeds before and after the collision.

Figure 3.6 ▲

Use the data in the table to determine the speed of car B after the collision and whether this bumper car collision is an elastic collision.

	Mass/kg	Initial velocity u/m s^{-1}	Final velocity v/m s^{-1}
Car G	180	+2.5	4.0
Car B	135	+4.0	v

Table 3.2 ▲

3.3 Collisions in two dimensions

Up to now all collisions we have considered have been between two objects moving in the same straight line: they were all one-dimensional collisions. A very brief look at a game of snooker will show you that collisions on a snooker table are rarely one-dimensional; nor are collisions between fast-moving protons and other nuclear particles.

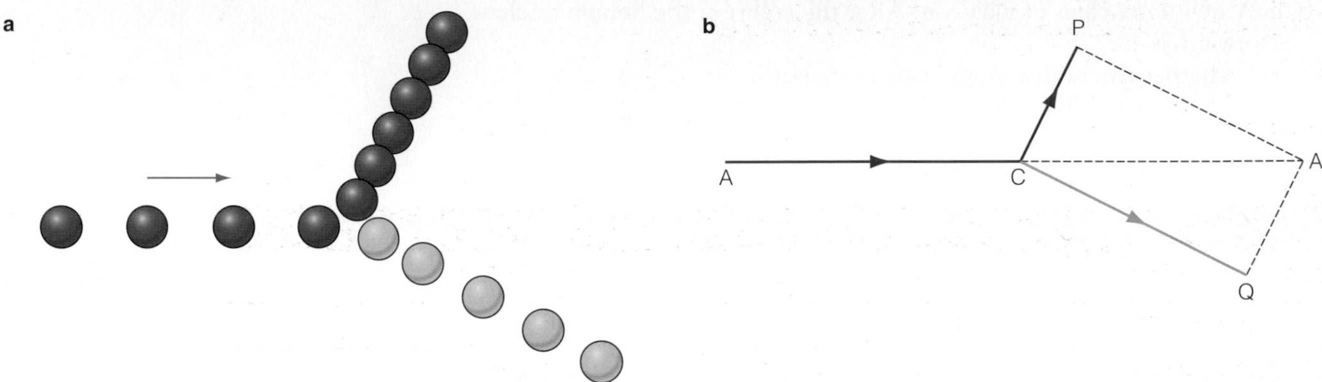

Figure 3.7 ▲

Figure 3.7a shows an off-centre collision between two balls of equal mass m, one moving from the left to strike a ball that is stationary. The positions of the balls are taken from a stroboscopic photograph of the collision.

Comparing the speeds of the balls by measuring the distances they each travel in, for example, three camera flashes, enables Figure 3.7b to be drawn. In this diagram the lengths of the arrows represent the vector momentum of each ball. Check for yourself that the ratios of the three arrows are about right for the three speeds. As each ball has the same mass, they are also correct for the momentums.

The **parallelogram of vectors** – here momentum vectors – now shows that the addition of the momentums of the two balls after the collision, CP + CQ = CA′, is equal to the momentum of the single ball before the collision, AC, i.e. momentum has been conserved.

Another way of using Figure 3.7a to show that momentum is conserved in this collision is to **resolve** each of the momentum vectors after the collision into their **components** parallel and perpendicular to the direction of motion of the incident ball. This will show that the momentum 'up' equals the momentum 'down' after the collision and thus that the 'vertical' momentum both before and after the collision is zero. The resolved momentums to the right will also add to be the same as the momentum of the incident ball.

Worked example

The photograph in Figure 3.8a shows an alpha particle making a collision in a cloud chamber filled with helium gas. Figure 3.8b shows the velocities of the incoming alpha particle before the collision and of the alpha particle and recoiling helium nucleus after the collision. The alpha particle and the helium nucleus each have a mass of 6.65×10^{-27} kg.

a) Make a table showing the calculated values of the momentum of each of the particles before and after the collision.
b) By resolving the vector momentums, show that the initial momentum of the incoming alpha particle is conserved after the collision.

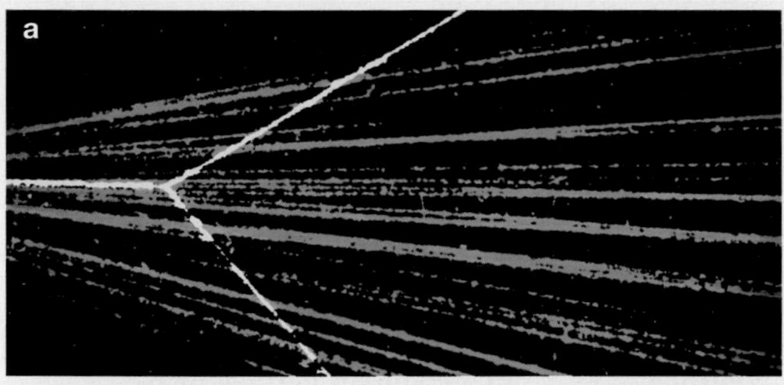

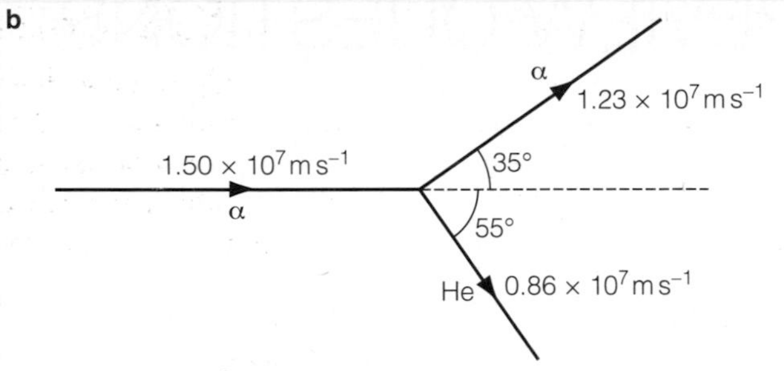

Figure 3.8 ▲

Answer

a)

	Mass times velocity *mu*/kg m s^{-1}	Momentum *p*/kg m s^{-1}
alpha particle before	$6.65 \times 10^{-27} \times 1.50 \times 10^{7}$	10.0×10^{-20}
helium nucleus before	0	0
alpha particle after	$6.65 \times 10^{-27} \times 1.23 \times 10^{7}$	8.18×10^{-20}
helium nucleus after	$6.65 \times 10^{-27} \times 0.86 \times 10^{7}$	5.72×10^{-20}

Table 3.3 ▲

b) Sum of components of momentum after collision parallel to initial path of alpha particle

$= (8.18 \times 10^{-20}\,\text{kg m s}^{-1}) \cos 35° + (5.72 \times 10^{-20}\,\text{kg m s}^{-1}) \cos 55°$

$= 6.70 \times 10^{-20}\,\text{kg m s}^{-1} + 3.28 \times 10^{-20}\,\text{kg m s}^{-1}$

$= 9.98 \times 10^{-20}\,\text{kg m s}^{-1}$

which is $10.0 \times 10^{-20}\,\text{kg m s}^{-1}$ to 3 sig. fig., i.e. equal to the initial momentum of the incoming alpha particle. Vector momentum is conserved.

In both Figures 3.7a and 3.8a, of two identical balls colliding and two equal-mass nuclear particles colliding, the angle between the two particles after the collision is 90°. It can be shown that the 90° angle only happens when the collision between objects of equal mass is an elastic one. (Snooker players know this and take steps to avoid possible 'in-offs' after making a pot.) It was photographs such as that in Figure 3.8a) that confirmed the fact that alpha particles are identical to helium nuclei.

You could check that the nuclear collision is elastic using $KE = \frac{1}{2}mv^2$ but, as the particles have the same mass, it is enough to show that:

$u^2(\text{alpha before}) = v_1^{\,2}(\text{alpha after}) + v_2^{\,2}(\text{helium after})$

Let's consider two other situations:

- A heavy ball or particle strikes a much lighter one that is at rest. The heavy ball 'carries on' and pushes the lighter one forward. The angle between the two after the collision is *less than* 90° if the collision is elastic.
- A light ball or particle strikes a much heavier one that is at rest. The light ball 'bounces back' and the heavy one moves forward. The angle between the two after the collision is *more than* 90° if the collision is elastic.

REVIEW QUESTIONS

1 Which of the following is **always** true?

A Kinetic energy is conserved in collisions.

B Recoil is the result of explosive charges.

C Impulsive forces only act for short times.

D Linear momentum is conserved in collisions.

2 A compressed spring is placed between two trolleys of mass m and $4m$ and the trolleys are let go. Describe what happens and explain where the kinetic energy of the trolleys comes from.

3 Explain the difference between an elastic and an inelastic collision.

4 A loaded railway wagon of mass 21 tonnes collides with an empty stationary wagon of mass 7 tonnes. They couple together and move off at $3.5\,\mathrm{m\,s^{-1}}$.

a) Show that the speed of the loaded wagon before the collision was $4.7\,\mathrm{m\,s^{-1}}$.

b) Calculate the percentage loss of kinetic energy in this collision.

5 Two ice dancers are moving together in a straight line across the ice at $5.8\,\mathrm{m\,s^{-1}}$. The man has a mass of 75 kg and the woman has a mass of 65 kg. They push each other apart (see for example Figure 2.7 on page 8). After they separate, they are both still moving in the same direction. The man's velocity is now $3.8\,\mathrm{m\,s^{-1}}$.

a) What is the woman's velocity after they separate?

b) What has happened to their kinetic energy?

6 In Figure 3.9, which diagram represents an elastic collision between a heavy moving body and a light stationary body?

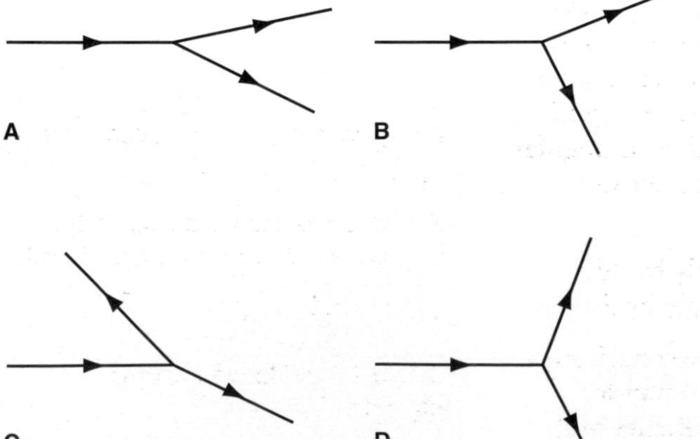

Figure 3.9 ▲

7 Taking data from the Worked example involving an alpha particle colliding with a helium nucleus (page 18), make calculations to decide whether there was any loss of kinetic energy in this two-dimensional collision.

8 A shell of mass 12 kg moving horizontally at $320\,\mathrm{m\,s^{-1}}$ explodes into three fragments A, B and C, which continue to move in a common vertical plane.

Fragment A: mass 2.0 kg continues at $450\,\mathrm{m\,s^{-1}}$, 45° above the horizontal.
Fragment B: mass 6.0 kg continues at $400\,\mathrm{m\,s^{-1}}$, horizontally.
Fragment C: mass m continues at speed v, in a direction θ to the horizontal.

a) Calculate the size of the momentums of the shell before it explodes and of fragments A and B afterwards.

b) Draw a vector diagram showing these momentums and hence deduce values for m, v and θ.

4 Motion in a circle

We all spend a great deal of time going round in circles; but we don't notice the motion. This is because the circles – around our latitude line on Earth once a day, around the Sun once a year – are very big and the time to complete a single journey is very long. When you complete a journey very quickly, for example on a playground roundabout or a fairground wheel, things are very different. In this chapter you will learn how to describe the motion of bodies moving in a circle and about the forces involved in keeping an object – perhaps you! – moving in a circle.

4.1 The language of circular motion

Figure 4.1 ▲
A long-exposure photograph of stars near the Pole Star

Over the five hours' exposure of the photograph, each star in the night sky near the Pole Star has rotated through the same angle, the same fraction of a circle. The length of each star track (the arc, s) divided by its distance from the centre of rotation (the radius, r) is the same – here about 1.3 – try it! This number is the angle θ through which each star has rotated – its **angular displacement**. This measurement is in **radians**, abbreviation rad, not in degrees.

- To convert from degrees to radians, divide by 2π and multiply by 360.
- To convert from radians to degrees, divide by 360 and multiply by 2π.

When describing how rapidly an object is rotating in a circle, we could use revolutions per second or revolutions per minute (r.p.m.), but in physics problems we use radians per second. We call this the **angular velocity** and give it the symbol ω (omega).

Average and instantaneous angular velocities are sometimes represented by the equations:

$$\omega_{av} = \frac{\Delta\theta}{\Delta t} \qquad \omega_{inst} = \frac{\delta\theta}{\delta t} \text{ or } \frac{d\theta}{dt}$$

Definition

$\theta = \frac{s}{r}$ gives angular displacement in radians.

The circumference of a circle is $2\pi r$, so there are 2π rad in a circle of 360°.

$1 \text{ rad} \equiv \frac{360}{2\pi} = 57.3°$

so the 1.3 rad found from the photograph is 75°.

Definition

$\omega = \frac{\theta}{t}$ gives angular velocity in rad s^{-1}

Worked example

Calculate the angular velocity of

a) the London Eye, which rotates once every 30 minutes,

b) a CD, which when first switched on rotates at 200 r.p.m.

Answer

a) $\omega = \dfrac{\Delta\theta}{\Delta t} = \dfrac{2\pi\,\text{rad}}{30 \times 60\,\text{s}} = 0.0035\,\text{rad}\,\text{s}^{-1}$

b) $\omega = \dfrac{\Delta\theta}{\Delta t} = \dfrac{200 \times 2\pi\,\text{rad}}{60\,\text{s}} = 21\,\text{rad}\,\text{s}^{-1}$

Putting together $v = \Delta s/\Delta t$ and $\omega = \Delta\theta/\Delta t$, we get an important link between v and ω:

$$v = r\omega$$

This is almost common sense. Think of a playground roundabout: everyone on it has the same angular velocity, ω. But the further you stand away from the centre, r, the faster you are travelling, v.

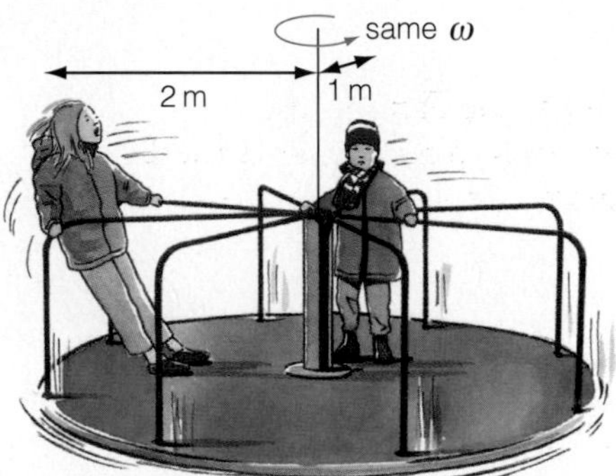

Figure 4.2 ▲
Stand near the centre and you'll go slower

Worked example

The information on a CD is recorded as a series of bumps in concentric circles. The information is 'read' by a laser beam. Calculate the speed of the bumps moving past the laser head at 4.0 cm from the centre of a CD rotating at 430 r.p.m.

Answer

430 r.p.m $= 430 \times 2\pi\,\text{rad}/60\,\text{s} = 45\,\text{rad}\,\text{s}^{-1}$

So $v = r\omega = 0.040\,\text{m} \times 45\,\text{rad}\,\text{s}^{-1} = 1.8\,\text{m}\,\text{s}^{-1}$

Another useful relationship relates the time taken for something to complete one circle. If an object is rotating at $1.5\,\text{rad}\,\text{s}^{-1}$, you can see that it will complete one circle in a time

$$T = \frac{2\pi\,\text{rad}}{1.5\,\text{rad}\,\text{s}^{-1}} = 4.2\,\text{s}$$

In general: $T = \dfrac{2\pi}{\omega}$

Reversing this to $\omega = 2\pi/T$ enables you to calculate the angular velocity of the spinning Earth, and then $v = r\omega$ lets you calculate the speed of any point on the Earth's surface provided you know the radius of the circle in which it is moving. (It would be zero at the poles!)

Worked example

The city of Birmingham is at latitude 52.5°. The Earth has a radius of 6400 km.

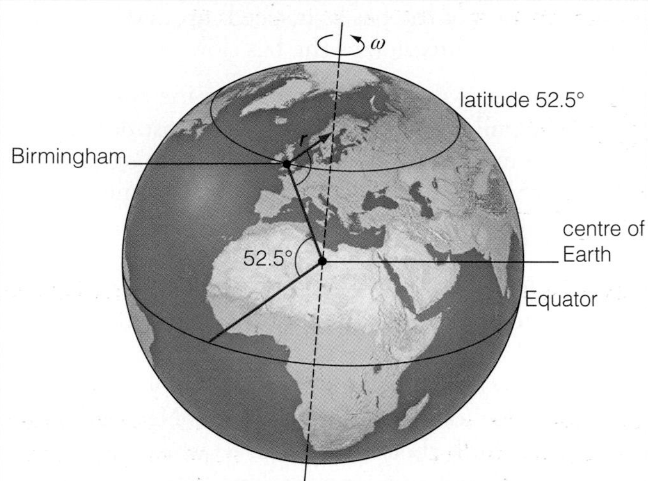

Figure 4.3 ▲

Calculate:

a) Birmingham's angular velocity as the Earth spins,
b) the radius of the circle in which Birmingham moves, and
c) the speed at which Birmingham's inhabitants are moving.

Answer

a) The Earth spins once every 24 hours.

$$\therefore \omega = \frac{2\pi}{T} = \frac{2\pi \text{ rad}}{24 \times 3600 \text{ s}} = 7.27 \times 10^{-5} \text{ rad s}^{-1}$$

b) At latitude 52.5°, $r = (6.4 \times 10^6 \text{ m}) \cos 52.5° = 3.90 \times 10^6 \text{ m}$

c) Using $v = r\omega \Rightarrow v = (3.9 \times 10^6 \text{ m})(7.3 \times 10^{-5} \text{ rad s}^{-1}) = 280 \text{ m s}^{-1}$

Tip

Be careful that your calculator is turned to degrees if, for example, you want to find the sin/cos/tan of an angle like 45°, and to radians if you want to find the sin/cos/tan of an angle like 0.79 rad.

4.2 Centripetal forces

Think of yourself as a passenger standing on a bus, holding on to a single vertical post. The bus first accelerates at $+a$ from the stop, then turns a sharp left-hand corner at a steady speed v, and finally brakes to a halt at the next stop with acceleration $-a$ (Figure 4.4). In what directions do you feel the force of the post on you during each part of this short journey?

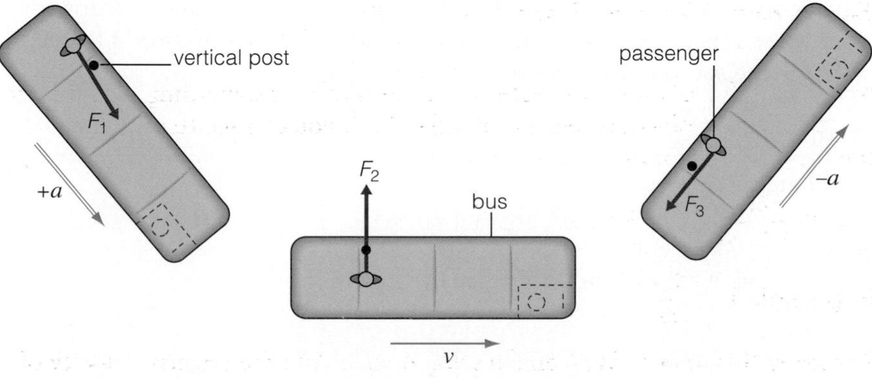

Figure 4.4 ▲

The forces on you for each part of the journey are shown in Figure 4.4 on the previous page. (Notice that you would have to move round the post to prevent yourself being thrown forward as the bus slows.) The forces are:

- a *forward* pull F_1 in the direction of the bus as it speeds up, and
- a *backward* pull F_3 in the reverse direction as the bus slows down.

The interesting part of the journey is the middle part when the bus turns the corner at a steady speed v. The pull you need, F_2, you would instinctively find is *perpendicular* to the direction in which the bus is travelling. It is towards the centre of the curve along which the bus is travelling. We say you need:

- a *centripetal* pull F_2 as the bus corners at a steady speed.

Without this force you would feel as if you were being 'thrown' away from the centre. This is a result of **Newton's first law of motion** – your body 'wants' to continue in a straight line.

The force on you while the bus turns the corner means that you must be accelerating in the direction of the force F_2. In this case it is **Newton's second law of motion** that you need to think about. The force on you as the bus turns is called a **centripetal force**, which means a force 'seeking' the centre of the circle that the bus is following at that instant. For all bodies or objects moving in the arc of a circle at a constant speed, the resultant force acting on them is a centripetal force, which causes a **centripetal acceleration**. The children on the roundabout in Figure 4.2 hang on and are pulled towards the centre.

To summarise:

- constant speed along the arc – Newton's first law applies
- centripetal acceleration – Newton's second law applies.

'Sideways' acceleration

Centripetal acceleration is 'sideways' to the direction of travel. Put more precisely, in the language of physics, a centripetal acceleration is radially inwards, perpendicular to the tangential velocity. The size of this acceleration is v^2/r where v is the velocity and r the radius of the circle, or, because $v = r\omega$, this can be expressed as $r\omega^2$.

$$\textbf{centripetal acceleration} = \frac{v^2}{r} = r\omega^2$$

For example, a runner moving at $8.0\,\text{m}\,\text{s}^{-1}$ round the end of a track that forms part of a circle of radius 35 m has a centripetal acceleration of $(8.0\,\text{m}\,\text{s}^{-1})^2/35\,\text{m} = 1.8\,\text{m}\,\text{s}^{-2}$.

Figure 4.5 ▲ A hammer thrower

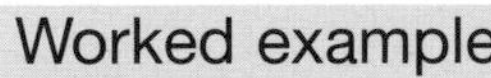

Worked example

The centripetal acceleration of the mass on the end of a hammer thrower's wire when it is being rotated in a circle of radius 1.5 m is $500\,\text{m}\,\text{s}^{-2}$ ($\approx 50g$). What is the speed of the mass as it is whirled in a circle?

Answer

Rearranging $a = \dfrac{v^2}{r}$

gives $v = \sqrt{(ar)}$

$\Rightarrow v = \sqrt{(500\,\text{m}\,\text{s}^{-2} \times 1.5\,\text{m})} = 27\,\text{m}\,\text{s}^{-1}$

Experiment

Investigating centripetal force

The centripetal force equation $F = mr\omega^2$ can be investigated experimentally as shown in Figure 4.6. Teachers should note that the glass tube should be about 20 cm long. For safety, both its ends should be smoothed by heat treatment in a Bunsen flame.

With a bit of practice, the rubber bung can be made to describe a horizontal circle and its period of rotation T can be found by a colleague timing 10 rotations. The length r of the string can be found once the timing has been completed by pinching the string between your fingers to stop the motion and then measuring the distance to the centre of mass of the bung.

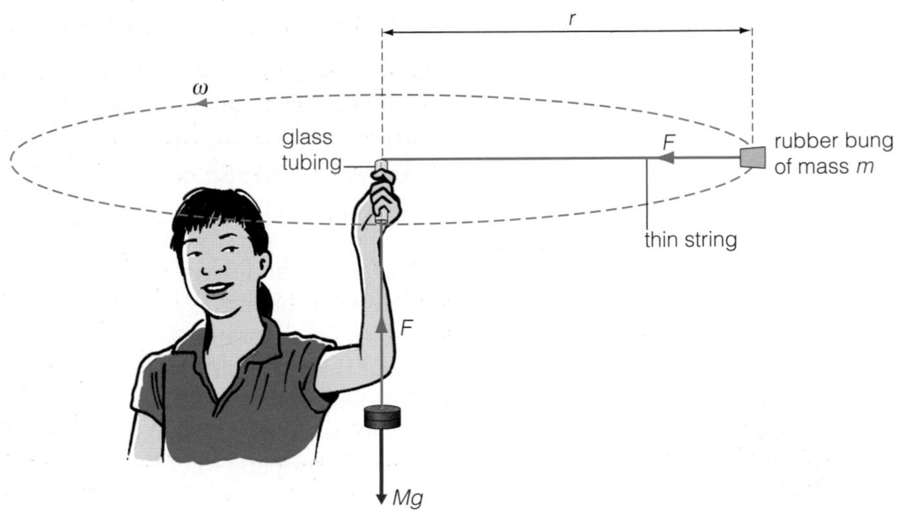

Figure 4.6 ▲

If we make the simplifying assumptions that, when the mass M is in equilibrium:

- there is negligible friction between the glass tube and the string, and
- the string is horizontal

then the pull, $F = Mg$, of the string on the mass M is equal to the pull of the string on rubber bung (of mass m). This provides the centripetal force $mr\omega^2$ for the bung.

Safety note

Eye protection should be worn. The whirling mass should be limited to a small/medium sized rubber bung. If the laboratory is crowded this should be done outside or in a hall.

Exercise

Two students carried out such an experiment and obtained the results shown in the table below. The mass m of the bung was 75 g.

M/kg	F/N	$10T$/s	ω/rad s^{-1}	r/m	$mr\omega^2$/N
0.050		21.9		0.83	
0.100		16.2		0.80	
0.150		13.5		0.93	
0.200		11.2		0.79	
0.250		10.7		0.99	

Table 4.1 ▲

a) Set up Table 4.1 as a spreadsheet to investigate the extent to which the experimental data confirms that $F = mr\omega^2$.

The students' teacher suggested that an alternative way to analyse their results would be to draw a graph, from which they could find a value for the mass m of the bung.

If $F = mr\omega^2$, a graph of F on the y-axis against $r\omega^2$ on the x-axis should give a straight line through the origin. The gradient should be equal to the mass m of the rubber bung.

b) Use the data provided to check this out for yourself.

Resultant centripetal force

Newton's second law tells you that the *resultant* force on a body of mass m accelerating at v^2/r is:

$$F_{res} = \frac{mv^2}{r}$$

In nearly all applications of the second law for objects moving in a circle of radius r at a uniform speed v, there is only a single external force acting on the object towards the centre of the motion. So the use of the above 'formula' is pretty easy. It can also be used in the form $F_{res} = mr\omega^2$ as in the Experiment on the previous page.

Think about water-skiers: they often swing away from the line the towing boat is taking, and move in the arc of a circle. A free-body force diagram of a water-skier as he moves in a circle would look like Figure 4.7a. The water pushes up on the water-skier with a force R (equal and opposite to his weight W), and the water also pushes him inwards (centripetally) with a force F_c. (It also pushes him backwards with a force to balance the pull of the tow rope on him. But as we are just looking at forces in the plane containing F_c, the resultant centripetal force, we can disregard this.)

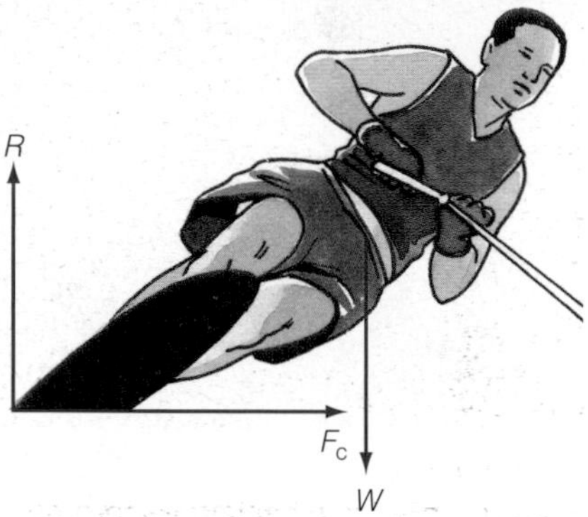

R
F_c
W

Figure 4.7a ▲

When you see a water-skier at the extreme of an outward curve, you see a huge plume of water that sprays outwards from the skier's path (Figure 4.7b). This is the result of **Newton's third law**: there is a force on the water equal and opposite to F_c. This force doesn't act on an object that is moving in a circle – here it acts on the water.

Tip

A free-body force diagram shows all the external forces acting on the body you have chosen. Sometimes, as here, you must choose forces in one plane.

Figure 4.7b ▲

Table 4.2 lists some more examples of objects moving in circles and the centripetal force(s) acting on them (you will find lots more at a fairground).

Note that the *resultant* centripetal force may involve other forces, such as the weight of the object moving in a circle.

Object moving in a circle	Centripetal force
Cyclist on a banked velodrome corner	Inward component of the force of the track on the cycle's wheels
Single sock in a spin drying machine	Inward force of the drum of the spin dryer on the sock
Fairground car at the bottom of a roller coaster dip	Upward force of the track on the car
Moon	Gravitational pull of the Earth on the Moon
Gymnast on the high bar	Inward force of the bar on the gymnast's hands

Table 4.2 ▲
Examples of circular motion

Figure 4.8 ▲
Racing cyclist on a banked velodrome track

Worked example

A bobsleigh of total mass 380 kg is travelling at a speed of $31\ \text{m s}^{-1}$ along an ice channel. At the moment shown in Figure 4.9 it is turning a corner that is part of a horizontal circle of radius 46 m and is tilted to the vertical at an angle of 65°.

Assume that friction is negligible.

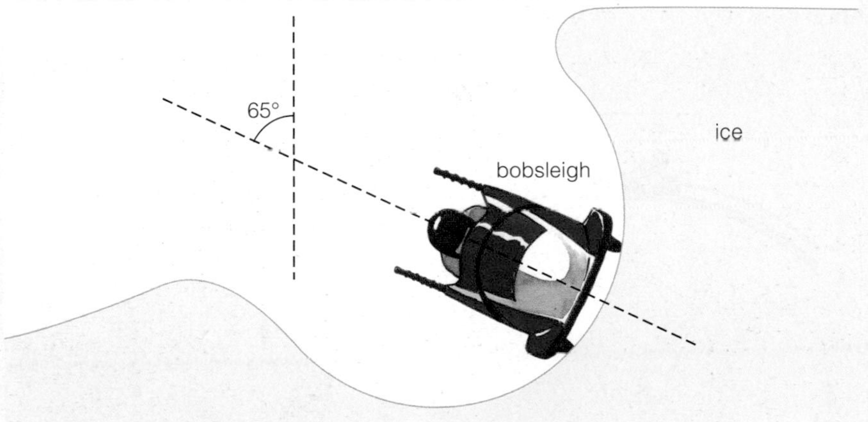

Figure 4.9 ▲

a) Calculate the centripetal acceleration of the bobsleigh and hence find the horizontal centripetal force acting on it.
b) What force provides this centripetal acceleration?

Answer

a) Centripetal acceleration $a = v^2/r = (31\ \text{m s}^{-1})^2/46\ \text{m} = 21\ \text{m s}^{-2}$
Resultant centripetal force $= ma = 380\ \text{kg} \times 21\ \text{m s}^{-2} = 8000\ \text{N}$
b) This centripetal force is the horizontal component R_h of the normal reaction force R, the push of the ice on the bobsleigh. R acts perpendicular to the bobsleigh's runners. So the angle between R_h and R is 25°.

$$\therefore R\cos 25° = 8000\ \text{N}$$
$$\Rightarrow R = 8800\ \text{N}$$

(Note that the vertical component R_v of R is equal to the weight of the bobsleigh, 3700 N.)

Tip

Whenever possible, use the cosine of the angle between a vector and the direction of its component.

4.3 Apparent weightlessness

We don't feel the pull of the Earth on us. What tells us that we have weight is the upward push of a seat or the ground on us. If for a moment we feel no such force, our brain thinks that we have no weight. A person might *feel* weightless when he or she:

- jumps off a trampoline
- treads on a non-existent floor in the dark
- travels in an aeroplane that hits an air pocket.

In each case the person is briefly in 'free fall': that is, he or she accelerates downwards at $9.8\,\mathrm{m\,s^{-2}}$ (g). Anyone diving off a ten-metre board might feel weightless for over a second – the time taken to dive from the board to the water surface, t, given by:

$$t = \sqrt{\left(\frac{2g}{h}\right)} = \sqrt{\left(\frac{2 \times 9.8\,\mathrm{m\,s^{-2}}}{10\,\mathrm{m}}\right)} = 1.4\,\mathrm{s}$$

Worked example

A rollercoaster at an amusement park includes a 'hump' which is part of a vertical circle of radius $r = 18.0\,\mathrm{m}$. A person in a car travelling over the hump feels momentarily weightless. Draw a free-body force diagram of the person and calculate the speed v at which the car is travelling.

Answer

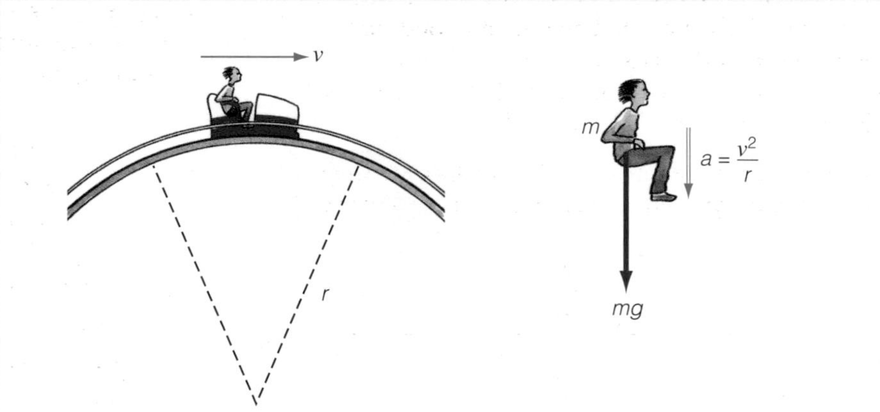

Figure 4.10 ▲

When the person feels weightless, there is no reaction force – the only force acting on them is their weight, mg.

As they are travelling in a circle, their centripetal acceleration $= v^2/r$. Applying Newton's second law gives:

$$\frac{mv^2}{r} = mg$$

$$\Rightarrow v^2 = gr = 9.8\,\mathrm{m\,s^{-2}} \times 18.0\,\mathrm{m}$$

So $v = 13\,\mathrm{m\,s^{-1}}$

REVIEW QUESTIONS

1 An angle of $\pi/3$ radians is the same as an angle of:

A 120° B 90° C 60° D 30°

2 450 revolutions per second is equivalent to:

A 2800 rad s^{-1} B 1800 rad s^{-1}
C 1400 rad s^{-1} D 290 rad s^{-1}

3 A playground platform is rotating once every 4.0 s. A young girl is hanging on 2.0 m from the centre of rotation. She is moving in a circle at a constant speed v of:

A 6.3 m s^{-1} B 3.1 m s^{-1} C 2.0 m s^{-1} D 1.0 m s^{-1}

4 Going round a tight corner at speed when sitting in a car, you feel a strong sideways force. This is because:

A you are being flung out by a centrifugal force

B your body tries to continue in a straight line

C you are pushing sideways against the seat

D your body is accelerating away from the centre

5 A laboratory centrifuge operates at a working speed of rotation of 200 rad s^{-1}. The end of a rotating test tube is 12 cm from the axis of rotation. What is the acceleration of matter at the end of the test tube? Express your answer as a multiple of g.

6 A proposed space station is designed in the shape of the circular tyre. Figure 4.11 shows a cross section.

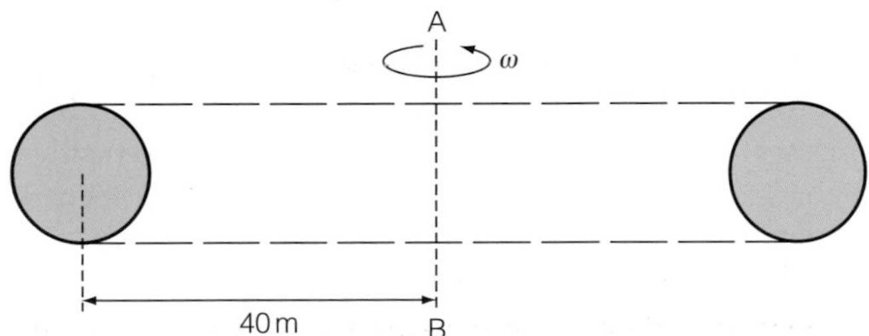

Figure 4.11 ▲

a) At what angular velocity must the station be rotated about its axis AB in order that a person living in it may experience a centripetal acceleration equal to g?

b) What advantage does such a space station have over existing orbiting space stations?

7 A 'car' of mass 1800 kg in an amusement park moves through the lowest part of a loop of radius 8.5 m at 22 m s^{-1}.

Draw a free-body force diagram for the car at this moment and calculate the upward force of the rails on the car.

8 The girl in Figure 4.12 has a mass of 56 kg. She is at the Earth's equator.

a) What is her centripetal acceleration? Take the Earth's radius to be 6400 km.

b) Show that the force needed to produce this acceleration is about 2 N.

c) Sketch the diagram and add force(s) showing how this resultant centripetal force arises.

Figure 4.12 ▲

9 What provides the centripetal force needed by a car cornering on a level road?

10 A teacher whirls a bucket of water in a vertical circle of radius 1.2 m. What is the minimum speed at which the bucket must be whirled if the teacher is not to get a soaking?

11 Figure 4.13 shows a 'conical pendulum' in motion, and a free-body force diagram for the pendulum bob. Use Newton's laws of motion to prove that $\tan\theta = v^2/rg$.

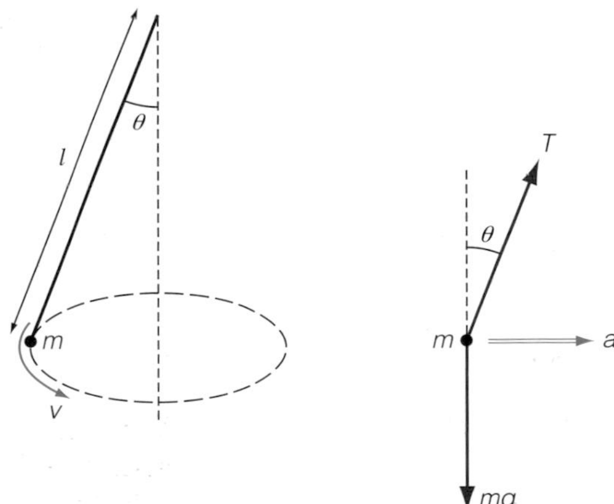

Figure 4.13 ▲

12 A metal ball of mass 2600 kg on the end of a 6 m cable is used to demolish buildings. The ball is pulled back until it is 3.0 m above its lowest position and it then swings down in an arc to strike the building.

Calculate the tension in the cable at the moment when the cable is vertical.

5 Electric fields

Figure 5.1 ▲
Michael Faraday

Michael Faraday was a brilliant scientist but not a strong mathematician and, perhaps because of this, he developed the idea of 'lines of force' to describe what you already know about magnetic fields. (Chapter 7 develops these ideas.) In this chapter you will learn about the forces between electric charges in electric fields, and how they too can be described using lines of force – another result of Michael Faraday's genius.

5.1 Fields in physics

Matter contains atoms and molecules. When you put a very large number of them together to make up a planet like the Earth, their **mass** produces what is called a **gravitational field**. When an apple (which also has mass) is released in this field, it falls to the ground because it feels a force – a gravitational force (Figure 5.2a). We do not 'see' what is causing the force, but we know it is there because the apple accelerates towards the Earth. Gravitational fields are studied further in Chapter 15.

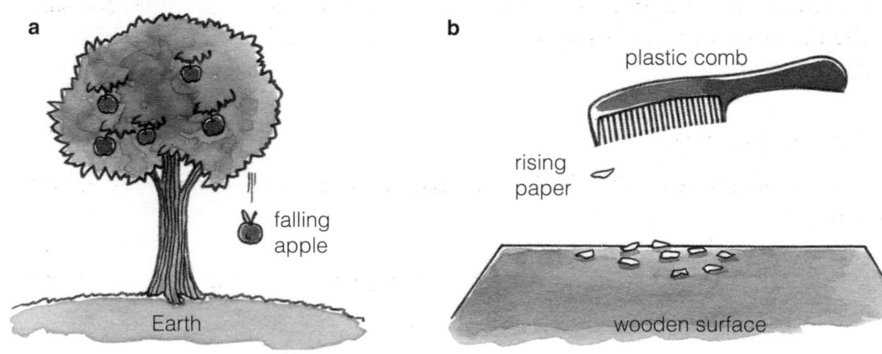

Figure 5.2 ▲
Invisible forces

The atoms and molecules in matter contain protons and electrons that carry electric **charge**. Normally the positive and negative charges in an object, like a plastic comb, exactly cancel each other out so that the comb is electrically neutral, i.e. has a total charge of zero. But sometimes a comb that has been rubbed on a cloth has a net charge and can pick up tiny pieces of paper (Figure 5.2b). We do not 'see' the force picking up the pieces of paper, but we know it is there because the paper accelerates towards the comb. The charge on the comb produces what is called an **electric field** and the force on the paper is called an electric or electrostatic force.

Both gravitational force and electric force are mysterious because there is no visible link between the two bodies that accelerate towards one another. The word **field** in physics is a general word used to describe regions in which these 'invisible' forces act.

Another type of field is produced by magnets and by electric currents. These **magnetic fields** are best understood by drawing magnetic **lines of force**, a technique you may have used in your earlier science studies. In exactly the same way, gravitational and electric fields can also be described using lines of force. Figure 5.3 is a familiar diagram of a magnetic field, shown by magnetic lines of force with arrows going from N to S on a bar magnet – a magnetic dipole. The arrows show the direction of the force on a N pole. The closeness (density) of the lines shows the strength of the field.

Tip

When asked: 'What is meant by a field in physics?', the answer is: 'A field is a region of space in which an object experiences a force'.

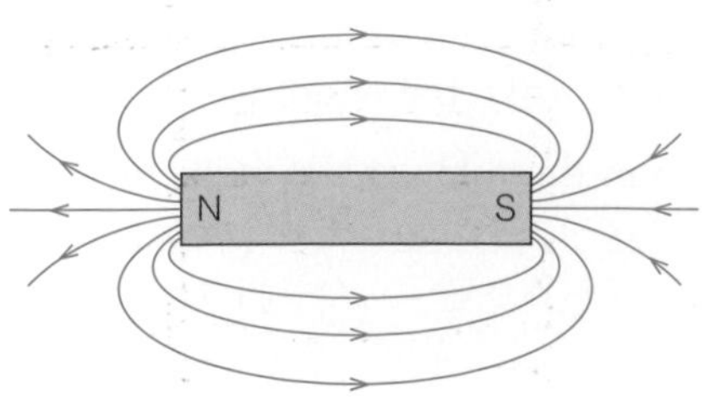

Figure 5.3 ▲
Magnetic lines of force

5.2 Electric forces

How does the comb pick up the pieces of paper in Figure 5.2b? The plastic comb is electrically charged: if it has extra electrons it is negatively charged. This negative charge repels negatively charged electrons in each tiny piece of paper, electrons that are forced away into the wooden surface. You may be surprised that both paper and wood can conduct electricity: they are poor conductors but not good insulators unless very, very dry. So the paper becomes positively charged and then, bingo! – the **unlike charges attract**. To confirm this fact, and to show that **like charges repel**, use charged insulating rods as shown in Figure 5.4.

There is an electric field in the region around a charged rod. If an object carrying a small electric charge Q feels a force F when placed close to the charged rod, we say the **electric field strength** produced there by the rod has a size F/Q.

If a piece of paper in Figure 5.2b was lifted by an electric force of 1.5×10^{-5} N and carried a charge of 5.0×10^{-8} C, then it must have been in an electric field of 300 N C^{-1}. Notice that a charge of 5.0×10^{-8} C means that (since the electronic charge $e = -1.6 \times 10^{-19}$ C) about 3×10^{11} electrons moved off the tiny piece of paper!

Electric field strength E is a **vector** quantity. The direction of E is the same as the direction of the electric force F, which is defined as the force on a *positive* charge.

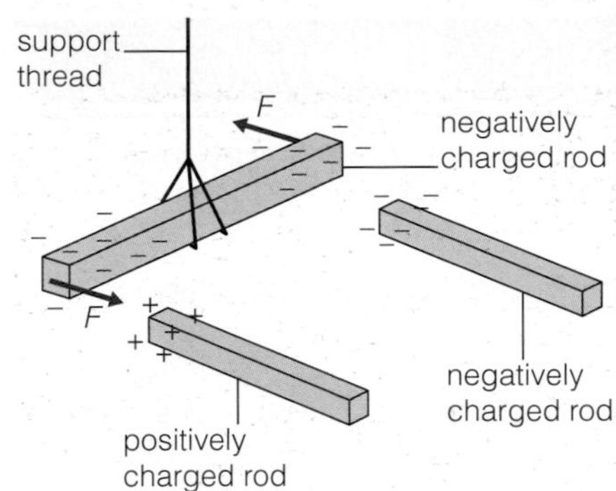

Figure 5.4 ▲
Unlike charges attract, like charges repel

Definition

Electric field strength

$$E = \frac{\text{electric force } F \text{ on a small charge}}{\text{small charge } Q}$$

$$E = \frac{F}{Q} = \text{force per unit charge}$$

The unit for electric field strength is the newton per coulomb, N C^{-1}.

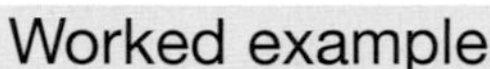

Worked example

A proton of charge $+1.6 \times 10^{-19}$ C is moving in an electric field of strength 500 N C^{-1}.

a) What electric force acts on it?

b) How does this force compare with the weight of a proton (proton mass = 1.7×10^{-27} kg)?

Answer

a) Rearranging $E = F/Q$ to find the force, we get:

$$F = EQ = 500\,\text{N C}^{-1} \times 1.6 \times 10^{-19}\,\text{C} = 8.0 \times 10^{-17}\,\text{N}$$

b) The weight of the proton $= mg$

$$= 1.7 \times 10^{-27}\,\text{kg} \times 9.8\,\text{N kg}^{-1}$$

$$= 1.7 \times 10^{-26}\,\text{N}$$

So the electric force here is 4.7×10^9 times bigger, that is nearly 5 billion times bigger, than the gravitational force. You can always ignore gravitational forces when dealing with electric forces on charged elementary particles.

5.3 Uniform electric fields

Two oppositely charged plates placed as shown in Figure 5.5 produce a **uniform** electric field between them. In a uniform electric field, the lines of force are equally spaced and parallel. (At the edges this is not quite true, but we will usually deal with the region of the field in which it *is* true.) When you move a small charged object around in a uniform electric field, the force on it remains constant. Because $E = F/Q$, this means that the value of E is the same everywhere.

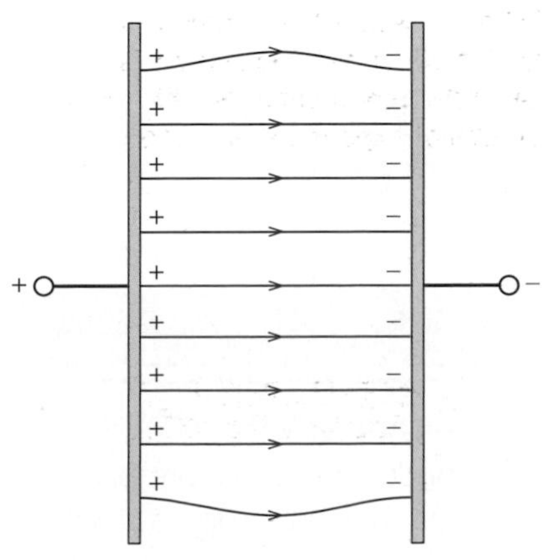

Figure 5.5 ▲
A uniform field

Experiment

To demonstrate that the electric field between two oppositely charged plates is constant

- Cut a test strip of aluminium foil, 20 mm by 5 mm.
- Attach it to the bottom of an uncharged insulating rod or plastic ruler.
- Hold the ruler vertical and lower the end with its foil into the space between the two charged plates.
- Move the ruler so that the aluminium foil is at different places between the charged plates.

If the field is uniform, the force on the foil will be constant and the aluminium foil will hang at the same angle to the vertical wherever it lies in the field. If the charges on the plates are reversed, the test strip swings to the same angle the other way.

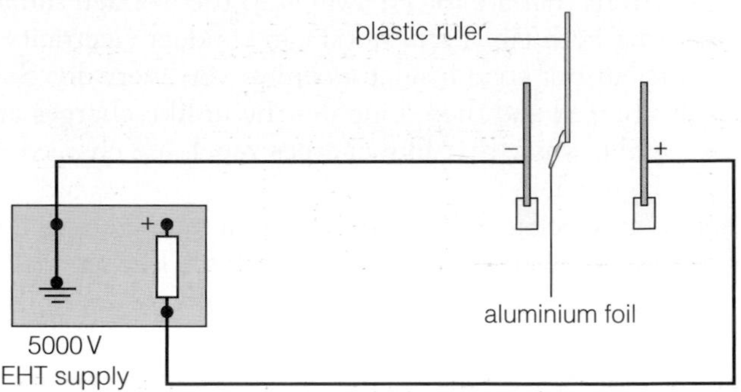

Figure 5.6 ▲
For safety the EHT supply should be strictly limited to less than 5 mA.

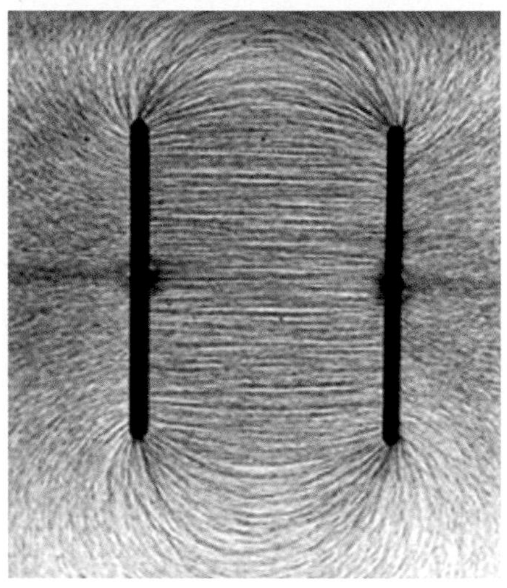

Figure 5.7 ▲
Demonstrating an electric field

Tip

If you can set up the sort of demonstration shown above, it is interesting to then replace the charged plates with point charges and other arrangements.

A more direct way of showing the uniform electric field is shown in Figure 5.7. In this photograph a potential difference has been applied to two metal plates that dip into a thin layer of insulating liquid. Lots of short pieces of fine thread are then sprinkled onto the liquid, and they line up end-to-end to show the shape of the field, rather as iron filings do in magnetic fields.

In a uniform electric field, the size of the electric field strength can also be expressed as

$$\mathbf{E} = \frac{\mathbf{V}}{\mathbf{d}}$$

where V is the (electric) potential difference between the oppositely charged plates or surfaces producing the electric field and d is the separation of the surfaces. V is a scalar quantity so this equation gives just the size of the E vector. The unit for E from this equation will be $\mathrm{V\,m^{-1}}$, apparently different from $\mathrm{N\,C^{-1}}$. However, the two units are equivalent; to show this, start with a volt:

$$\mathrm{V \equiv J\,C^{-1} \equiv N\,m\,C^{-1}}$$

so dividing the first and last by the metre gives $\mathrm{V\,m^{-1} \equiv N\,C^{-1}}$.

If the plates in Figure 5.7 were 30 mm apart and had a p.d. of 12 V across them, the strength of the electric field between them would be $12\,\mathrm{V}/0.030\,\mathrm{m} = 400\,\mathrm{V\,m^{-1}}$.

There seem to be two ways of charging objects and so two ways of producing electric fields. One is to use the frictional contact between two bodies, a nylon shirt and a woollen jumper or a Perspex ruler and a cloth; the other way is to use a simple battery or a d.c. supply. In both cases electric charge is produced, and it is electric charge that generates the electric field.

Experiment

Investigating the electric field between two charged plates

Two conducting plates are placed parallel to one another in a shallow solution of copper sulphate (Figure 5.8a). A metal probe is connected to the negative terminal of the supply via a digital (very high resistance) voltmeter. The plates are connected to a 12 V d.c. supply via a switch S.

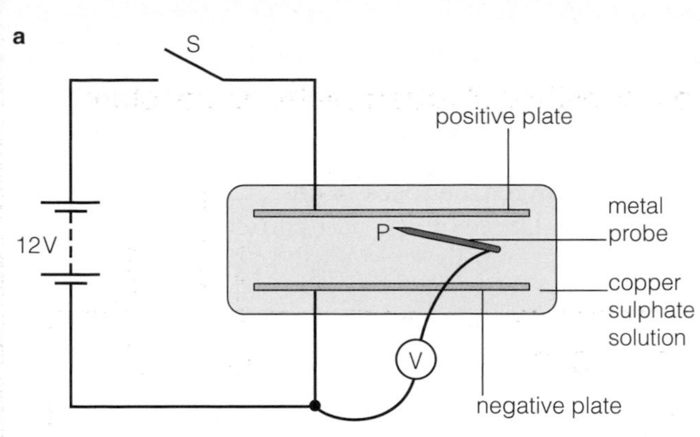

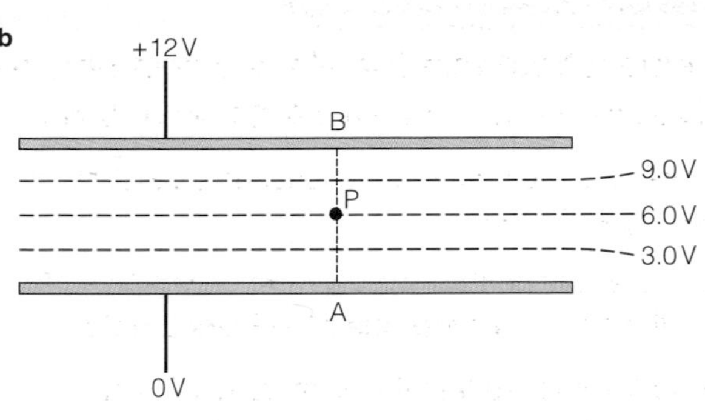

Figure 5.8 ▲

When the switch is closed there is a uniform electric field in the liquid between the plates. When the probe is dipped into the liquid at P the voltmeter registers the potential difference (p.d. or voltage) between the negative plate and the point in the liquid where the probe is placed. By dipping the probe at different places in the liquid, a 'map' of potential difference such as that in Figure 5.8b can be produced.

Safety note

Avoid contact with the copper sulphate solution and wash your hands if you happen to touch it.

Exercise

A careful set of voltmeter readings taken along the line AB, recording both the voltmeter reading and the distance AP, might produce the following results.

Here AB = 15.0 cm and V_{AB} = 12.0 V.

Distance AP *x*/cm	2.5	5.0	7.5	10.0	12.5
Voltage AP *V*/V	1.9	4.1	5.8	8.0	10.2

Table 5.1 ▲

Plot a graph of *V* (up) against *x* (along). The gradient of this graph you will see has units V m^{-1}. This **potential gradient** equates to the field strength. Find this gradient and hence work out the strength of the uniform electric field between the plates.

In the experiment above, the electric field is perpendicular to the lines along which the voltmeter readings are constant. These lines (as shown in Figure 5.8b) are called **equipotential lines**. As the volt is a name for a joule per coulomb, V ≡ J C^{-1}, the work done in moving a charged object *along* an equipotential line is zero. The work done *W* in moving an object of charge *Q* *between* two equipotential lines is

$$W = QV$$

where V is the difference in potential or voltage between the lines.

There is an electric field near to the Earth's surface. The upper atmosphere carries a permanent positive charge and the Earth's surface a permanent negative charge. It is not of a constant size day-to-day, but averages about 120 N C^{-1}. The field is directed downwards towards the Earth's surface and can be considered uniform. In thunderstorms the field builds up to much higher values, and lightning can result.

Figure 5.9 ▲
A sudden lightning discharge

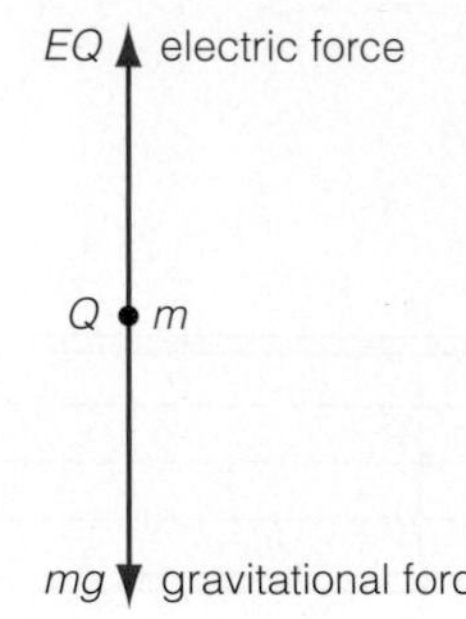

Figure 5.10 ▲

Worked example

A speck of dust of mass 2.0×10^{-15} g carries a negative charge equal to that of one electron. The dust particle is above a flat desert where the Earth's electric field is $150\,\mathrm{N\,C^{-1}}$ and acts downwards. Ignoring any forces on the dust particle resulting from air movements:

a) draw a free-body force diagram for the speck of dust,

b) calculate the resultant vertical force on the dust particle.

Answer

a) See Figure 5.10.

b) Weight of the dust particle $= mg = 2.0 \times 10^{-18}\,\mathrm{kg} \times 9.8\,\mathrm{N\,kg^{-1}}$
$= 1.96 \times 10^{-17}\,\mathrm{N}$

From $E = F/Q$ we get $F = EQ = 150\,\mathrm{N\,C^{-1}} \times 1.6 \times 10^{-19}\,\mathrm{C} = 2.40 \times 10^{-17}\,\mathrm{N}$

Resultant force $= 2.40 \times 10^{-17}\,\mathrm{N}$ *up* $- 1.96 \times 10^{-17}\,\mathrm{N}$ *down* $= 4.4 \times 10^{-18}\,\mathrm{N}$ *up*

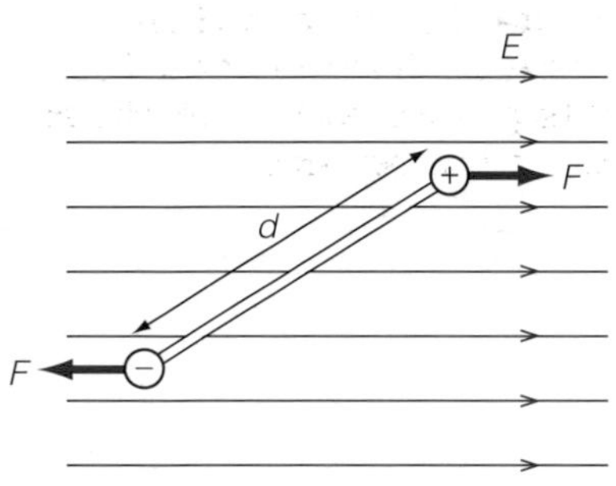

Figure 5.11 ▲
An electric dipole in a uniform field

Dipoles in uniform electric fields

An electric dipole consists of two equal but opposite charges $\pm Q$ separated by a distance d. When placed in a uniform electric field E as in Figure 5.11, the two charges experience equal-sized but oppositely directed forces.

The resultant force on the dipole is therefore zero, but the two forces will exert a twisting action on the dipole, which will try to swing it round until it lies along the electric field. This twisting effect lies at the heart of two widespread modern technologies: the microwave oven and liquid crystal displays.

A **microwave oven** operates by generating an electric field that reverses in direction several billion times per second. The water molecules in the food in the microwave oven are all tiny electric dipoles. These respond to this changing field by trying to align themselves with it, but the field is reversing so rapidly (at about 2500×10^6 Hz) that the molecules jostle against one another. The energy gained by the water molecules from the field is dissipated as internal energy (heat) in the surrounding food material.

Many electronic devices display alphanumeric information using **liquid crystals**. In a normal liquid such as water the dipole-like molecules are randomly orientated. Those in a liquid crystal tend to line up in response to one another's electric fields. An imposed external electric field can rotate the lines of liquid crystal dipoles. Figure 5.12 shows how this effect is exploited by influencing the behaviour of the seven small segments used in alphanumeric displays.

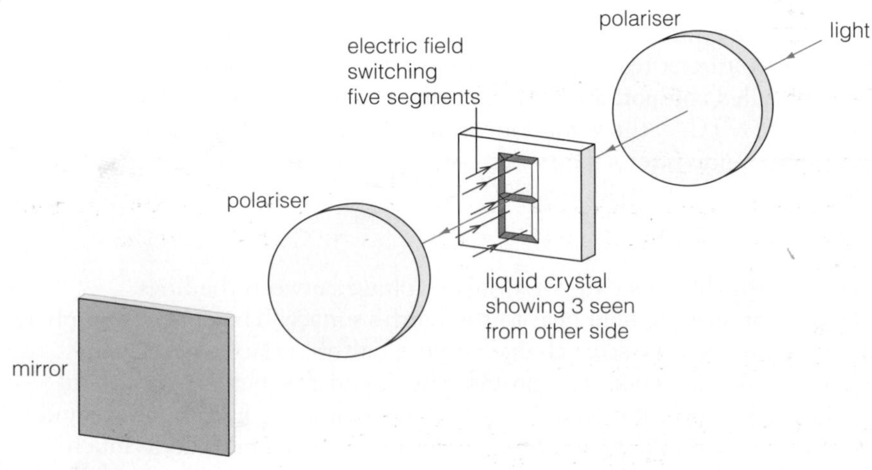

Figure 5.12 ▲
How a liquid crystal display works

An electric field can be switched on or off for each segment. The resulting change in alignment of molecules changes the response of the liquid crystal material in that segment to polarised light. When the electric field is on, the effect is to make the segment black, and we see the number or letter resulting from the pattern of black segments. Liquid crystal displays like this consume very little power, so your calculator can operate quite easily from the power generated by a small solar cell.

5.4 Radial electric fields

We've seen that the electric field between two parallel charged plates is a uniform field; what shape of electric field does a small 'point' charge like a single proton produce?

An isolated positive charge P produces an electric field like that shown in Figure 5.13. The radial lines of force show the direction of the field around P.

Tip

Of course, the electric field lines spread out in three dimensions from the charge – the diagram only shows them in two dimensions.

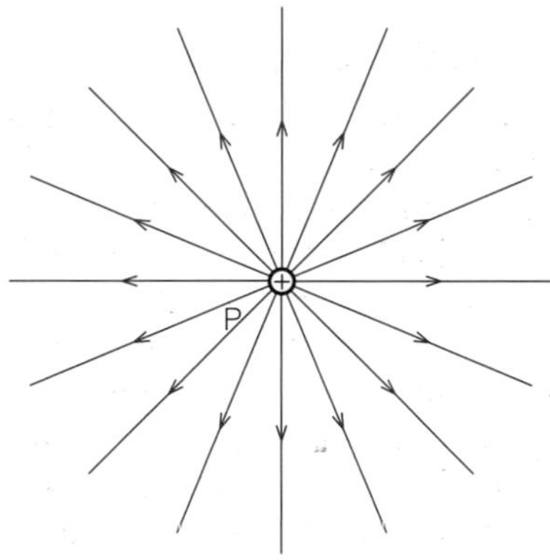

Figure 5.13 ▲
A radial electric field

The arrows on the lines of force would be reversed if the charge were negative. Equipotential lines in this case will be drawn as circles, but in three dimensions they will be spherical surfaces.

A charged sphere, or even an almost-spherical charged head, also produces a field that is radial. Figure 5.14 confirms this as the boy's hair shows clearly the direction of the electric field around his head.

Mathematically, the size of the electric field E a distance r from a point charge Q is:

$$E = \frac{kQ}{r^2}$$

The constant $k = 8.99 \times 10^9\,\mathrm{N\,m^2\,C^{-2}}$.

Figure 5.14 ▲
It makes your hair stand on end!

Worked example

Calculate the size of the electric field 1.0×10^{-10} m from the single proton at the nucleus of a hydrogen atom, the radius at which we find the electron.

Answer

The proton charge is 1.6×10^{-19} C, so

$$E = \frac{(8.99 \times 10^9\,\mathrm{N\,m^2\,C^{-2}}) \times (1.6 \times 10^{-19}\,\mathrm{C})}{(1.0 \times 10^{-10}\,\mathrm{m})^2} = 1.44 \times 10^{11}\,\mathrm{N\,C^{-1}}$$

Tip

The use of brackets around very large or very small numbers often helps to prevent you forgetting that, as here, you are squaring one of them.

5.5 Coulomb's law

The basic law describing the size of the force F between two point charges Q_1 and Q_2 is an inverse-square law that depends on the distance r between the charges and the size and sign of the charges Q_1 and Q_2:

$$F = \frac{kQ_1Q_2}{r^2}$$

with the constant k as above. This is called Coulomb's law after the scientist who discovered it.

You can see that if Q_1 is an isolated charge and Q_2 is a small charge placed near it, then

$$\frac{F}{Q_2} = \frac{kQ_1}{r^2} \quad \text{or} \quad E = \frac{kQ_1}{r^2}$$

That is, the way a radial electric field is described depends on Coulomb's force law between two point charges.

Experiment

Testing Coulomb's inverse-square law for the force between two charges

The forces are going to be very small, so a sensitive measuring device is required to register changes in the force as the distance between the charges is changed. Here an electronic top-pan balance that registers changes in mass to 0.01 g (vertical force changes of 0.0001 N) is used.

Figure 5.15 shows how the top-pan balance is used. Two conducting spheres, shown in blue, are charged, e.g. by flicking the negative lower sphere with a woollen cloth and using the positive terminal of a d.c. supply set at about 30 V to charge the upper sphere. Note the way in which the charged spheres are insulated so that they do not lose their charge during the experiment.

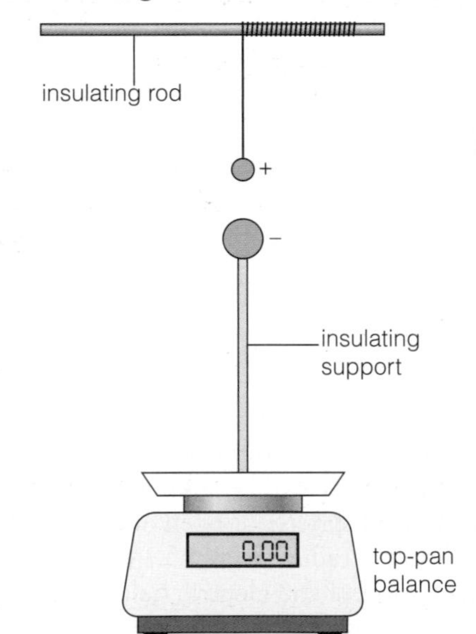

Figure 5.15 ▲

Readings m on the top-pan balance are now taken for different distances between the centres of the spheres. One way of recording how this distance changes without touching the charged spheres is to project a shadow of them onto a nearby screen. Calling the distance between the centres of the shadows y, a set of results like those in Table 15.2 might be collected.

m/g	0.11	0.38	0.62	0.83	1.14
y/cm	12.0	6.4	5.0	4.3	3.7

Table 5.2 ▲

As $m \propto F$, the force between the charged spheres, and $y \propto r$, the distance between the centres of the spheres, then if it is found that $m \propto 1/y^2$ we can say that $F \propto 1/r^2$.

Exercise

Look at the data collected in the experiment above to test Coulomb's law.

1 Show, by a non-graphical method, that m is approximately proportional to $1/y^2$.

2 Suggest difficulties that might arise in this experiment.

Worked example

Two point charges, of $+0.50\,\mu C$ and $-0.50\,\mu C$, form an electric dipole of length 0.12 m.

a) Calculate the size of the force between the charges. Comment on its size.
b) Calculate the electric field strength at the midpoint between the charges.
c) Sketch the shape of the electric field in the region of the dipole.

Answer

a) Substituting in $F = kQ_1Q_2/r^2$:

$$\Rightarrow F = \frac{(9.0 \times 10^9\,\mathrm{N\,m^2\,C^{-2}}) \times (30 \times 10^{-9}\,\mathrm{C}) \times (30 \times 10^{-9}\,\mathrm{C})}{(0.25\,\mathrm{m})^2}$$

$$= 1.3 \times 10^{-4}\,\mathrm{N}$$

This is a small force; it would support a mass of only $1.3 \times 10^{-4}\,\mathrm{N}/10\,\mathrm{N\,kg^{-1}} = 1.3 \times 10^{-5}\,\mathrm{kg}$ or 0.013 g.

b) Each charge will produce an equal-sized electric field at the midpoint. Both fields act from the positive towards the negative charge.

$$\text{So total } E = \frac{2 \times (9.0 \times 10^9\,\mathrm{N\,m^2\,C^{-2}}) \times (30 \times 10^{-9}\,\mathrm{C})}{(0.25\,\mathrm{m})^2}$$

$$= 8600\,\mathrm{N\,C^{-1}}$$

c)

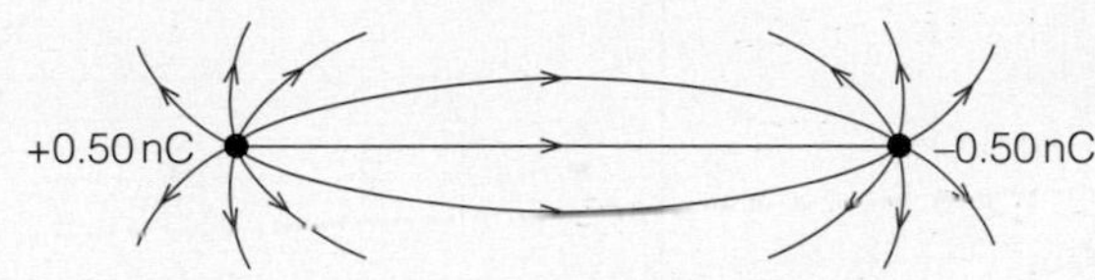

Figure 5.16 ▲

Tip

When commenting it is often a good idea to be quantitative, i.e. to offer comments that include numbers or even give calculated values.

REVIEW QUESTIONS

1 Which of the following is **not** a vector quantity?

A Electric field strength
B Electric charge
C Gravitational field strength
D Gravitational force

2 The electric field between two charged plates is known to be $650\,\mathrm{N\,C^{-1}}$.

If the distance between the plates is 12 cm, the potential difference between the plates must be:

A 54 V B 78 V C 4500 V D 7800 V

3 The field in Figure 5.17:

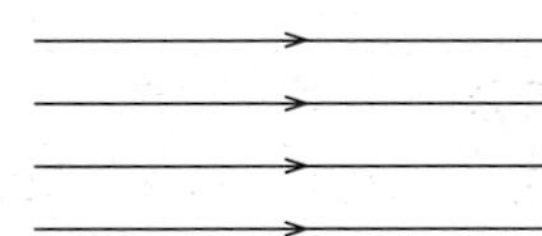

Figure 5.17 ▲

A could only be a gravitational field
B could only be a magnetic field
C could only be an electric field
D could be any one of gravitational, magnetic or electric fields

4 A balloon is rubbed on a woollen jumper and then 'stuck' to a classroom wall. Explain how it sticks to the wall.

5 A positively charged rod is lowered carefully into a cylindrical metal can without touching it. The can is standing on the ground, a conducting surface. Explain how the can becomes charged and sketch the electric field between the rod and the can.

6 a) Write out the full meaning of both $F = mg$ and $F = QE$.

b) State one difference between gravitational and electric fields.

7 A tiny oil drop of mass 4.0×10^{-15} kg is observed to remain stationary in the space between two horizontal plates when the potential difference between the plates is 490 V and their separation is 8.0 mm.

a) What is the electric field between the plates?

b) Draw a free-body force diagram of the drop showing the gravitational and electric forces acting on it.

c) Calculate the size of the forces and, by equating them, deduce the charge Q on the drop.

d) Comment on your value for Q.

8 A proton of mass 1.7×10^{-27} kg is moving to the right at a speed of $4.2 \times 10^5\,\mathrm{m\,s^{-1}}$. It enters an electric field of $45\,\mathrm{kN\,C^{-1}}$ pointing to the left.

How far will the proton travel before it comes momentarily to rest?

9 The maximum electric field there can be near the Earth's surface is about $3 \times 10^6\,\mathrm{N\,C^{-1}}$. For greater fields, sparking occurs.

What is the order of magnitude of the acceleration of a singly ionised nitrogen molecule of mass 4.6×10^{-26} kg in an electric field of $2.8 \times 10^6\,\mathrm{N\,C^{-1}}$?

10 In one type of ink-jet printer a tiny ink droplet of mass m carrying charge $-q$ is steered to its place on the paper by a uniform electric field. Figure 5.18 shows the path of this droplet.

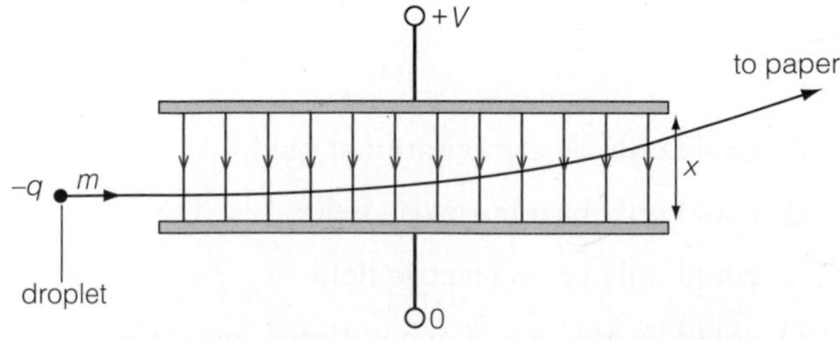

Figure 5.18 ▲

a) Give an expression for the acceleration of the droplet due to this electric field.

b) Describe the path of an uncharged droplet in this printer.

c) Suggest how different droplets can be steered to different places on the paper.

11 a) Show that the graph in Figure 5.19, of E against r in the region of an isolated proton, follows an inverse-square law.

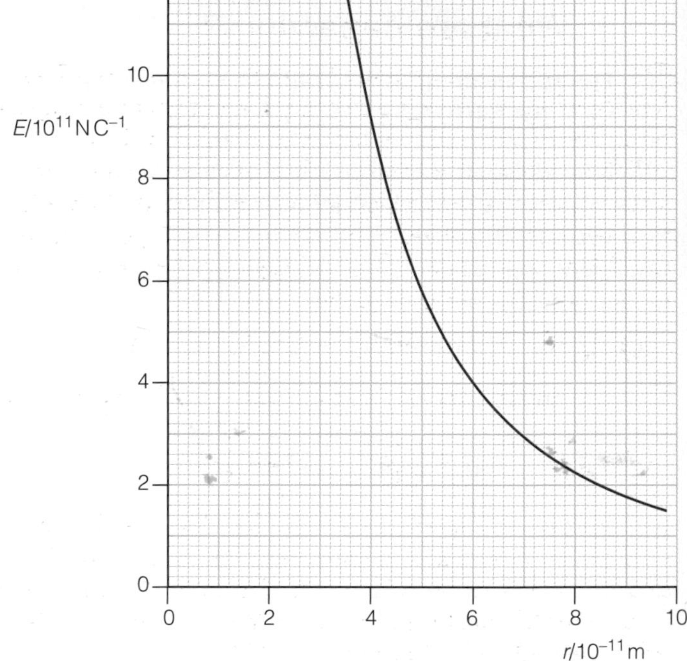

Figure 5.19 ▲

b) Calculate a value for the electric field constant k.

12 In the spark plug of a petrol-driven car there are two electrodes separated by a gap of about 0.67 mm. What p.d. must be applied across the electrodes to cause a spark in air? Air begins to ionise when the electric field is about $3 \times 10^6\,\mathrm{V\,m^{-1}}$.

13 The nucleus of a gold atom can be regarded as a sphere of radius 7.2×10^{-15} m. It contains 79 protons. What is the repulsive force on a proton that just 'touches' the gold nucleus? Comment on your answer.

6 Capacitance

You may not know much about capacitors, yet you probably make use of them quite often without realising it. You can find out about some of their uses in Section 6.3 'Capacitors in the real world'.

In this chapter you will also learn how capacitors work – how they store charge and how they can discharge through a resistor. This discharge is called an 'exponential decay' of charge. Exponential decays are very significant in many real-life situations, as are changes that involve exponential growth, for example in energy use and in population statistics.

6.1 What are capacitors?

The electrical circuit symbol for a capacitor is two parallel lines of equal length, which suggests (correctly) that connecting a capacitor in series with other electrical components produces a gap in the circuit.

When you apply a potential difference to the terminals of the capacitor, the sides or plates of the capacitor become charged. As a result there is an electric field between the plates – see Figure 5.5 on page 31. If the potential difference (p.d.) is V and the charge that this p.d. displaces from one side of the capacitor to the other is Q, then

$$Q \propto V \qquad \text{or} \qquad Q = (\text{a constant}) \times V$$

The constant, C, is called the **capacitance** of the capacitor.

$$Q = CV \qquad V = \frac{Q}{C} \qquad C = \frac{Q}{V}$$

For many capacitors the charge displaced or stored for a p.d. of 12 volts, for example, is very small, perhaps only a few microcoulomb (μC), so C is very small, of the order of μC per volt. The unit C V^{-1} is called a farad (symbol F) – yes, after Michael Faraday – so commonly capacitors have capacitances of μF or even nF (10^{-9} F) or pF (10^{-12} F).

Capacitors of more than 1000 μF usually have a plus sign marked on one end and a maximum voltage written on them, e.g. 25 V – see Figure 6.8 on page 43. It is very important that the + end is connected to the positive terminal of the supply and that the maximum p.d. is not exceeded.

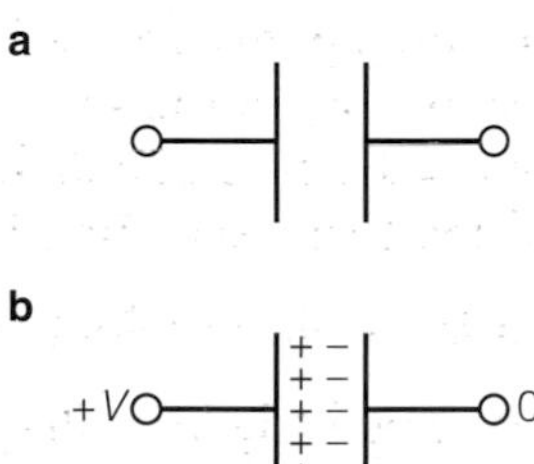

Figure 6.1 ▲
a) Capacitor symbol and b) displacement of charge when a p.d. is applied

Tip

'Queue equals sea vee' may be an easy way to remember this relationship, but make sure you remember what the symbols stand for: charge, capacitance and potential difference.

Worked example

A potential difference of 30 V displaces 4.1×10^{16} electrons from one plate of a capacitor to the other.

a) How much charge does this represent?

b) Calculate the capacitance of the capacitor.

Answer

a) The charge on an electron is -1.6×10^{-19} C.

$$\therefore \text{total charge displaced} = -1.6 \times 10^{-19}\,\text{C} \times 4.1 \times 10^{16} = -6.6 \times 10^{-3}\,\text{C}$$

b) $C = \dfrac{Q}{V} = \dfrac{6.6 \times 10^{-3}\,\text{C}}{30\,\text{V}}$

$= 2.2 \times 10^{-4}$ F or 220 μF or 0.22 mF

Experiment

Measuring the capacitance of a capacitor

Measuring C involves measuring both V and Q. The p.d. is easy – a digital voltmeter does the job. The charge is more difficult; you need to measure the charge that moves from one plate to the other. This is best done using the circuit shown in Figure 6.2, where the combination of a stopclock and a microammeter enables the charge to be found.

The charge is equal to the area under the current–time graph, Figure 6.3. If the current is constant, then the charge is just the product It, e.g. a trickle of 60 μA for 12 s means that 720 μC or 7.2×10^{-4} C of charge has been displaced from one plate to the other.

To keep the current constant, begin with the rheostat at its maximum value of 100 kΩ. Close the switch, start the clock and note the reading on the microammeter, then keep the current at that initial value by gradually reducing the resistance of the rheostat. You will find that you need to reduce this slowly at first and then more quickly. This may take a bit of practice. Another way of keeping the current constant is simply to use a constant current source, symbol ⦶⦶, in place of the cell and rheostat.

An alternative method of finding the charge on a capacitor by discharging it through a fixed resistor is described on pages 48 and 49. A coulombmeter is another option but this can only be used to measure very small charges – see page 44.

Safety note

If you are using a capacitor with a + sign marked on one end, be sure to connect this to the positive of the cells or battery.

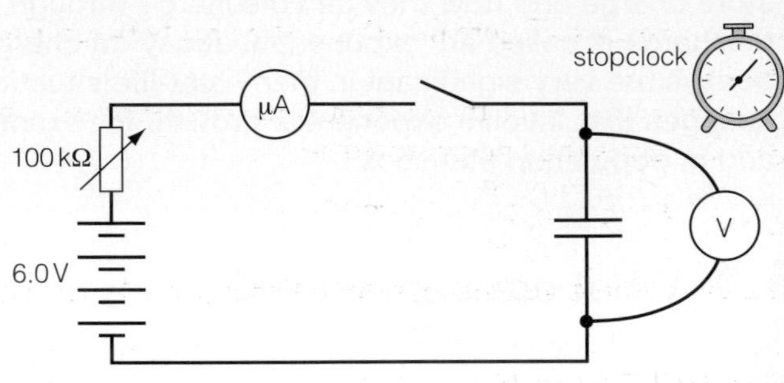

Figure 6.2 ▲

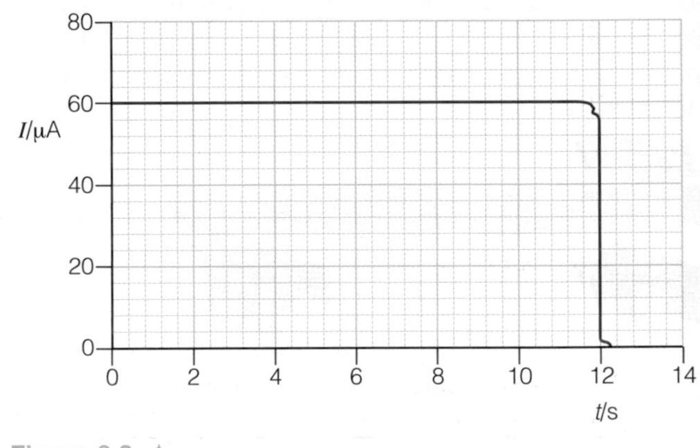

Figure 6.3 ▲

6.2 Energy storage by capacitors

Tip

Remember a volt is a joule per coulomb.

Capacitors store energy. The energy transferred, or work done, when a charge Q moves across a potential difference V is given by $W = VQ$.

When charging capacitors there is a problem in calculating the energy transferred from this formula: the p.d. varies as the capacitor charges!

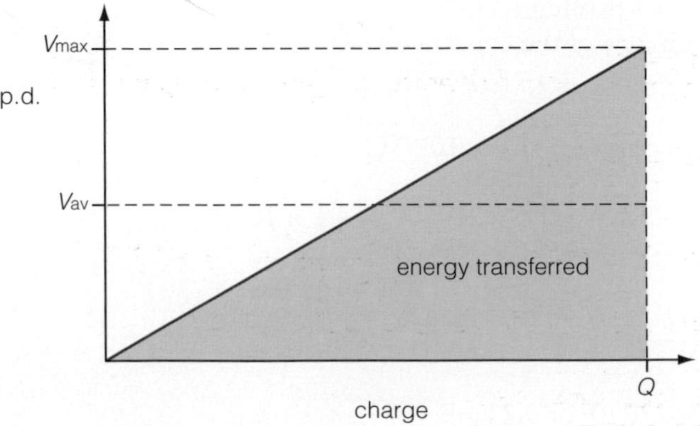

Figure 6.4 ▲

Fortunately, V is proportional to Q (Figure 6.4), so the *average* p.d. is exactly half the final p.d., and we can use

$$V_{av} = \frac{1}{2}V_{max}$$

when calculating the energy transferred.

So $W = V_{av}Q = \frac{1}{2}V_{max}Q$

Referring back to Figure 6.4, this is the area between the graph line and the Q-axis.

Using simply V for the maximum p.d. and substituting for Q or V using $Q = CV$ gives the energy stored as:

$$W = \frac{1}{2}VQ = \frac{1}{2}\frac{Q^2}{C} = \frac{1}{2}CV^2$$

The capacitor stores this transferred energy as electric potential energy.

Tip

Try to remember formulas like $W = V_{av}Q = ½V_{max}Q$ in words as 'energy transferred equals average potential difference times charge stored' and 'energy stored equals half maximum potential difference times charge stored'. This will help your understanding.

Worked example

Figure 6.5 shows how the potential difference across a capacitor changes as it is charged.

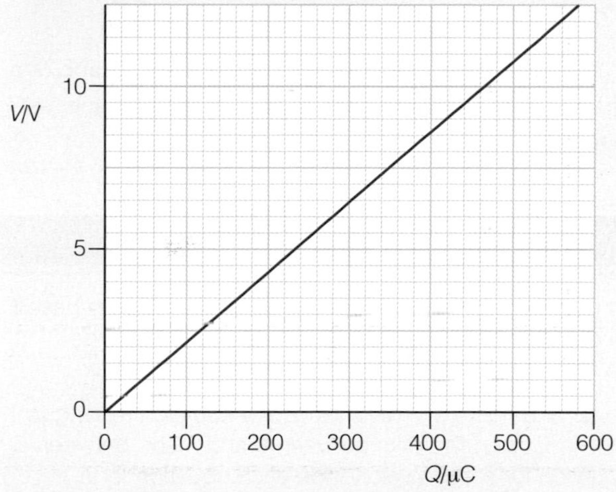

Figure 6.5 ▲

a) What is the capacitance of the capacitor?

b) Calculate the energy stored in the capacitor when it has a charge of Q/μC: 100, 200, 300, 400, 500.

c) Sketch a graph of energy stored against charge.

d) What shape would you expect a graph of stored energy against potential difference to have?

Answer

a) $C = \frac{Q}{V} = \frac{460 \times 10^{-6}\,\text{C}}{10\,\text{V}} = 46\,\mu\text{F}$ (*Not* the gradient of the graph!)

b) Energy stored $W = \frac{1}{2}QV = \frac{1}{2}\frac{Q^2}{C}$

For 100 μC the energy stored $= \frac{1}{2}\frac{(100 \times 10^{-6}\,\text{C})^2}{(46 \times 10^{-6}\,\text{F})} = 1.1 \times 10^{-4}\,\text{J}$

Other values worked out similarly are shown in Table 6.1.

Q/μC	100	200	300	400	500
W/mJ	0.11	0.43	0.98	1.74	2.72

Table 6.1 ▲

Tip

Don't try to use milli- or micro- *inside* calculations – always go to standard form.

And check that the units work out, remembering that $\text{V} \equiv \text{J C}^{-1}$.

c)

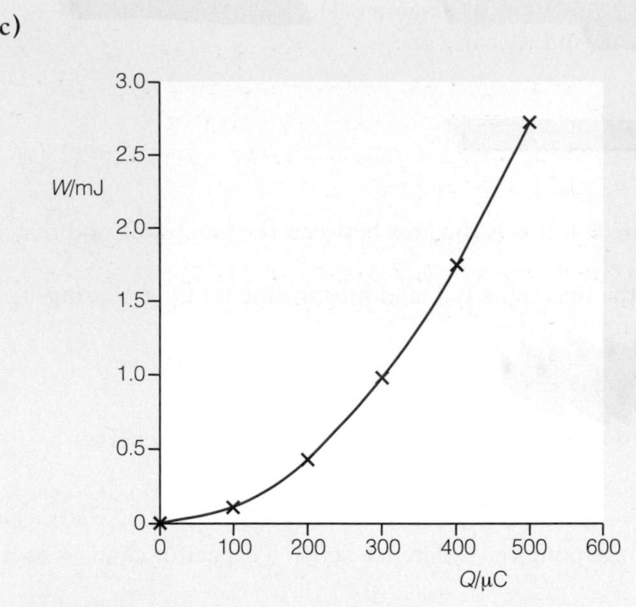

Figure 6.6 ▲

d) As W is proportional to Q^2, the graph shows the characteristic shape of a y against x^2 graph. A graph of W against V will show the same shape, as W is also proportional to V^2.

Exercise

There is an analogy between the behaviour of springs and capacitors. This is based on the similarity between the equation defining capacitance, $V = Q/C$, and the equation describing the behaviour of a spring that obeys Hooke's law, $F = kx$.

Look through this chapter so far and make a vertical list of any equation relating to the properties of a capacitor. Then, using the links between V and F and between Q and x, write an analogous equation for a spring alongside each equation for a capacitor. Be careful, as the analogy links $1/C$ to k, not C to k.

Experiment

Investigating the efficiency of energy storage in a capacitor

The energy stored in a 10 000 μF capacitor can be transferred to gravitational potential energy (*GPE*) by discharging the capacitor through a small electric motor. As the capacitor discharges, the motor raises a small mass. The set-up in Figure 6.7a, with the circuit shown in Figure 6.7b, is used.

The mass to be lifted must be quite small – a ball of plasticine of mass $m = 8.0$ g is suitable. Wind the

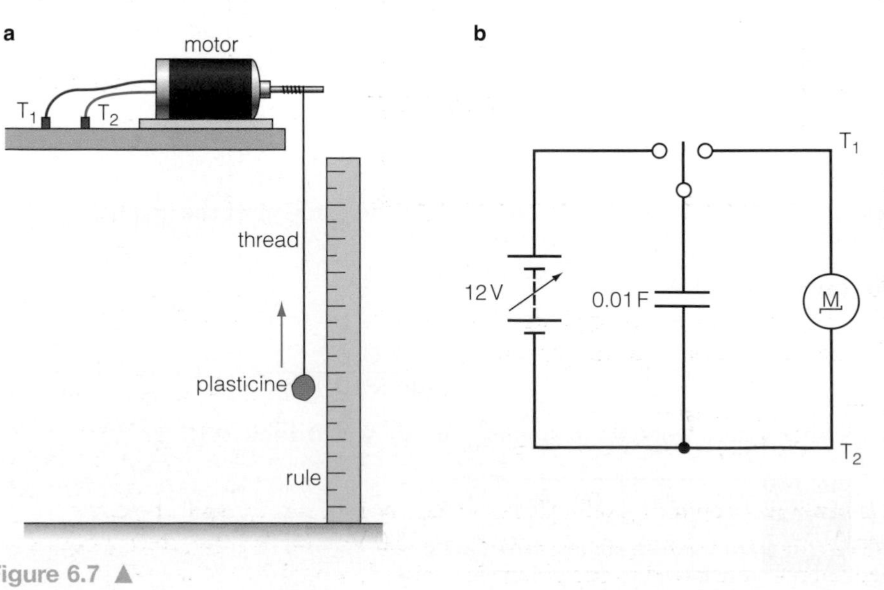

Figure 6.7 ▲

cotton thread around the spindle of the motor and arrange a vertical rule to measure how far (h) the plasticine is lifted. The capacitance of the capacitor must be large – here it is an electrolytic capacitor of 10 000 μF or 0.01 F – so that $\frac{1}{2}CV^2$ is large enough to provide a large enough gain in *GPE* (*mgh*) of the plasticine to give a measurable height h.

Take a series of readings of V, the charging voltage, and h as the capacitor is first charged from the d.c. power supply, and then as it is discharged through the motor. Table 6.2 shows a typical set of readings.

V/V	7.5	9.0	10.5	12.0
h/m	0.27	0.40	0.53	0.71

Table 6.2 ▲

Plot a graph of *mgh* (in joules) against $\frac{1}{2}CV^2$ (in joules) and comment on the form of the graph.

Calculate values for the efficiency η of the energy transfer and suggest where the 'lost' energy has gone.

Safety note

Be sure to connect the + marked on the capacitor to the positive terminal of the supply, and that the largest voltage you use is less than the maximum voltage marked on the capacitor.

6.3 Capacitors in the real world

Figure 6.8 shows a 2200 μF capacitor, which may be used in a camera flash unit.

Figure 6.8 ▲
A capacitor

Worked example

A capacitor is fully charged by connecting it to a 6.0 V d.c. supply and the stored energy is then transferred to a discharge tube, creating a flash. The capacitance of the capacitor is 2200 μF. Calculate the energy stored in the capacitor and the energy given to the charge by the d.c. supply. Sketch a Sankey diagram to show the operation of the flash unit.

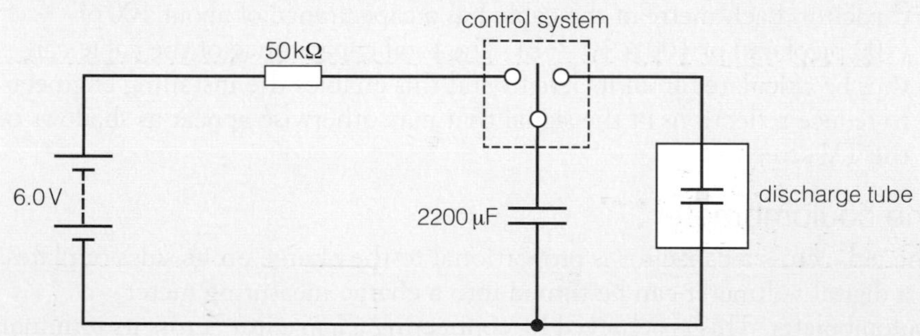

Figure 6.9 ▲

Answer

Using $Q = CV$, the stored charge is $6.0\,\text{V} \times 2200 \times 10^{-6}\,\text{F} = 0.0132\,\text{C}$

So the energy stored in the capacitor is

$$W = \tfrac{1}{2}VQ = \tfrac{1}{2} \times 6.0\,\text{V} \times 0.0132\,\text{C}$$
$$= 0.0396\,\text{J or } 0.040\,\text{J to 2 sig. fig.}$$

The energy given to the 0.0132 C of charge as it passes through the d.c. supply is 6.0 joules for each coulomb – the 6.0 V or $6.0\,\text{J}\,\text{C}^{-1}$ of the supply. So the energy transferred to the charge is $0.0132\,\text{C} \times 6.0\,\text{J}\,\text{C}^{-1} = 0.0792\,\text{J}$.

This energy transfer is exactly twice 0.0396 J, the energy stored. The other half of the energy is transferred to internal energy in the connecting wires.

The Sankey diagram is therefore as shown in Figure 6.10.

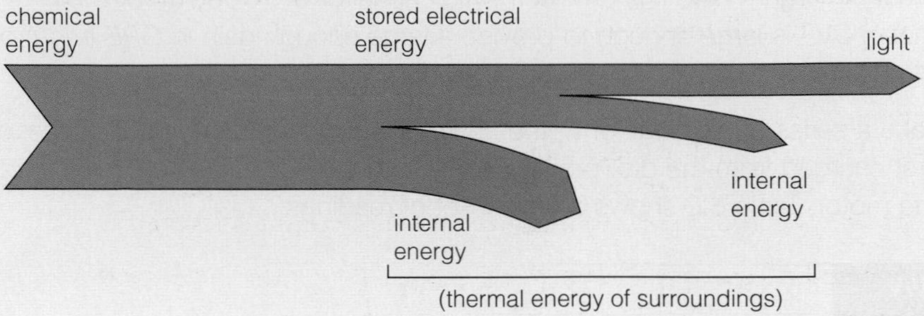

chemical energy
stored electrical energy
light
internal energy
internal energy
(thermal energy of surroundings)

Figure 6.10 ▲

For the camera flash unit in the Worked example above, if the discharge occurs in 0.10 ms, the average power transfer is $(0.040\text{ J})/(0.10 \times 10^{-3}\text{ s}) = 400\text{ J s}^{-1}$ or 400 W. This is a very bright short flash.

Ordinary capacitors cannot be used to store large quantities of electrical energy: it would take a whole room full of them to supply a hundred watts for an hour. But recent developments have produced **super-capacitors** that can store as much as 8000 J, and a bank of such super-capacitors can be used, for example, to back up electrical systems in hospitals in the event of a power cut.

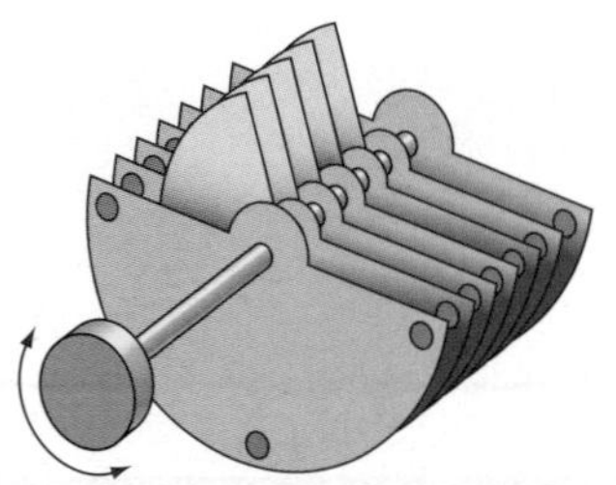

Figure 6.11 ▲
A variable capacitor

Figure 6.11 shows a common device – a **variable capacitor**. One like this is turned when you search for a new frequency on a radio with a tuning knob.

Two other places where capacitors are found in the home are below the keys on a PC and inside the cable from an aerial to a TV set.

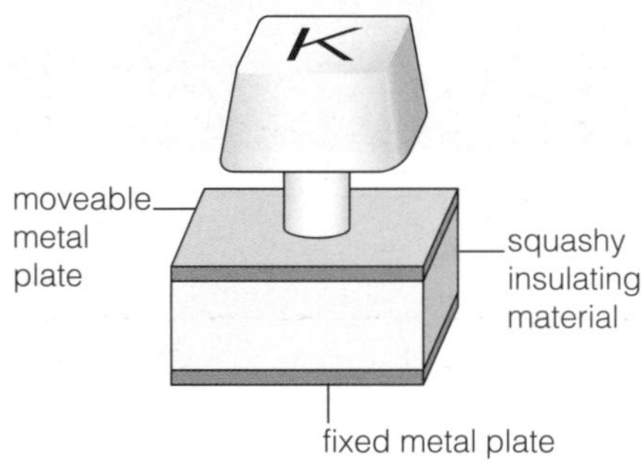

Figure 6.12 ▲
The capacitor key

- There is a capacitor under each key on the keyboard of a PC, as shown in Figure 6.12. When a key is pressed the two sides or plates of the capacitor become a little closer together. In so doing the capacitance changes slightly; the computer scans all the keys and deduces which one has been pressed.
- The cable to a TV set is called a coaxial cable. It acts as a cylindrical capacitor. Each metre of the cable has a capacitance of about 100 pF (100 picofarad or 100×10^{-12} F). The total capacitance of the cable can thus be calculated from its length and this enables the installing engineer to reduce reflections of the signal that may otherwise appear as shadows on the TV screen.

The coulombmeter

The p.d. across a capacitor is proportional to the charge on its sides or plates, so a digital voltmeter can be turned into a charge-measuring meter – a coulombmeter. This is achieved by connecting a capacitor across its terminals (Figure 6.13).

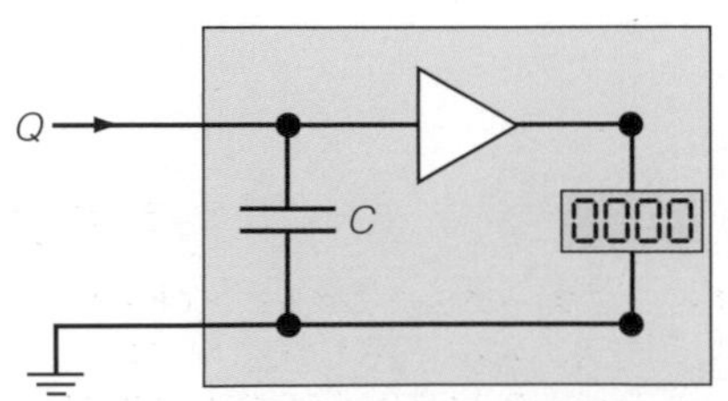

Figure 6.13 ▲
Converting a digital voltmeter to a coulombmeter

Assume the capacitance of the capacitor is 4.7 μF. If this capacitor is initially uncharged, charges from 0 up to 940 nC (940 nanocoulomb or 940×10^{-9} C) delivered to it will produce potential differences from 0 to 0.20 V. The scale of the meter can, using suitable amplification, be adjusted to read nanocoulombs directly. Such meters can only measure very small charges – up to about 2000 nC or 2 μC.

6.4 Exponential change

What do we mean by an exponential change?

Exponential growth is when something gets larger by the same fraction or proportion in each time interval.

For example, the number of bacteria in an opened tin of meat at room temperature doubles about every 6 hours. Graphs illustrating exponential growth such as this show very rapid increase – see Figure 6.14a. In real situations, for example the bacteria growth, something will intervene to stop the growth, i.e. the exponential change does not continue forever.

Exponential decay is when something gets smaller by the same fraction or proportion in each time interval.

For example, the number of radioactive nuclei in a sample of sodium-24 halves every 15 hours. Graphs illustrating true exponential decay get closer and closer to the time axis, but never quite reach it – see Figure 6.14b. In real situations, such as the decay of sodium-24, there will come a time when only one and then no radioactive nucleus remains.

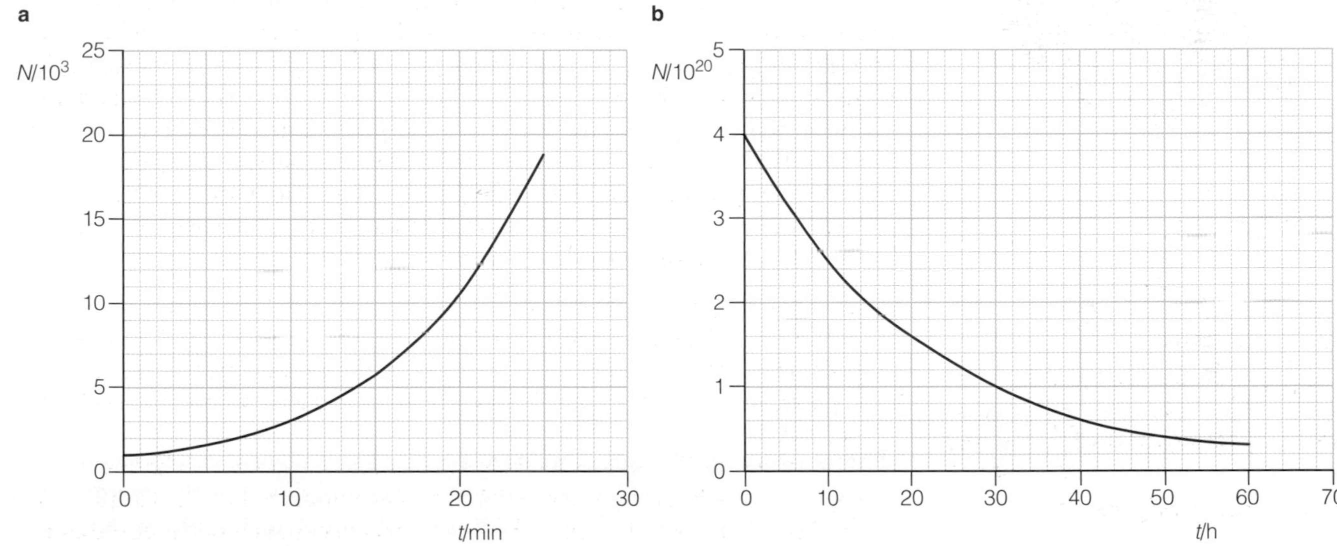

Figure 6.14 ▲
a) Exponential growth and b) exponential decay

Worked example

Show that the graph in Figure 6.14b is showing exponential decay.

Answer

The graph will be showing exponential decay if the **ratios** of values of N after successive equal time intervals are constant. From the graph, the values of $N/10^{20}$ at 10-hour intervals are as given in Table 6.3.

t/h	0	10	20	30	40	50
$N/10^{20}$	4.0	2.5	1.6	1.0	0.6	0.4

Table 6.3 ▲

Tip

It is sometimes tempting to look for successive ratios that are 2 : 1, i.e. 'half-lives', but very often only one or two half-lives are shown. This approach can still be used, however, by measuring the time for the y-value to halve starting at different initial y-values and looking to see if the time to halve is always (about) the same.

The ratios of successive values of N in Table 6.3 are:

$$\frac{4.0}{2.5}, \frac{2.5}{1.6}, \frac{1.6}{1.0}, \frac{1.0}{0.6}, \frac{0.6}{0.4}$$

To 2 sig. fig. these are:

1.6, 1.6, 1.6, 1.7, 1.5

The curve is showing exponential decay.

Exercise

Another way of testing to see that a set of data is showing exponential change is to plot a graph with a logarithmic y-axis and a linear x-axis. (This is explained further in Section 6.6.). It is usual to plot natural logarithms – ln on your calculator, not log. If data show exponential decay, the graph will be a straight line with a negative gradient.

Plot a suitable graph using the data from the Worked example above to show that the decay is exponential, and determine the gradient.

6.5 Capacitor discharge

When a capacitor of capacitance C discharges through a resistor of resistance R, the current I in the resistor is proportional to the charge Q remaining on the capacitor, i.e. the rate of flow of charge is proportional to the remaining charge.

Expressed algebraically:

$$I \propto Q \qquad \text{or} \qquad \frac{-\Delta Q}{\Delta t} \propto Q$$

$$I = \text{constant} \times Q \qquad \frac{-\Delta Q}{\Delta t} = \text{constant} \times Q$$

This is a very special type of relationship: *the less you have, the more slowly you lose it*. The discharge of a capacitor gets slower and slower as the charge left on the capacitor gets less and less, and it is never complete. Graphs of capacitor discharge turn out to be **exponential decay** graphs. The constant in the above relationship is $1/RC$ and has the unit s^{-1}. So that

$$I = \frac{Q}{RC}$$

The quantity RC is called the **time constant** for the decay. The larger the time constant, the longer the decay takes (see Figure 6.18 on page 48). The 'half life' $t_{\frac{1}{2}}$ of the decay – the time for the charge to decrease to one half of its value – is:

$$t_{\frac{1}{2}} = RC \ln 2 = RC \times 0.693$$

Figure 6.15 shows a typical graph of how the charge Q varies with time in a discharge of this kind. Testing the graph to show that it represents exponential decay is easy here, as three half-lives can be read off: 128 to 64; 64 to 32; and 32 to 16, each taking 28 ms. Try to read off some other half-lives starting at different points.

Here $RC = t_{\frac{1}{2}}/0.693 = 28 \times 10^{-3}\,\text{s}/0.693 = 40.4 \times 10^{-3}\,\text{s}$ or about 40 ms to 2 sig. fig. So, for example, if $C = 47\,\mu\text{F}$ then R must be $860\,\Omega$.

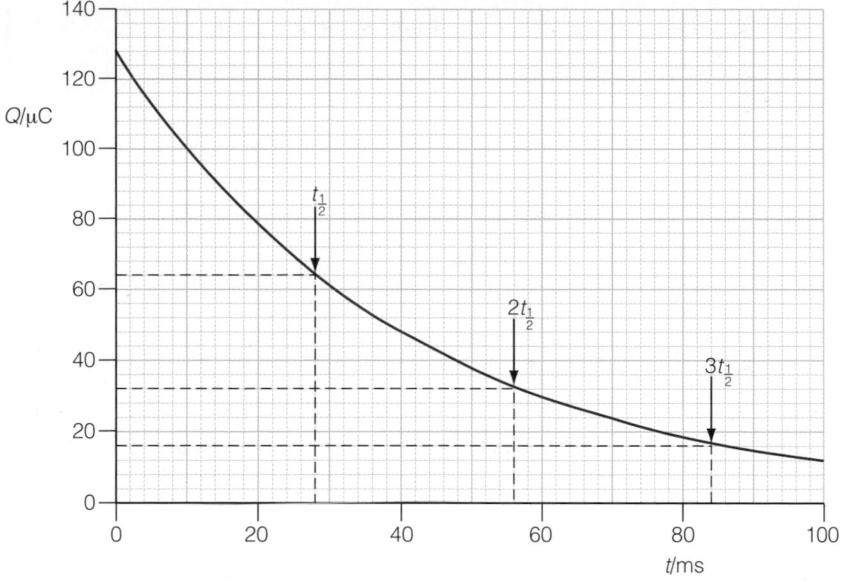

Figure 6.15 ▲
Exponential capacitor discharge

Tip

Amazingly, and we won't attempt any higher mathematics, a tangent drawn at $t = 0$ on a curve like Figure 6.15 cuts the time axis at RC. Does it work here?

Worked example

Show that the unit of RC – the ohm times the farad – is the second.

Answer

An ohm is a volt per ampere: $\Omega \equiv \mathrm{V\,A^{-1}}$ (from $V = IR$)
A farad is a coulomb per volt: $\mathrm{F} \equiv \mathrm{C\,V^{-1}}$ (from $Q = CV$)

$$\therefore\ \Omega \times \mathrm{F} \equiv (\mathrm{V\,A^{-1}}) \times (\mathrm{C\,V^{-1}}) = \mathrm{A^{-1}} \times \mathrm{C}$$

But $\mathrm{C} \equiv \mathrm{A\,s}$ (from $Q = It$), so the unit of $RC = (\mathrm{A\,s}) \times (\mathrm{A^{-1}}) = \mathrm{s}$, the second.

Experiment

Studying the discharge of a capacitor

Charge a 470 μF capacitor to 6.0 V, and discharge it through a resistor and a microammeter of total resistance R.

Take readings every 20 s. Plot a graph of the discharge current I against time t and deduce a value for R.

The readings in one experiment of this kind were as in Table 6.4.

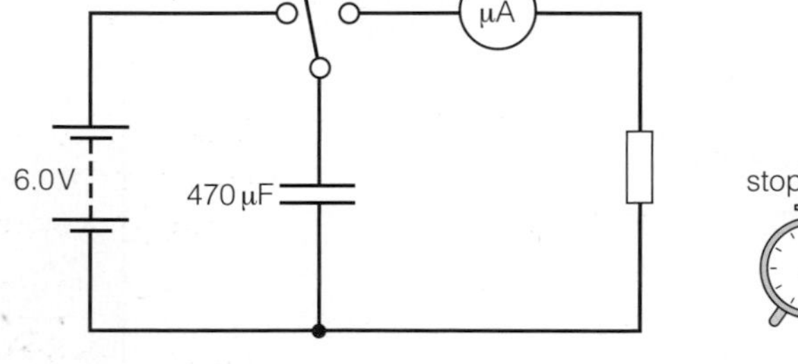

Figure 6.16 ▲

t/s	0	20	40	60	80	100
I/μA	59	39	26	17	11	7

Table 6.4 ▲

The graph of I against t is shown in Figure 6.17.

The half life $t_{\frac{1}{2}}$ can be read off starting at several places, e.g. as the current falls from 50 μA to 25 μA, or from 30 μA to 15 μA.

From these two readings

$t_{\frac{1}{2}} = (42 - 8)\,\text{s} = 34\,\text{s}$ and

$t_{\frac{1}{2}} = (67 - 33)\,\text{s} = 34\,\text{s}$

$$\therefore RC = \frac{t_{\frac{1}{2}}}{0.693} = \frac{34\,\text{s}}{0.693} = 49\,\text{s}$$

Hence, as $C = 470\,\mu\text{F}$,

$$R = \frac{49\,\text{s}}{470 \times 10^{-6}\,\text{F}} = 104\,000\,\Omega \text{ or } 104\,\text{k}\Omega.$$

Notice how the graph of I against t has exactly the same shape as one of Q against t: you should have expected this as $Q \propto I$.

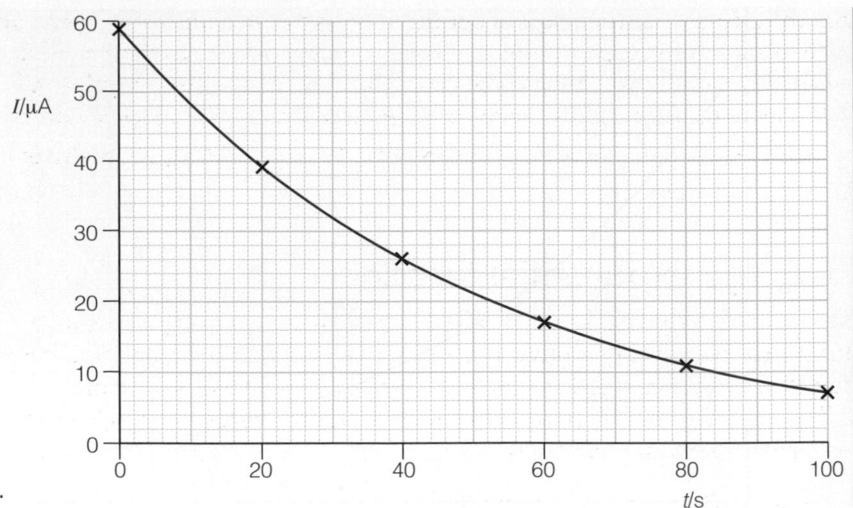

Figure 6.17 ▲

Worked example

Explain what the discharge curve of I against t would be like in the above experiment if the circuit resistance were doubled, and sketch the curve.

Answer

Referring to the circuit diagram, Figure 6.16, the capacitor is initially charged to a p.d. of 6.0 V. When the switch is closed to connect the capacitor across the resistor, this p.d. of 6.0 V is across the resistor, and so the initial current is given by $I = V/R$, i.e. 6.0 V divided by R.

If R is doubled the initial current will be halved, so in this case it will be about 30 μA.

But the total charge that must flow from the capacitor as it discharges is unchanged.

This charge is represented by the area under the I–t graph, so in order to get the same area the discharge must be slower and last longer.

The new discharge curve is shown in blue in Figure 6.18.

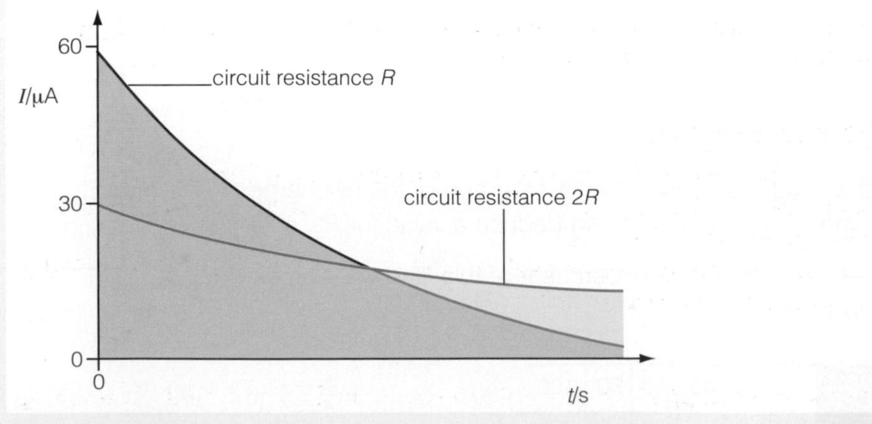

Figure 6.18 ▲

Tip

In situations like this, think of the physics. Don't get bogged down in the mathematics.

Estimating areas

The total initial charge on a capacitor that is discharging is equal to the total area under the I–t curve. For the decay curve in Figure 6.17, this is best estimated by drawing a triangle of height 60 μA. You now must draw the hypotenuse so that the area of your triangle is about the same as the area under the curve. You have to allow for that part of the curve that is not shown off to the right, but because this is only an estimate you can't expect to get the area exactly right.

In Figure 6.17 we can estimate that the total charge will be somewhere between $\frac{1}{2}(60 \times 10^{-6}\,\text{A} \times 90\,\text{s})$ and $\frac{1}{2}(60 \times 10^{-6}\,\text{A} \times 100\,\text{s})$, i.e. between 2.7×10^{-3} C and 3.0×10^{-3} C.

Tip

Notice that the two estimates differ by about 10%, which is quite acceptable when estimating a quantity. Counting squares takes a lot longer and is usually **not worth it** during a timed test or examination.

6.6 The exponential function

Exponential decay graphs are described mathematically using a special 'function'. You have met other functions, like sin x and cos x, and you know that they have characteristic graphs – they are wavelike or sinusoidal.

The exponential function is written as e^x, and is pronounced as 'ee to the ex' (strictly 'ee to the power ex').

e is a number: e = 2.718 to 4 sig. fig.

A graph of $y = e^x$ is an exponential graph. If x is a positive number the graph shows exponential growth like Figure 6.14a, and if x is a negative number the graph shows exponential decay like Figure 6.14b.

A graph of $y = y_0 e^{-kt}$ (where k is a constant) is an exponential decay with time t. It starts on the y-axis at $y = y_0$ when $t = 0$, and is a 'concave' curve that approaches but never reaches the t-axis.

The natural logarithm of e^x is simply x, i.e. $\ln e^x = x$, so if we take logarithms of each side of the equation for exponential decay, we get $\ln y = \ln y_0 - kt$. This means that a graph of $\ln y$ against t will be a straight line with a gradient $-k$. For capacitor discharge, $k = 1/RC$.

Summarising the mathematical relationships for capacitor discharge, we therefore have:

$$I = \frac{Q}{RC}$$

$$Q = Q_0 e^{-t/RC} \Rightarrow \ln Q = \ln Q_0 - \frac{t}{RC}$$

$$I = I_0 e^{-t/RC} \Rightarrow \ln I = \ln I_0 - \frac{t}{RC}$$

$$V = V_0 e^{-t/RC} \Rightarrow \ln V = \ln V_0 - \frac{t}{RC}$$

All these logarithmic graphs have a gradient $-1/RC$ and it can be shown that $t_{\frac{1}{2}} = RC \ln 2$ ($= 0.693RC$, as before).

Worked example

Refer to the Experiment on page 47. Using the data in Table 6.4, plot a graph of the natural logarithm of the current, $\ln(I/\mu\text{A})$, against time t and deduce a second value for R.

Discuss whether this second value provides a more reliable value for R than that calculated on page 48.

Answer

The values of $\ln(I/\mu\text{A})$ to 2 sig. fig. are given in Table 6.5, and the resulting graph is shown in Figure 6.19.

t/s	0	20	40	60	80	100
ln (I/μA)	4.1	3.7	3.3	2.8	2.4	1.9

Table 6.5 ▲

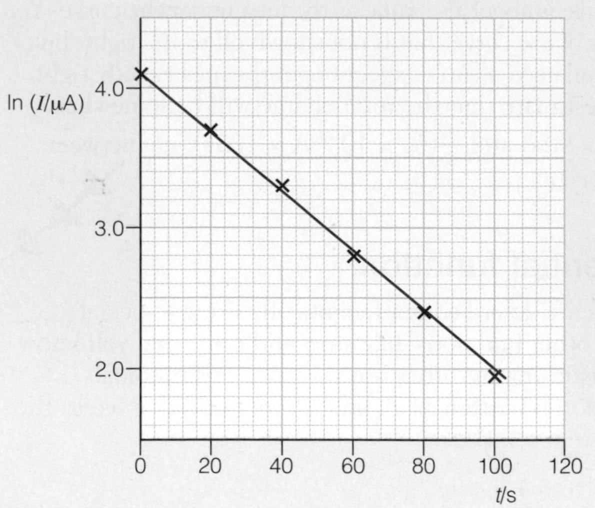

Figure 6.19 ▲

Here $\ln I = \ln I_0 - t/RC$ because $I = I_0 e^{-t/RC}$, and so the gradient of the graph against t is $-1/RC$.

$$\text{Gradient} = \frac{(2.0 - 4.1)}{100\,\text{s}} = -0.021\,\text{s}^{-1}$$

$$\therefore RC = \frac{1}{0.021\,\text{s}^{-1}} = 47.6\,\text{s}$$

Substituting $C = 470 \times 10^{-6}\,\text{F} \Rightarrow R = 101\,000\,\Omega$ or $101\,\text{k}\Omega$.

This value is more reliable than that on page 48, as it has been deduced from a straight line graph using all the experimental results, rather than from only a few points on a curved graph.

6.7 Charging a capacitor

Figure 6.20a shows how you would charge a capacitor C through a resistor R from a fixed d.c. supply of e.m.f. $\mathcal{E}$. When the switch is closed the p.d. across the capacitor V_C is zero as it has not yet received any charge. The initial p.d. across the resistor V_R is therefore $\mathcal{E}$ and the initial current is high.

At every stage of the charging process: $V_C + V_R = \mathcal{E}$. But as the capacitor charges, $V_C = Q/C$ increases and so V_R decreases. Figure 6.20b shows the general shape of how the p.d.s. vary until the capacitor is almost fully charged and the current in the circuit $I = V_R/R$ is nearly zero.

Just as the rate of flow of charge is proportional to the remaining charge, giving the equation $Q = Q_0 e^{-t/RC}$ when **discharging** a capacitor, so when **charging** a capacitor the rate of flow of charge is proportional to how close to fully charged the capacitor is, giving $Q = Q_0(1 - e^{-t/RC})$. While describing the discharge as exponential decay, we can describe the charging curve as one that 'approaches a final value exponentially'.

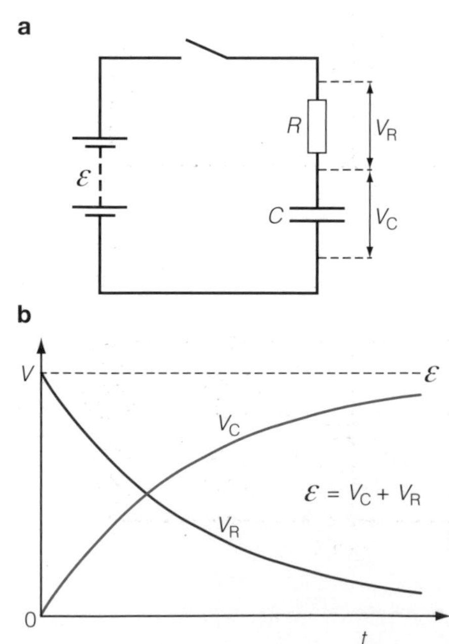

Figure 6.20 ▲

REVIEW QUESTIONS

1 A capacitor holds 1.3 μC when charged to 60 V. What is its capacitance?

2 Two capacitors of 47 μF and 470 μF are connected to a d.c. supply as shown.

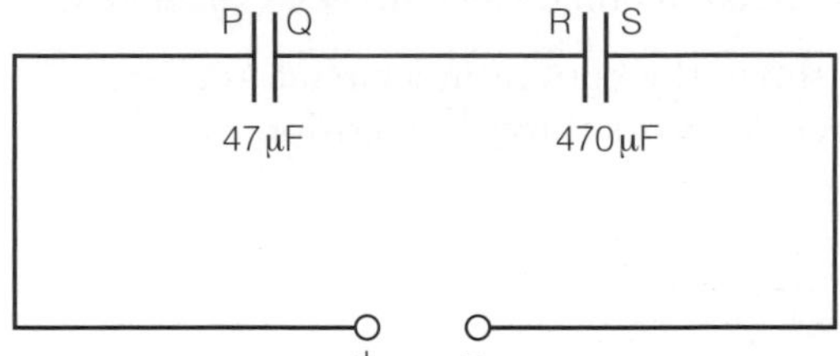

Figure 6.21 ▲

The ratio of the sizes of the charges on plates Q and R is:

A 1 : 1 B 1 : 10 C 1 : 11 D not calculable without knowing the e.m.f. of the d.c. supply

3 Two capacitors P and Q are storing equal amounts of energy, yet P has twice the capacitance of Q.

The p.d. across capacitor P is V. The p.d. across capacitor Q must be:

A 0.50V B 0.71V C 1.4V D 2.0V

4 A 100 μF capacitor is charged at a steady rate of 80 μC s^{-1}. The potential difference across the capacitor will be 12 V after a time of:

A 6.7 s B 9.6 s C 10.4 s D 15.0 s

5 A 220 mF capacitor stores 4.0 J of energy.

a) Calculate the p.d. across the capacitor.

b) Calculate the charge on each of the capacitor plates.

c) How are your answers to a) and b) related?

6 A 100 μF capacitor can be charged to a maximum p.d. of 20 V, while a 1.0 μF capacitor can be charged to a maximum p.d. of 300 V.

Which row in Table 6.6 correctly shows which capacitor can store the most charge *and* which can store the most energy?

	Stores most charge	Stores most energy
A	100 μF	100 μF
B	100 μF	1.0 μF
C	1.0 μF	100 μF
D	1.0 μF	1.0 μF

Table 6.6 ▲

7

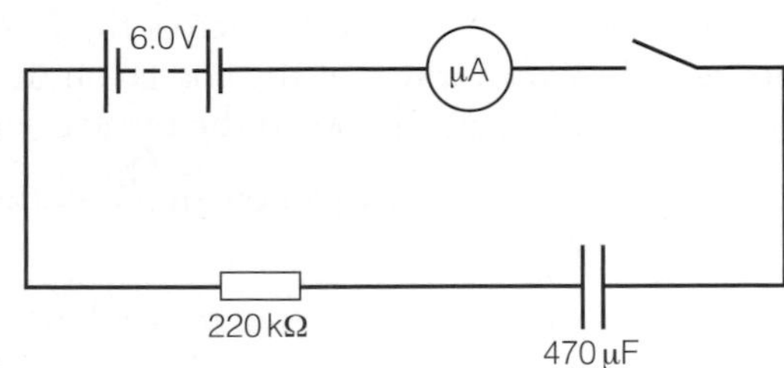

Figure 6.22 ▲

At a certain moment after the switch was closed in the above circuit, the current registered by the ammeter was 8.5 μA. Calculate, for this moment,

a) the p.d. across the resistor,

b) the p.d. across the capacitor,

c) the charge on each of the capacitor plates.

8 In circuit 1 of Figure 6.23, the capacitor is charged to 3.0 V and discharged through the lamp. The brightness and the length of the flash are noted.

The same procedure is followed with circuit 2, and the brightness and the length of the flashes are observed to be exactly similar to those in circuit 1.

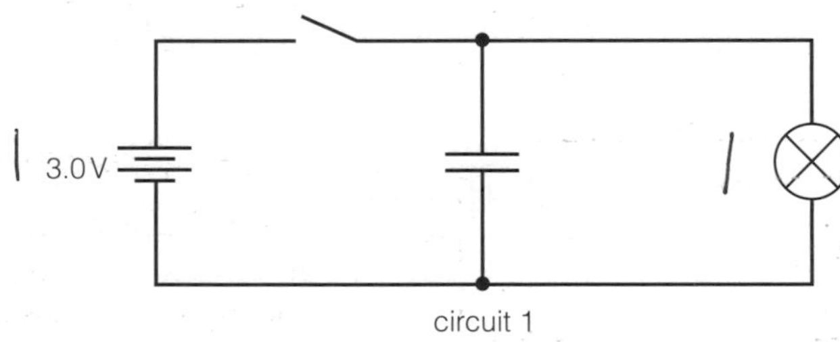

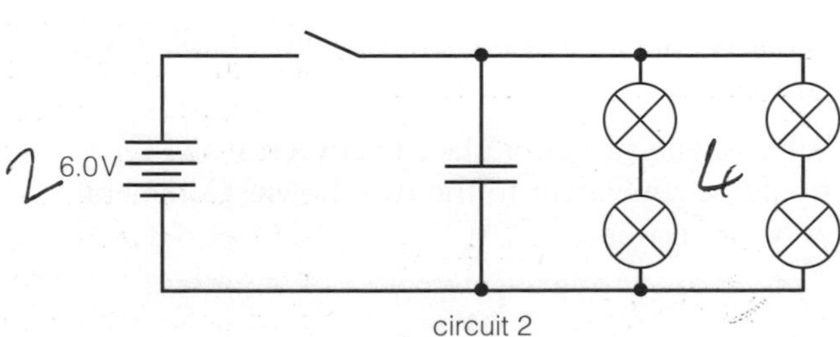

Figure 6.23 ▲

a) Explain these observations. (All the lamps are identical.)

b) Describe in words the arrangement of lamps you would set up to provide similar results when the capacitor is charged from a 9.0 V supply.

9 a) A 22 mF capacitor is connected across a 10 V d.c. power supply. What is the charge on the capacitor plates, and what energy is stored?

b) The power supply is then disconnected and an identical capacitor is connected across the 22 mF capacitor, in such a way that the charge on the plates is shared equally between the two capacitors.

i) What is the total energy now stored in the two capacitors?

ii) Suggest where the 'lost' energy may have gone.

10 A bank of capacitors of total capacitance 0.26 F at the Lawrence Livermore Laboratory in California can deliver an energy pulse of 6×10^{16} W. The pulse lasts for about a nanosecond.

Estimate the p.d. to which the capacitor bank is charged and explain why your answer is only an estimate.

11 In Figure 6.24 a potential difference of 12 V is applied between P and Q.

a) Calculate the charges on each capacitor.

b) What single capacitor placed between P and Q would be equivalent to the two shown? Comment on your answer.

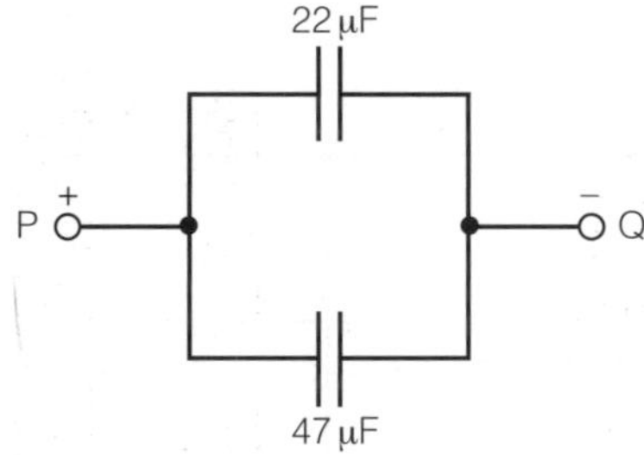

Figure 6.24 ▲

12 In Figure 6.25 a potential difference of 12 V is applied between P and Q.

a) What is the p.d. across each capacitor? Explain your answer.

b) What single capacitor placed between P and Q would be equivalent to the two shown? Comment on your answer.

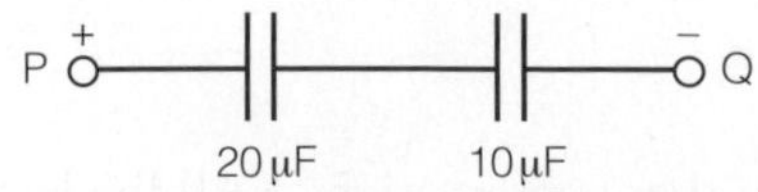

Figure 6.25 ▲

13 Refer to Figure 6.14a on page 45. Show that this curve is exhibiting exponential growth.

14 Refer to the graph in Figure 6.15 on page 47. Use the graph to find the current at $t_{\frac{1}{2}}$ and at $2t_{\frac{1}{2}}$. Comment on your answer.

15 When discharging a capacitor of unknown capacitance C through a resistor of 22 kΩ, it is found that the current falls to half its final value in 34 s. Calculate C.

16 Figure 6.26 shows the current in a circuit as charge from a capacitor decays through a resistor.

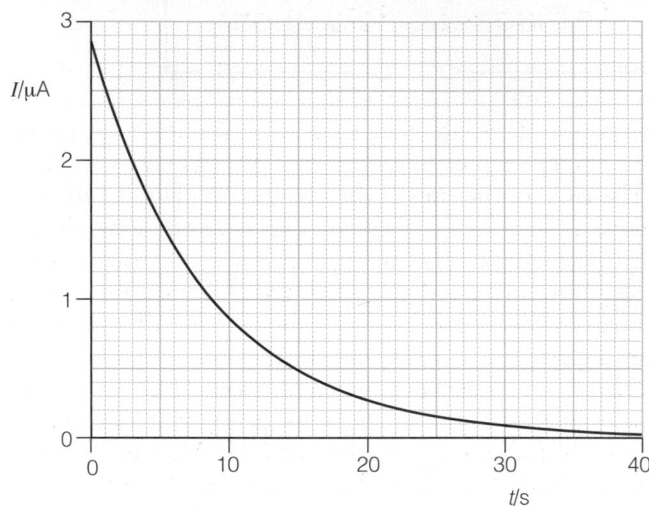

Figure 6.26 ▲

Use the graph to find the initial charge on the capacitor.

17 Refer to the circuit in Figure 6.20a on page 50 and to the two graphs in Figure 6.20b (V_C and V_R against time t). Draw, on separate axes, graphs of:

a) the current I in the circuit against t,

b) the charge Q on the capacitor against time t.

Explain why your graphs have the shape they have.

7 Magnetic fields

You have met magnetic fields before; but do you know how the strength of a magnetic field is measured? Electric currents produce magnetic fields and so forces are exerted on current-carrying wires placed in a magnetic field. These are the forces that drive electric motors.

Michael Faraday (see page 30) found out how to produce electric currents from the motion of wires in magnetic fields. This is how all the electrical energy in your home is generated.

7.1 Magnetic field lines

Around a magnet there is a region where objects made of iron or containing iron, a 2p coin for example, 'feel' a force pulling them towards the magnet. We say that there is a **magnetic field** around the magnet. It is invisible, just as gravitational and electric fields are invisible; we only know it is there because of the effects it produces.

Figure 5.3 on page 30 shows, in two dimensions, the magnetic field around a bar magnet – a magnetic dipole. The red lines are called **magnetic lines of force**. You can investigate the shape of a magnetic field like this using a tiny compass or by sprinkling iron filings on a sheet of paper or card placed over the magnet.

There is a simple convention for the N/S labels on the magnets and the arrows on the lines of force. N and S are used for magnets rather like the + and − that are used for electric charges. Magnetic lines of force then flow from N to S, which are called the poles of the magnet, just as electric lines of force flow from positive charge to negative charge. Like electric (and gravitational) fields, the magnetic field is strongest, that is exerts the largest forces, where the magnetic field lines are most closely bunched.

Where the magnetic lines of force from two magnets overlap, their effects add to give a single magnetic field shape. Figure 7.1 shows the resultant field for two bar magnets placed side by side. Somewhere between the two N poles there will be a place where the two magnetic fields cancel out. Such a place is called a **neutral point**. It will show up experimentally as a place where a tiny compass does not know which way to point. There is another neutral point between the two S poles. Composite fields like this show us the vector nature of magnetic fields.

Electric currents produce magnetic fields. Magnetic lines of force in these magnetic fields always produce closed loops. The simplest field is that produced by a steady current in a long straight wire.

Note

Unlike positive and negative electric charges, a magnetic N or S pole has never been found by itself.

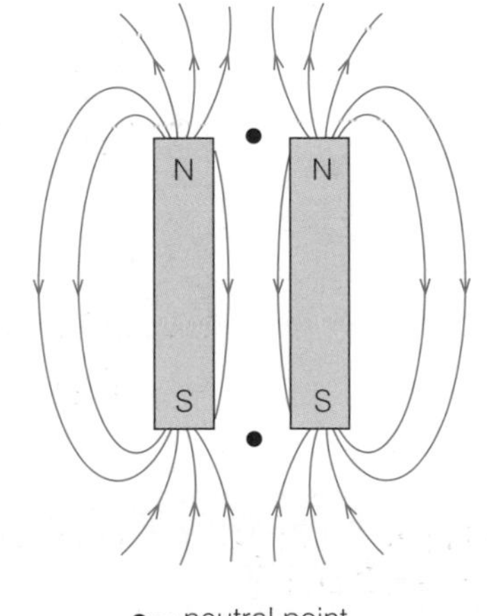

Figure 7.1 ▲

a current into page

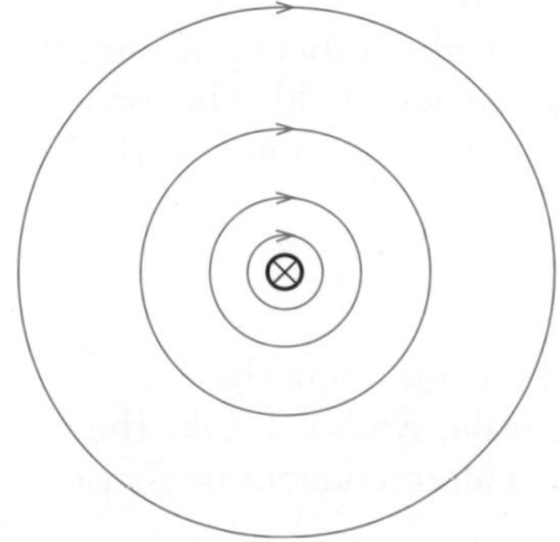

b current out of page

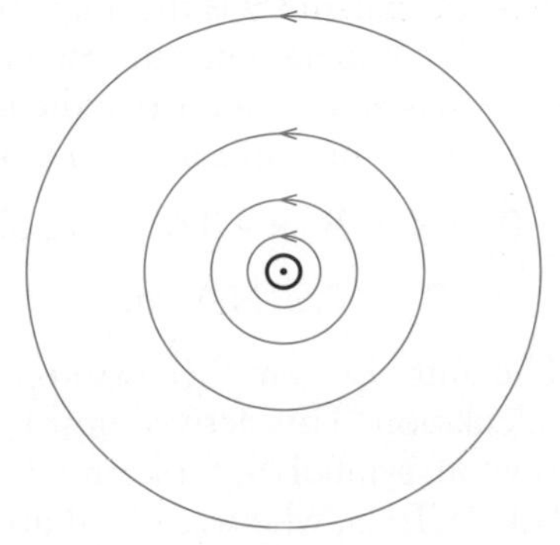

Figure 7.2 ▲
Magnetic field of a wire carrying a current

Figure 7.2 shows in two dimensions the magnetic field lines as a series of concentric circles, the field getting weaker further from the wire. The directional arrows on the magnetic lines of force are clockwise when the current is going away from you 'into the paper'. To remember this, imagine turning a screw in the direction of the current: the way you turn the screw gives the clockwise sense. The symbols × and • inside the circle of the wire are used to indicate that the current is into or out of the paper.

When current-carrying wires are twisted into coils or spring-like coils called solenoids, the magnetic fields are like those shown in Figure 7.3.

a simple coil

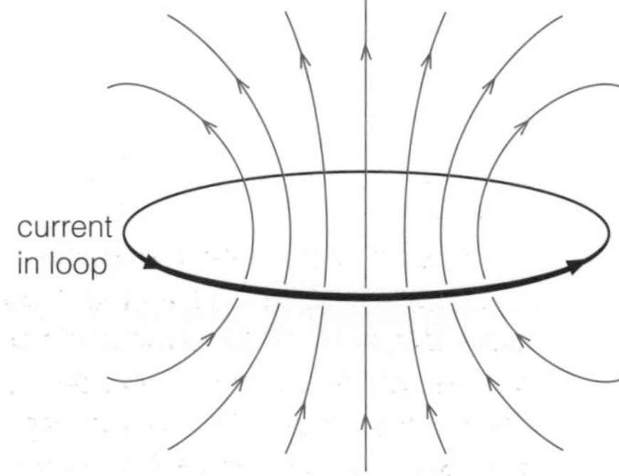

b solenoid

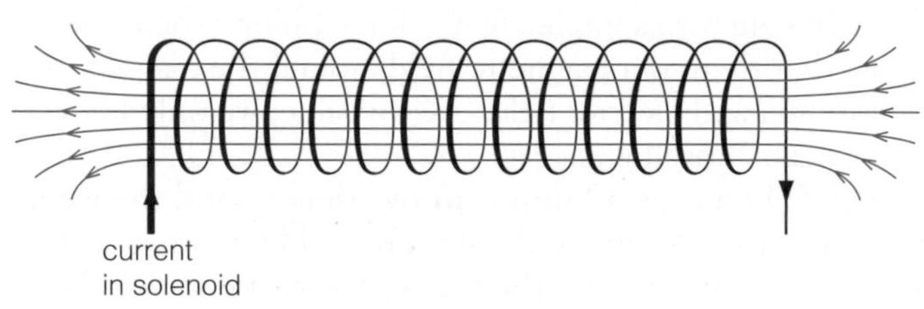

Figure 7.3 ▲
Magnetic fields of current-carrying coils

7.2 How strong are magnetic fields?

When a current-carrying wire is placed at right angles to a uniform magnetic field, for example the field between the N and S poles of two ceramic magnets (they are the black ones that have their poles across each flat face), the magnetic fields interact, resulting in a force F on the wire. This force depends on the current I in the wire and the length ℓ of the wire that lies in the field:

$$F \propto I\ell$$

Like electric or gravitational fields, the magnetic field is strongest, that is exerts the largest forces, where the magnetic field lines are most closely bunched.

Definition

Magnetic field strength is B in the equation

$F = B I \ell \sin \theta$

The symbols are defined in the text.

The strength of the field, which is called the **magnetic flux density** B, a vector quantity, is defined as the constant of proportionality, so:

$$\mathbf{F = B_{\perp} I\ell}$$

The $\perp$ sign indicates that the wire carrying the current must be perpendicular to the magnetic field. This equation can be written as $F = BI\ell \sin \theta$, where $B_{\perp} = B \sin \theta$ and θ is the angle between B and the wire.

Let's try some numbers. Suppose 12 cm of wire in which there is a current of 4.0 A lies across, i.e. perpendicular to, a uniform magnetic field. The force on the wire is measured and found to be 0.24 N. Using the equation $F = B_{\perp} I\ell$,

$$0.24\,\text{N} = B_{\perp} \times 4.0\,\text{A} \times 0.12\,\text{m}$$

$$\Rightarrow B_{\perp} = 0.50\,\text{N}\,\text{A}^{-1}\,\text{m}^{-1}$$

The unit $\text{N A}^{-1}\,\text{m}^{-1}$ (newton per ampere metre) emerges from the calculation. This derived unit is given the name **tesla**, symbol T. Like the newton, symbol N, for kg m s^{-2}, the tesla is named after a famous physicist, Nikola Tesla, who was Croatian.

A magnetic flux density or magnetic field strength of $0.50\,\text{N A}^{-1}\,\text{m}^{-1}$ or 0.50 T is much larger than the Earth's magnetic field in the UK. In fact it is

10 000 times stronger. The magnetic flux density of the Earth's magnetic field in London is about 50 μT and is inclined downwards at about 65° to the horizontal. As B is a vector, this means that a compass needle is responding to a horizontal magnetic field of only about $(50 \times 10^{-6}\,\text{T})\cos 65° = 21 \times 10^{-6}\,\text{T}$ or 20 μT.

The left-hand rule

Like gravitational field strength g and electric field strength E, magnetic field strength B is a vector quantity. But unlike g- and E-fields, where the forces are parallel to the fields, the magnetic force is perpendicular to both the field and the current-carrying wire. We are looking at a three-dimensional situation that can be represented by what is called Fleming's left-hand rule – Figure 7.4. It must be a left hand.

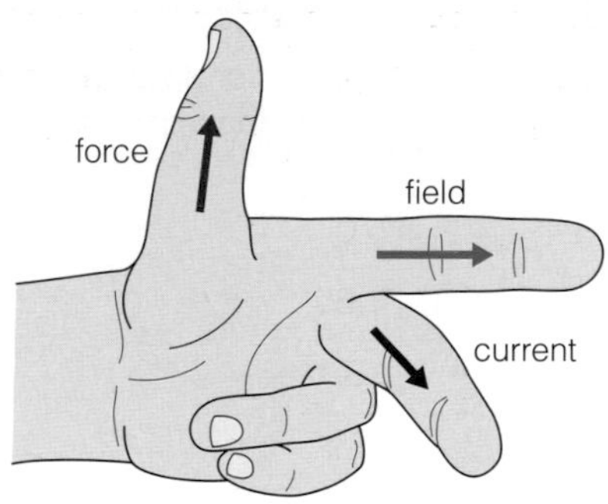

Figure 7.4 ▲
The left-hand rule

The thumb and first two fingers of the left hand are set at right angles to each other. With the **F**irst finger pointing in the direction of the **F**ield and the se**C**ond finger pointing in the direction of the **C**urrent, the **T**humb gives the direction of the force or **T**hrust.

Tip

Because of the three-dimensional nature of $F = B_{\perp}I\ell$, it is a great help to draw diagrams with the magnetic field into the paper so that the current and the force lie in the plane of the paper. For example, see Figure 7.5, where the magnetic field is represented by a pattern of crosses × × ×.

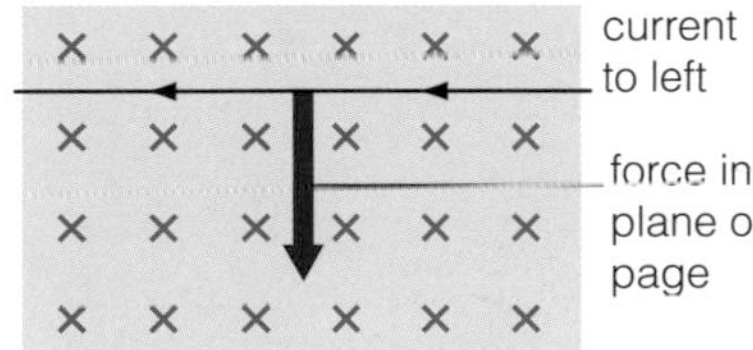

Figure 7.5 ▲

Experiment

Studying the force on a current-carrying conductor

The arrangement in Figure 7.6, with a sensitive electronic balance to measure forces, is used.

First zero the balance and then switch on the current. Make sure the horizontal piece of the wire carrying the current lies between the poles of the U-magnet (formed by two ceramic magnets on an iron 'yoke') and check that the wire is perpendicular to the magnetic field. A series of balance readings, in grams, can now be taken for a range of currents.

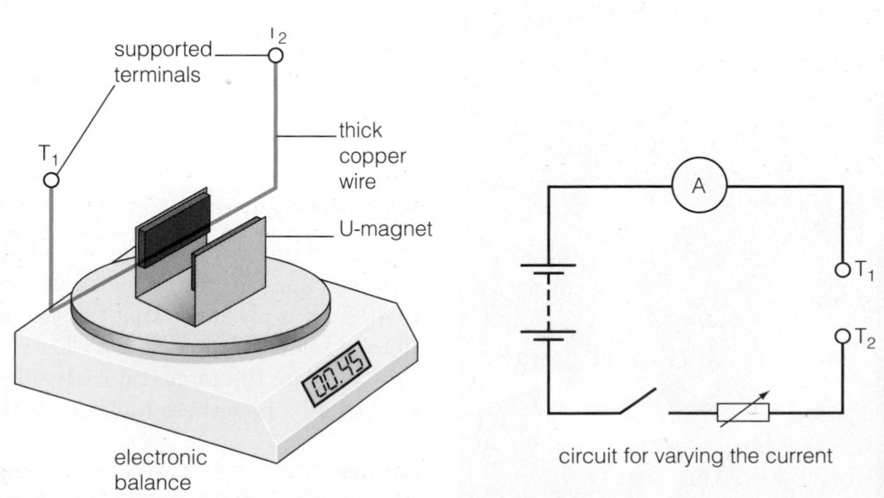

Figure 7.6 ▲
The power supply should be switched off between readings to avoid the copper wire getting too hot.

The magnetic force on the wire is, by Newton's third law, equal in size but opposite in direction to the magnetic force on the magnet – and it is this latter force that is registered by a change in the balance reading. The force is then calculated using $F = mg$, with $g = 9.8\,\mathrm{N\,kg^{-1}}$ and the mass in kg.

Plot a graph of F against I and deduce a value for the magnetic field between the poles of the U-magnet.

It is possible to alter the length ℓ of the horizontal wire in the magnetic field by using a second U-magnet, but such an experiment gives only a rough test of how the force varies with ℓ.

Exercise

Table 7.1 shows a typical set of observations for a set-up as in Figure 7.6, where the length of current-carrying wire between the magnetic poles is $\ell = 4.5\,\mathrm{cm}$.

I/A	1.2	2.5	3.4	4.2	5.0
m/g	0.45	0.90	1.25	1.50	1.85
$F/10^{-3}$ N	4.4	8.8	12.3	14.7	18.1

Table 7.1 ▲

As the currents are only measured here to 2 significant figures, the balance readings have been recorded to the nearest 0.05 g.

Plot a graph of these results and deduce a value for the magnetic field between the poles of the U-magnet. You should obtain a value of about $0.08\,\mathrm{N\,A^{-1}\,m^{-1}}$ or 80 mT.

7.3 D.C. electric motors

Figure 7.7 ▲ Electric mobility

The force on current-carrying wires in magnetic fields is made use of in electric motors. A motor using direct current (d.c.) enables you to start a car engine. Electric mobility buggies, golf buggies and electric vehicles such as milk floats run on d.c. motors. Many companies are now developing electric vehicles to replace the town car. High-powered electric motors, such as those used in trains and pumping stations, make use of the same electromagnetic force but their complex modern technologies are not described here.

Worked example

Figure 7.8 shows a rectangular coil abcd with the dimensions given below the diagram. The coil, which can rotate about the axis XY, carries a current of 40 A. It is placed so that the plane of the coil lies parallel to the magnetic lines of force of a field of 0.65 T.

a) Calculate the forces on each side of the coil: F_{ab}, F_{bc}, F_{cd}, F_{da}.
b) Describe how these forces vary as the coil rotates through 90°.

Answer

a) Using $F = B_{\perp}I\ell$:

$$F_{ab} = F_{cd} = 0.65\,\mathrm{N\,A^{-1}\,m^{-1}} \times 40\,\mathrm{A} \times 0.30\,\mathrm{m} = 7.8\,\mathrm{N}$$

$F_{bc} = F_{da} = 0$, as there is no component of B perpendicular to I ($\sin\theta = 0$)

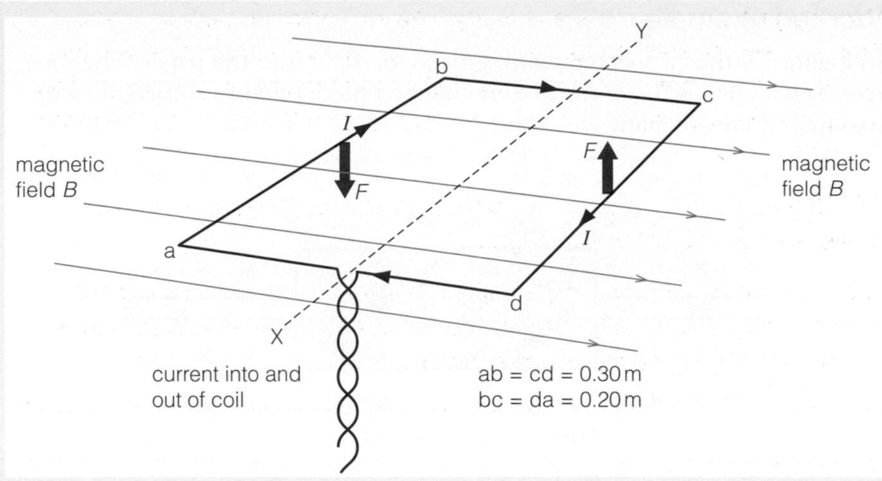

Figure 7.8 ▲

Such a pair of forces, down on ab and up on cd, each of 7.8 N, produce a strong twisting effect. The forces can be multiplied by 10 if there are 10 insulated turns to the coil.

b) As the coil rotates, the forces on sides ab and cd remain the same size. By the time it has rotated 90°, however, the two forces are no longer twisting the coil about XY but trying to stretch it. The forces on sides bc and da grow from zero, but again these forces do not try to twist or turn the coil about the axis XY. (You need to use the left-hand rule to convince yourself of these statements.)

7.4 Some useful algebra

The current I in a wire is the result of the drift of very large numbers of electrons. In your AS studies you will have met the relationship

$$I = nAQv$$

where n is the number of charge carriers (electrons) per unit volume, A is the cross-sectional area of the wire, Q is the charge on each charge carrier ($Q = -e$ for electrons) and v is the drift speed of the charge carriers.

As the force on a wire is given by $F = B_{\perp}I\ell$, inserting $I = nAQ$ gives $F = B_{\perp}(nAQv)\ell$, which can be rearranged as:

$$F = B_{\perp}Qv \times nA\ell$$

In this equation $A\ell$ is the volume of the wire in the magnetic field, so $nA\ell$ is the *total number* N of charge carriers in that piece of wire, and so $F \div nA\ell = F \div N$ is the force on *one* of these charge carriers (usually electrons).

The result is that the force F *on one charged particle* moving at speed v perpendicular to a magnetic field of flux density B is given by:

$$\mathbf{F = B_{\perp}Qv}$$

As the size of the charge on an electron or singly ionised atom is only 1.6×10^{-19} C, the force is very small. But the mass of an electron is of the order of 10^{-30} kg, so the accelerations that these tiny forces produce can be enormous!

Worked example

In Figure 7.9 the crosses represent a magnetic field into the paper. The long rectangle PQRS is the outline of an enlarged piece of current-carrying wire in which there are many electrons.

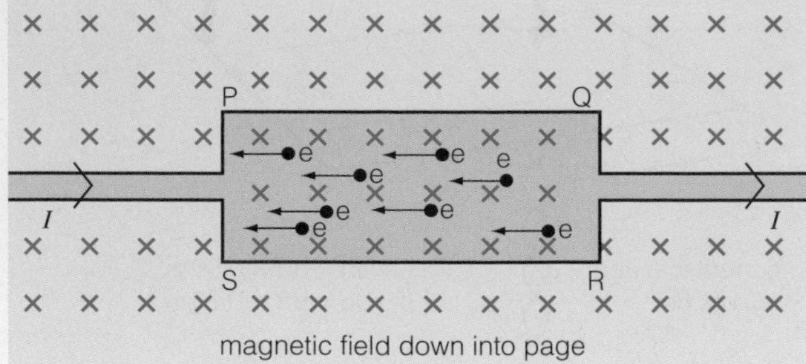

Figure 7.9 ▲

Explain why there is a potential difference across the wire, and predict which side of the wire is at the higher potential.

Answer
Although the electrons are moving from right to left, the direction of conventional current – the flow of positive charge – is from left to right. With the downward magnetic field, the left-hand rule produces a force on each electron, pushing it toward side PQ as it moves along the wire. There is no magnetic force on the protons in the nucleus of the atoms forming the wire as they are not moving.

Therefore the side PQ of the wire acquires a negative charge and the side RS, where there is a dearth of electrons, acquires a positive charge. These two charged zones mean that there is an electric field between the sides of the wire and thus a potential difference across it. The lower side is at the higher potential.

Note

This potential difference is called a Hall p.d. and forms the basis of the Hall probes you may have used to compare or measure magnetic fields.

7.5 Electron beams

Electrons carrying electric currents in wires are drifting very slowly, a fraction of a millimetre per second, but both oscilloscopes and X-ray tubes produce beams of electrons moving at high speeds in evacuated glass tubes. The electron guns inside such devices, operating at a potential difference V, give each electron a kinetic energy equal to the product eV.

Tip

Remember that a volt is a joule per coulomb, so eV comes out in joules.

Worked example

Calculate the speed of an electron of mass 9.1×10^{-31} kg after it has been accelerated across a potential difference of 25 V. State any assumption you make.

Answer
Kinetic energy of the electron $= 1.6 \times 10^{-19}\,\text{C} \times 25\,\text{V} = 4.0 \times 10^{-18}\,\text{J}$

As $KE = \frac{1}{2}mv^2$,

$$\frac{1}{2} \times 9.1 \times 10^{-31}\,\text{kg} \times v^2 = 4.0 \times 10^{-18}\,\text{J}$$

$$\Rightarrow v = 3.0 \times 10^6\,\text{m}\,\text{s}^{-1}$$

Assumption: all the energy becomes kinetic energy – the electron is moving in a vacuum and any change in gravitational potential energy is negligible.

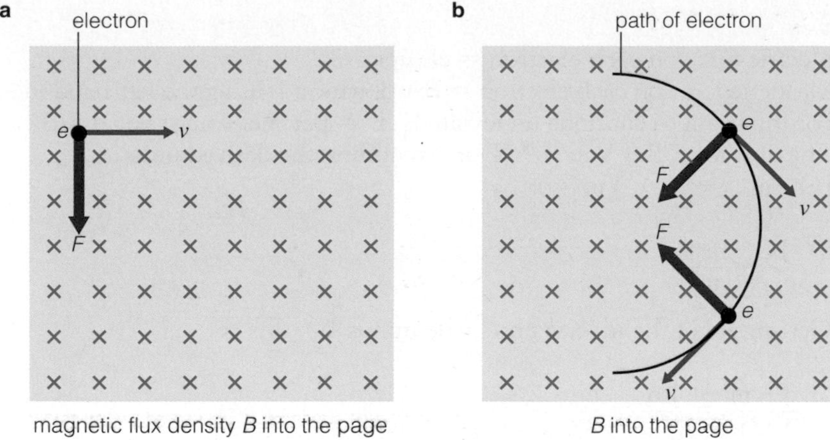

Figure 7.10 ▲

Now suppose that a beam of electrons is fired with a speed v in a vacuum perpendicular to a magnetic field of flux density B. The magnetic force F on each electron is given by Bev, and this force is perpendicular to both the magnetic field and the direction in which the electron is moving, by the left-hand rule. Figure 7.10a shows the directions of v and F with the magnetic field into the page. For the $3.0 \times 10^6\,\text{m}\,\text{s}^{-1}$ electron in the previous Worked example, $Bev \approx 10^{-14}\,\text{N}$ for magnetic flux densities of a fraction of a tesla, so the gravitational force $mg \approx 10^{-29}\,\text{N}$ on the electron can be completely ignored.

In Figure 7.10b the effect of the magnetic force on the electron is illustrated: it continues to act perpendicular to both B and v, making the electron move in a circle. In this situation the magnetic force Bev is the only force acting on the electron and forms the centripetal force needed for circular motion (see page 26). As Bev is always perpendicular to the path of the electrons, it does no work on them. Hence their kinetic energy and their speed as they move in a circle remain constant.

Tip

Remember that the left-hand rule applies to conventional current, or positive charge.

Worked example

A beam of electrons moving to the right in a vacuum enters a region where an electric field E (red) and a magnetic field B (green) act, as shown in Figure 7.11, at right angles to each other. The electrons continue to move in a straight line through the crossed fields.

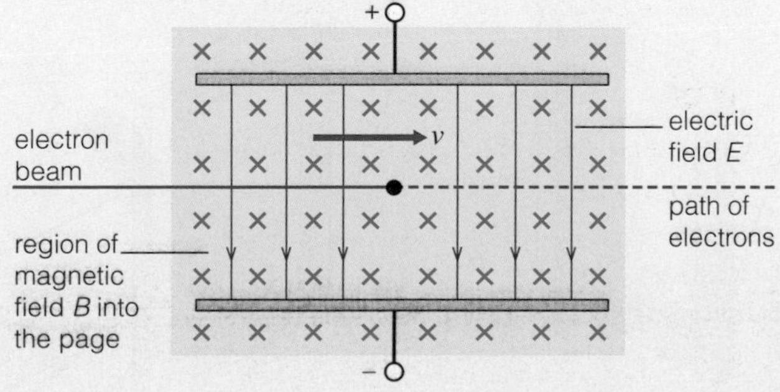

Figure 7.11 ▲

a) Write down the electric and magnetic forces acting on an electron in the beam.
b) Explain under what condition the electron beam continues in a straight line.
c) Check that your algebraic expression for this condition 'works' for units.

Answer

a) Electric force on each electron = eE upwards.
Magnetic force on each electron = Bev downwards using the left-hand rule.

b) For the beam to continue undeviated, $eE = Bev$, i.e. v must equal E/B.

c) The unit for E/B is $\mathrm{V\,m^{-1} \div T}$ or, expanding the derived units, $\mathrm{J\,C^{-1}\,m^{-1} \div N\,A^{-1}\,m^{-1}}$ or:

$$\frac{\mathrm{J\,C^{-1}\,m^{-1}}}{\mathrm{N\,A^{-1}\,m^{-1}}}$$

Getting rid of the inverse units (the minus 1s), this is:

$$\frac{\mathrm{J\,A\,m}}{\mathrm{N\,C\,m}} = \frac{\mathrm{J\,A}}{\mathrm{N\,C}} = \frac{\mathrm{m}}{\mathrm{s}}$$

because

$$\frac{\mathrm{J}}{\mathrm{N}} = \frac{\mathrm{N\,m}}{\mathrm{N}} = \mathrm{m}$$

and

$$\frac{\mathrm{A}}{\mathrm{C}} = \frac{\mathrm{C}}{\mathrm{s\,C}} = \frac{1}{\mathrm{s}}$$

So the unit for E/B is (as hoped) $\mathrm{m\,s^{-1}}$, the unit of v.

Tip

This sort of units exercise is excellent revision for lots of areas of physics from your earlier studies and, even if you are not asked to do it very often in examinations, it is well worth trying from time to time in examples or exercises such as this.

7.6 Changing magnetic flux

What is magnetic flux? Magnetic flux *density* B tells us how close together magnetic field lines are; it tells us the strength of the magnetic field. The product of magnetic flux density and the area through which it acts is called the **magnetic flux** through the area.

The magnetic flux through a window of area $1.0\,\mathrm{m^2}$ (using a value of $18\,\mu\mathrm{T}$ for the horizontal component of the Earth's magnetic field, and assuming that the window faces north or south) will only be $18\,\mu\mathrm{T} \times 1.0\,\mathrm{m^2} = 18 \times 10^{-6}$ weber.

Definition

The symbol for magnetic flux is Φ, so magnetic flux $\Phi = B_{\perp}A$

The $\perp$ stresses that the magnetic field must be measured perpendicular to the area. The meaning of 'density' in calling B the magnetic flux density now becomes apparent, as B is the magnetic flux per unit area. The unit of Φ must be $\mathrm{T\,m^2}$. This unit is sometimes given the name weber (Wb), but it is safe to use $\mathrm{T\,m^2}$ and important to remember that a tesla T is a name for $\mathrm{N\,A^{-1}\,m^{-1}}$.

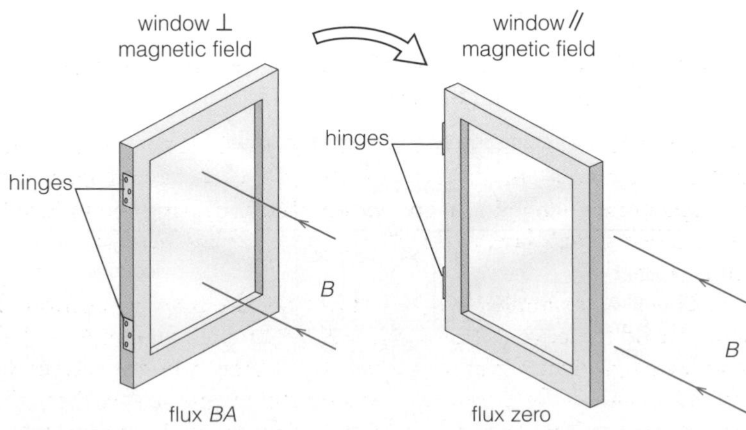

Figure 7.12 ▲

The importance of magnetic flux is not simply how strong it is but *how quickly it changes*. Suppose you open the window very quickly – say in 0.2 s – and that the magnetic flux through the window after you have opened it is zero (Figure 7.12). The average rate of change of magnetic flux through the window is given by

$$\frac{\Delta\Phi}{\Delta t} = \frac{18 \times 10^{-6}\,\mathrm{Wb}}{0.2\,\mathrm{s}} = 90 \times 10^{-6}\,\mathrm{Wb\,s^{-1}}$$

This looks very uninteresting, until you sort out the units. Let's try it.

$$\begin{aligned}\mathrm{Wb\,s^{-1}} &\equiv \mathrm{T\,m^2\,s^{-1}} \equiv \mathrm{N\,A^{-1}\,m^{-1}\,m^2\,s^{-1}} \equiv \mathrm{N\,m\,A^{-1}\,s^{-1}}\\ &\equiv \mathrm{J\,s^{-1}\,A^{-1}} \equiv \mathrm{W\,A^{-1}} \equiv \mathrm{V}\end{aligned}$$

Wow – the volt!

So the act of opening the window quickly seems to have produced something that can be calculated as 90 μV.

When a magnetic field passes through a coil of wire that has N turns, the magnetic flux through the coil, or the magnetic flux linking the coil, is $N\Phi$ or NBA. The rate of change of this **magnetic flux linkage** is the key to understanding where our electricity supplies come from.

7.7 Electromagnetic induction

When Michael Faraday first produced an electric current from changing magnetic fields in 1831, the second industrial revolution – the revolution involving electric generators and motors – was born. He called the effect **electromagnetic induction** and gave his name to the basic law governing induced e.m.fs:

$$\mathcal{E} = \frac{-N\,\Delta\Phi}{\Delta t} \text{ or } -N\frac{\mathrm{d}\Phi}{\mathrm{d}t}$$

Here $\mathcal{E}$ is the e.m.f. induced in a circuit when the average rate of change of magnetic flux through the circuit is $\Delta\Phi/\Delta t$ (or, instantaneously, $\mathrm{d}\Phi/\mathrm{d}t$). The value of N is greater than 1 when the circuit is a coil with a lot of turns, because the e.m.f. is induced in each coil. This is rather like having a number N of cells in series: if each has an e.m.f. of 1.5 V then 20 cells in series will produce 30 V.

In words, the above equation, **Faraday's law** of electromagnetic induction, reads as:

> the induced e.m.f. in a circuit is equal in size to the rate of change of magnetic flux linkage through the circuit

The minus sign in the equation above shows that the induced e.m.f. could send an induced current around the circuit that would set up a magnetic field that would *oppose* the change in magnetic flux that is causing the induced e.m.f. This is called **Lenz's law**, and is really an example of the law of conservation of energy.

Let's go back to our example of the window from the previous section. The metal frame around the edge of the window is our circuit, so here $N = 1$. The induced e.m.f. will produce a current around this frame, and the current will itself produce a magnetic field that tries to stop the magnetic flux through the frame from getting smaller. The induced e.m.f. is doing work sending the charge round the circuit and we must provide the energy for that by pushing on the window. Of course for the window we will never 'see' this electromagnetic induction, even though 90 μV could produce a noticeable induced current, as the resistance of the metal frame might be only a milliohm (mΩ). Nor will we ever notice that it is harder to open the window quickly than to do it slowly.

Tip

You need to know that the expression $\mathrm{d}\Phi/\mathrm{d}t$ is a shorthand way of saying 'the rate of change of magnetic flux at a particular instant'.

Figure 7.13 ▲
Lenz's law could be called the law of cussedness, as whatever you try to do produces a result that tries to stop you!

Experiment

Capturing an induced e.m.f. on a data logger

Drop a short bar magnet through a coil that is in series with a resistor. Connect a data logger across the resistor – Figure 7.14a.

The data logger records the e.m.f. induced in the coil at short time intervals and later draws a graph to show how the e.m.f. varies with time. It will look something like Figure 7.14b.

Repeat the experiment, dropping the magnet from different heights above the coil.

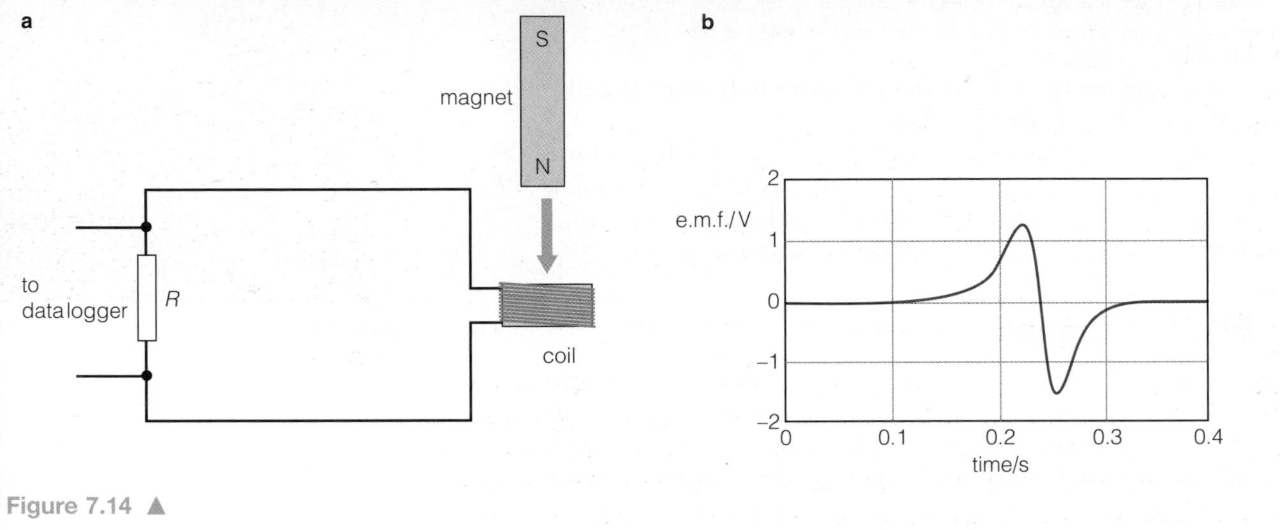

Figure 7.14 ▲

Worked example

a) Explain the shape of the graph produced by the data logger in the above experiment.
b) Explain why the acceleration of the falling magnet is not quite $9.8\,\text{m}\,\text{s}^{-2}$ (g).
c) Describe how the output of the data logger would differ from that above if
   i) the coil is replaced by a coil with double the number of turns, and
   ii) the magnet is dropped from a greater height.
d) The area under a graph of induced e.m.f. against time represents the change in magnetic flux linkage. Explain why, in Figure 7.14b, the area under the curve above zero e.m.f. should equal the area under the curve below zero e.m.f.

Answer

a) Where the graph is positive, the magnetic flux through the coil is increasing as the magnet approaches the coil. Where the graph is negative, the magnetic flux is decreasing as the magnet exits the coil. The maximum e.m.f. as the magnet approaches is a little less than the maximum e.m.f. as it exits because the magnet accelerates through the coil. Thus the rate of change of magnetic flux is bigger as it exits.
b) The induced e.m.f. produces a current in the coil–resistor circuit. This current in turn produces a magnetic field (see Figure 7.3a, page 54) that will be up out of the coil as the magnet approaches, so repelling the falling magnet – Lenz's law. Similarly, the falling magnet will be attracted back into the coil as the magnet exits the coil. Both these effects are producing upward forces on the magnet so its acceleration is not quite as big as g.
c) i) Double the number of turns means that the magnetic flux linking the coil is doubled. Therefore the induced e.m.f. is doubled at each stage of the fall. The peaks are doubled.

ii) When the magnet is dropped from a greater height the magnetic flux linking the coil changes at a greater rate. Therefore the e.m.f. is greater and the time taken for the magnet to pass through the coil is reduced. The peaks are higher and narrower.

d) The total change in flux linkage as the magnet enters the magnetic field of the coil should be equal to the total change in flux linkage as the magnet exits the coil (although the rate of change of flux linkage is not the same because the magnet speeds up).

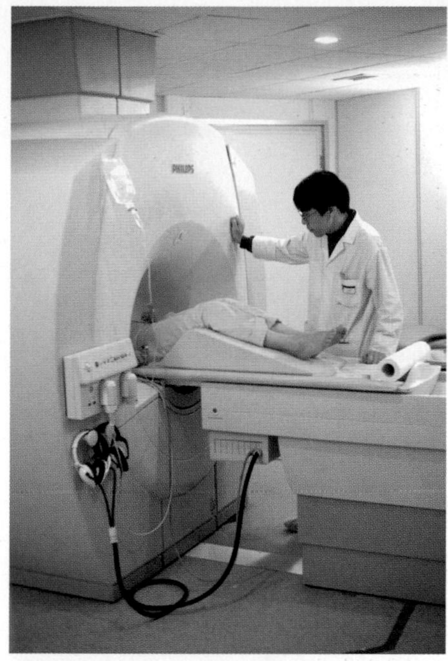

Figure 7.15 ▲ An MRI scanner

A **solenoid** is a long spring-like coil, of length many times its diameter, with many turns every centimetre. A current I in a solenoid produces a magnetic field B along its axis (see Figure 7.3b, page 54).

Huge magnetic field strengths, of the order of 1.5 T, are produced in the solenoids of nuclear magnetic resonance imaging (MRI) scanners (Figure 7.15). The solenoid is in the 'box' and the patient slides into the solenoid for examination. It uses superconducting coils at liquid helium temperature and carries very large currents. An MRI scanning system can cost a hospital over £1 million. There are few safety problems for the patient because no ionising radiations are involved, as there are with X-ray CT scans, though some patients do suffer from claustrophobia. Patients are asked to remove any body jewellery – can you think why?

Figure 7.16 shows a simple laboratory solenoid connected to a variable d.c. power supply. A small coil of cross-sectional area A with N turns is placed as shown inside the solenoid. Any change in the current in the solenoid will now produce, in accordance with Faraday's law, an induced e.m.f. in the small coil, because there will be a change in the magnetic flux linking the coil.

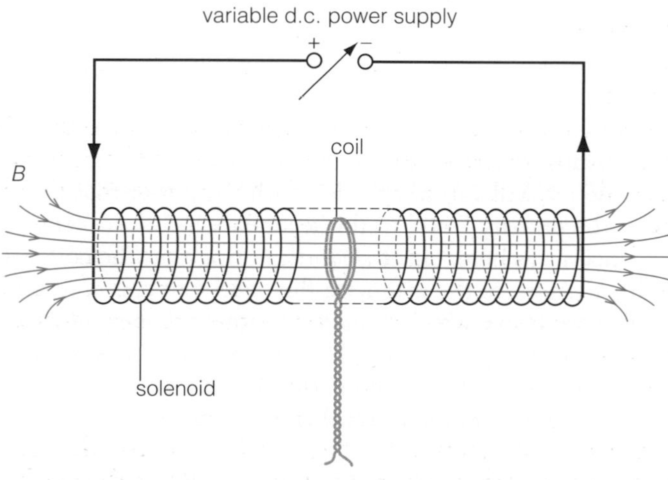

Figure 7.16 ▲

Suppose the current in the solenoid is reduced to zero in a time t. The *change* of magnetic flux linkage in the small coil is NAB and the (average) *rate of change* is NAB/t. Let's try some numbers – say

$N = 40$ $\quad B = 50\,\text{mT} = 50 \times 10^{-3}\,\text{N A}^{-1}\,\text{m}^{-1}$

$A = 8\,\text{cm}^2 = 8 \times 10^{-4}\,\text{m}^2$ $\quad t = 0.5\,\text{s}$

Putting in the numbers for NAB/t gives an induced e.m.f. of 3.2×10^{-3} V. So the result of changing the current in the solenoid is to produce an induced e.m.f. in the small coil, in this case, of 3.2 mV.

7.8 The transformer

Transformers are everywhere: they are in those black heavy plugs used in recharging devices such as that for your mobile; they provide safe voltages for experiments in your laboratories; they are the key to the efficient transfer of power over the national grid. What do they transform?

Transformers increase or decrease the voltage of alternating a.c. electrical power supplies. They do not work with direct d.c. power supplies.

- In all transformers a coil – the primary P coil – produces a magnetic field.
- This magnetic field links another coil – the secondary S coil.
- With alternating currents, the P magnetic field varies continuously.
- Therefore the magnetic flux linkage through the S coil also varies continuously.
- The change of this magnetic flux linkage induces a varying e.m.f. in the S coil.

Tip

Explanations of this type are often best laid out as a series of bullet points rather than as an 'essay'.

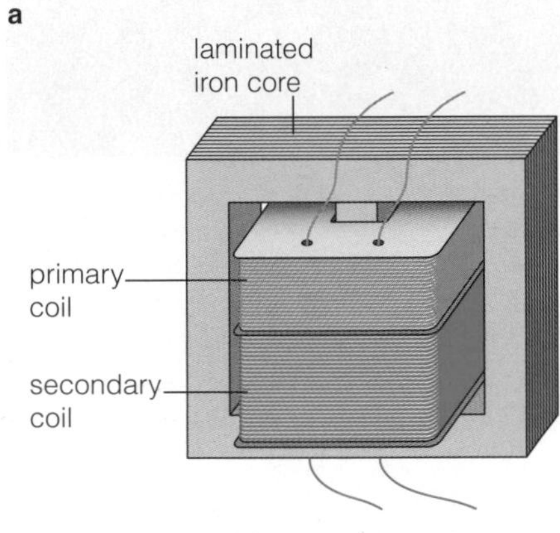

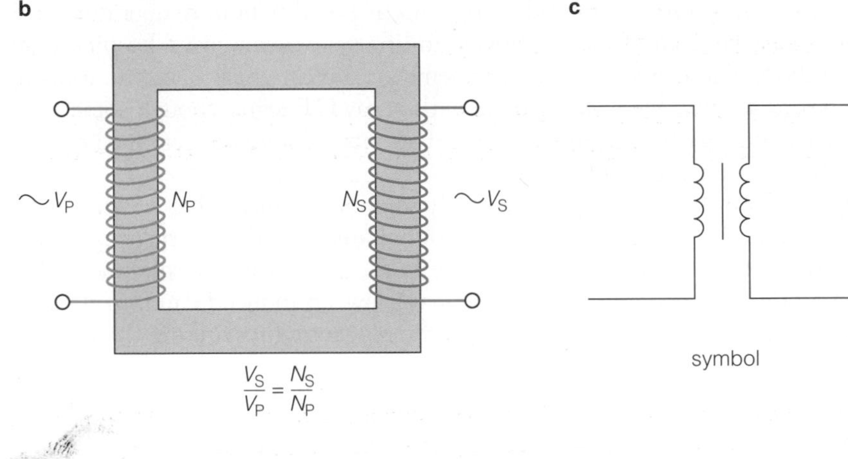

Figure 7.17 ▲

Figure 7.17 shows a) a realistic idea of the structure of a transformer, b) the principle of the transformer and c) the circuit symbol for a transformer. The detail of the structure will vary very much with the power that the transformer must pass from the primary P coil to the secondary S coil. It is the principle that mainly concerns us here, but it is worth noting that the coils are wound on an iron core to increase the magnetic flux, and that this core is formed of thin laminated iron sheets, which reduces the internal energy losses. (Most transformers become warm when switched on and large ones need special cooling systems to prevent them overheating.)

The transformer equation, for an **ideal transformer**,

$$\frac{V_S}{V_P} = \frac{N_S}{N_P}$$

states that the ratio of the number of turns on the secondary and primary coils determines the ratio of the voltages 'out of' and 'into' the transformer. It may seem that you can get something for nothing, for example get out 2400 V having only put in 240 V. Yes, you can get more volts! But you can't get more *power* out than you put in. The principle of conservation of energy tells us this. For an ideal transformer, the power output is equal to the power input, i.e.

$$I_S V_S = I_P V_P$$

In practice, there are always energy losses and so $I_S V_S < I_P V_P$.

Worked example

Figure 7.18 gives information about an ideal transformer.

a) Explain quantitatively why this transformer is said to be ideal.

b) Suggest values for the number of turns the transformer might have on its coils.

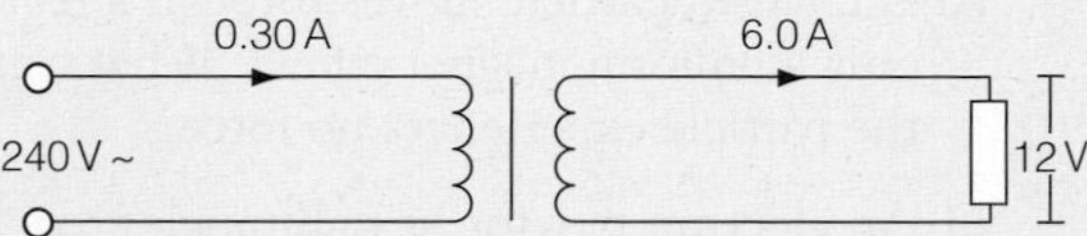

Figure 7.18 ▲

Answer

a) Input power $= I_P V_P = 0.30\,\text{A} \times 240\,\text{V} = 72\,\text{W}$
Output power $= I_S V_S = 6.0\,\text{A} \times 12\,\text{V} = 72\,\text{W}$
As there is no power loss, the transformer is ideal.

b) The ratio $\dfrac{V_S}{V_P} = \dfrac{240\,\text{V}}{12\,\text{V}} = 20$, so $\dfrac{N_S}{N_P} = 20$.

There must be 20 times as many turns on the primary coil as on the secondary. For example: N_P might be 2000 and then N_S would be 100.

REVIEW QUESTIONS

1 Two bar magnets are placed side by side with opposite poles facing. The number of neutral points produced will be:

A 0 B 1 C 2 D 4

2 Three parallel wires, each carrying equal currents down 'into' the paper, are arranged as shown in Figure 7.19. The resulting magnetic field at P (marked by red circle) will be in the direction of which arrow?

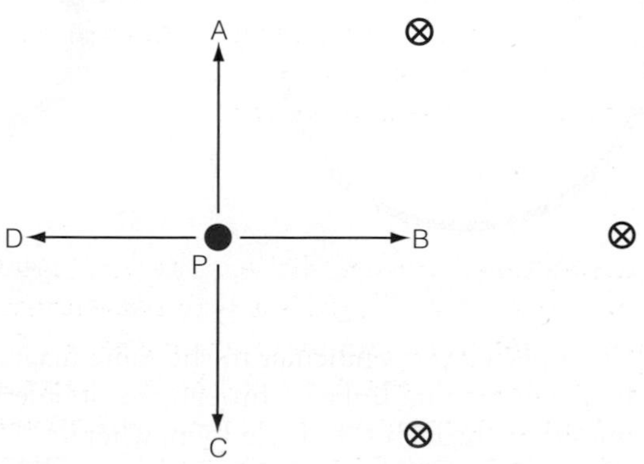

Figure 7.19 ▲

3 The tesla (T) expressed in base SI units is:

A $\text{kg}\,\text{m}^{-1}\,\text{A}^{-1}$ B $\text{kg}\,\text{s}^{-1}\,\text{C}^{-1}$
C $\text{kg}\,\text{s}^{-2}\,\text{A}^{-1}$ D $\text{kg}\,\text{m}^{2}\,\text{s}^{-2}\,\text{A}^{-1}$

4 What are:

a) the horizontal component, and

b) the vertical component

of the Earth's magnetic field at a place where the magnetic flux density is 86 μT and the field dips in to the Earth at an angle of 15° to the vertical?

5 The cross-channel d.c. cable between France and the UK carries a maximum current of 15 kA. The Earth's magnetic field has a flux density of 65 μT in a region where the cable is horizontal. Calculate the size of the force on 1.0 km of cable.

6 The magnetic flux density B varies with distance r from a long straight wire carrying a current. Numerically B is inversely proportional to r.

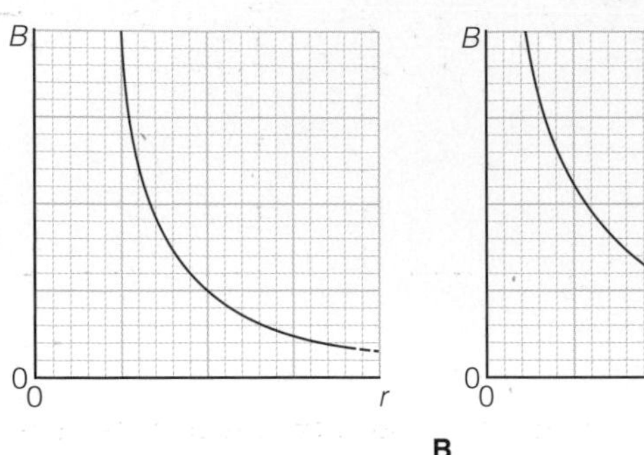

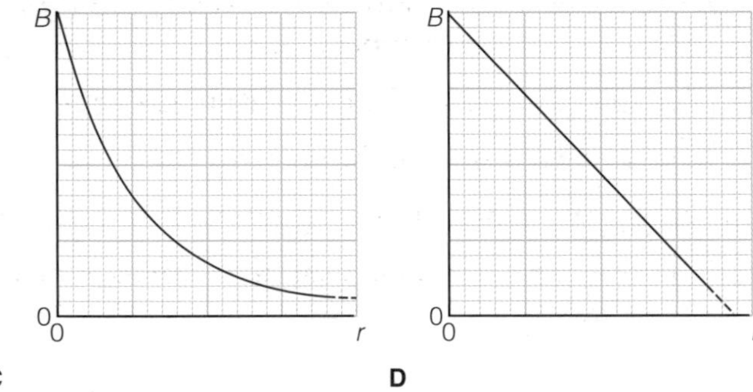

Figure 7.20 ▲

Which graph in Figure 7.20 correctly shows such a relationship between B and r?

7 There is a current of 2.0 A in each of the conductors OP, OQ and OR shown in Figure 7.21. The conductors are in a magnetic field of flux density 0.25 T parallel to the plane of the diagram.

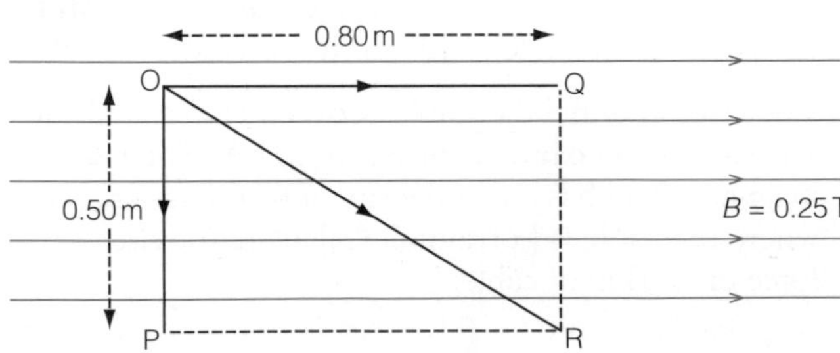

Figure 7.21 ▲

What is the size and direction of the force on each conductor?

8 Take one set of readings from Table 7.1 on page 56 and confirm that a B-field of 0.08 T is compatible with 4.5 cm of wire in the magnetic field.

9 A horizontal copper wire with a mass per unit length $\mu = 80\,\mathrm{g\,m^{-1}}$ lies at right angles to a horizontal magnetic field. When there is a current of 5.6 A in the wire it levitates, that is, it is supported against the pull of the Earth. Calculate the magnetic field strength.

10 A charged particle moves through a region containing only a uniform magnetic field. What can you deduce if the particle experiences no force?

11 An electron moving at right angles to a magnetic field of flux density 0.20 T experiences an acceleration of $3.0 \times 10^{15}\,\mathrm{m\,s^{-2}}$.

a) What is the electron's speed?

b) By how much does its *speed* change in the next 10^{-7} s?

12 An alpha particle (charge $+2e$) moving at a speed of $5.0 \times 10^{6}\,\mathrm{m\,s^{-1}}$ enters a region in which there is a uniform magnetic field of 0.15 T. What is the magnetic force on the alpha particle if the angle between its initial path and the magnetic field is
a) 90°, b) 45°, c) 30°?

13 Compare the size of the forces produced by the Earth's gravitational, electric and magnetic fields in England on a proton that is moving horizontally at close to the speed of light.

Look up or estimate any numerical values that you need.

14 A length of wire is formed into a loop and placed perpendicular to a uniform magnetic field. The circle is then cut and the wire formed into a double loop as shown.

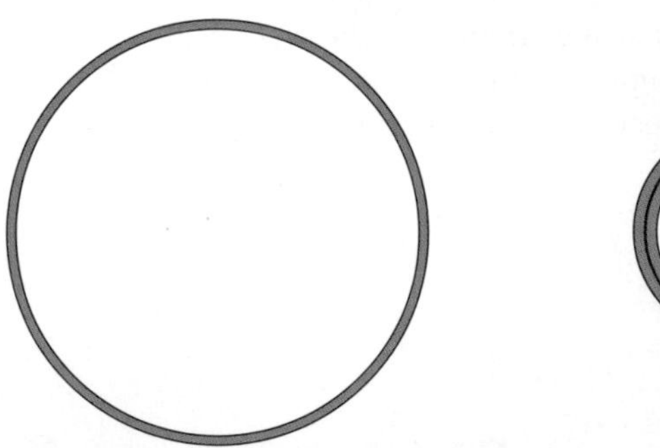

Figure 7.22 ▲

When placed perpendicular to the same magnetic field, the magnetic flux linkage through the double loop is N times that through the single loop, where:

A N = 0.25 B N = 0.32
C N = 0.40 D N = 0.50

15 Show that a weber per second is equivalent to a volt, i.e. $\text{Wb s}^{-1} \equiv \text{V}$.

16 Why might Lenz's law be called the 'law of cussedness'?

17 A U-shaped magnet is formed by placing two ceramic magnets on a steel yoke. The face of each ceramic magnet measures 3.5 cm by 5.0 cm. If the magnetic flux between the magnets is 2.5×10^{-4} Wb, calculate the size of the average magnetic flux density between them.

18 The magnetic flux density in an MRI scanner is 1.5 T. A woman enters the scanner wearing, unnoticed, a gold wedding ring with a diameter of 18 mm.

a) What are the maximum and minimum values of the magnetic flux through the ring as she moves her ring finger about?

b) Calculate the e.m.f. induced in her ring when she moves it from maximum to minimum flux linkage in 0.30 s.

19 A student, instead of dropping the magnet through the coil as in Figure 7.14 on page 62, holds the magnet and pushes and pulls it in and out of the coil twice per second.

a) Sketch what will be recorded by the data logger.

b) Explain the shape of your sketch.

20 The switch S is closed in the circuit below.

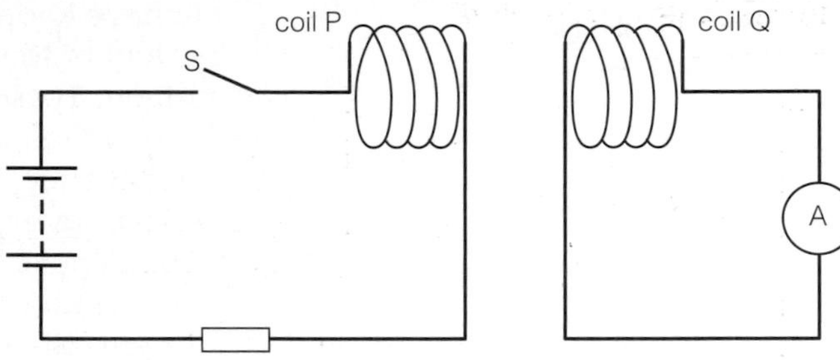

Figure 7.23 ▲

Explain why the ammeter registers a sudden current surge and then returns to zero.

21 A transformer is designed to reduce the 230 V a.c. supply in the laboratory to 12 V to operate a particular device. There are 200 turns on the secondary coil and the device takes a current of 1.5 A.

a) How many turns are there on the primary coil?

b) What is the current in the primary coil?

c) State any assumption you have made.

22 Make a list of where transformers might be found in a modern house.

12) $Q = 2(1.6 \times 10^{-19})$ C
$v = 5.0 \times 10^{6}$ m s^{-1}
$F = BI\ell$

8 Electrons and nuclei

We have known about electrons since the discovery in the late 19th century of thermionic emission – electrons leaving a hot metal surface. These electrons can be accelerated by electric and magnetic fields, which enables their charge and mass to be determined. The evidence for the very small positive nucleus at the centre of an atom came from experiments in 1909 in which alpha particles were fired through gold foil and found to be scattered.

In this chapter you will learn the language of the atom and about modern experimental techniques for accelerating and detecting charged particles.

8.1 The language of the atom

The **proton number** (or atomic number) Z denotes the number of protons in an atomic nucleus of a given element (and is also the number of electrons in a neutral atom of that element), e.g. for gold $Z = 79$.

The **neutron number** N denotes the number of neutrons in the nucleus of an atom, e.g. for gold $N = 118$.

The **nucleon number** (or atomic number) A denotes the total number of protons and neutrons in the nucleus of an atom, e.g. for gold $A = 197$.

mass of a proton	$m_p = 1.67262 \times 10^{-27}$ kg
mass of a neutron	$m_n = 1.67493 \times 10^{-27}$ kg
mass of an electron	$m_e = 0.00091 \times 10^{-27}$ kg = 9.1×10^{-31} kg

Table 8.1 ▲
Masses of sub-atomic particles

Atomic nuclei are represented by their symbol, with the proton number at bottom left and the nucleon number at top left, e.g. $^{197}_{79}$Au or $^{12}_{6}$C.

$$\begin{array}{l} \text{nucleon number} \rightarrow A \\ \qquad\qquad\qquad\qquad\quad X \leftarrow \text{symbol for element} \\ \text{proton number} \rightarrow Z \end{array}$$

These symbols and numbers are also used to represent **nuclides** – the nucleus plus its electrons. Most elements have at least one stable nuclide plus several unstable nuclides – see Chapter 13. Nuclides of the same element are called **isotopes**. All isotopes have the same Z but different N, and hence different A because $A = Z + N$. For example, for the only stable isotope of gold,

$$A(197) = Z(79) + N(118)$$

For one of the unstable isotopes of gold,

$$A(196) = Z(79) + N(117)$$

Unstable isotopes decay to stable ones with characteristic **half lives** – see Chapter 13. In the process they emit alpha (α), beta (β) and gamma (γ) radiations. The properties of these are described in Section 13.4 and summarised in Table 13.1 on page 134.

8.2 Alpha-particle scattering

The material of the nucleus is very, very dense – about 10^{16} kg m^{-3}! (Compare this with the density of a metal like gold – about 2×10^{4} kg m^{-3}.) How do we know this?

The fact that the atom is mainly 'empty' was first established when α-particles from a naturally radioactive source were fired at a very thin sheet of gold, in 1909. Most passed through undeflected from their path; others were deflected though small angles. A tiny fraction – less than 0.01% – of the α-particles however were deflected by more than 90°, and from this unexpected scattering the physicist Ernest Rutherford was able to describe the **nuclear model** of the atom with its tiny, massive, positively charged nucleus. He was further able to calculate a rough value for the diameter of the nucleus. Figure 8.1a shows a modern version of the experiment. The paths of three α-particles that pass just above and very close to a gold nucleus are shown in Figure 8.1b.

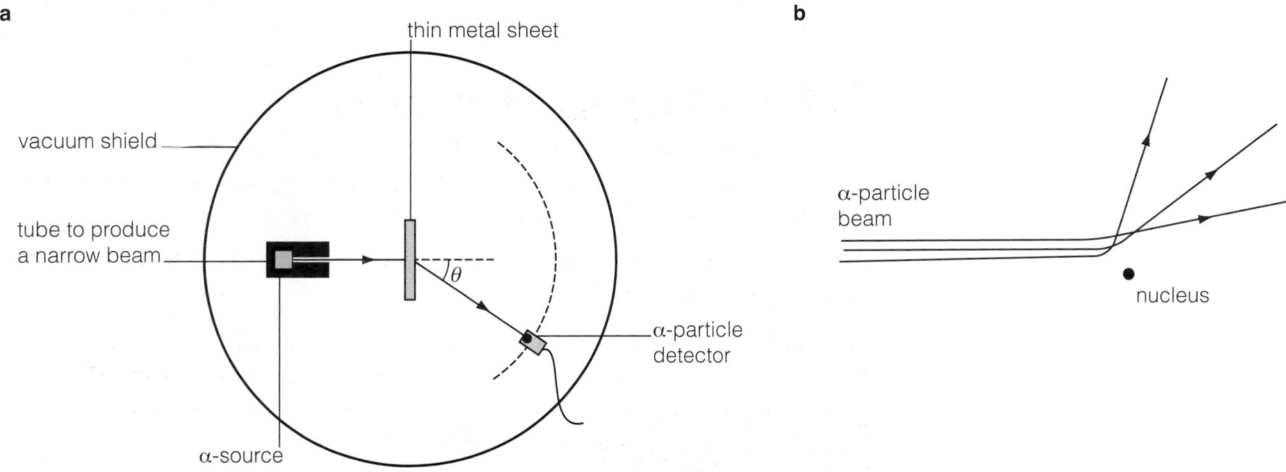

Figure 8.1 ▲

Worked example

An α-particle aimed directly towards the nucleus of a gold atom approaches it to within 5.0×10^{-14} m before its motion is reversed.

a) What are the charges on an α-particle and a gold nucleus?
b) Explain why the α-particle rebounds from the gold nucleus.
c) Suggest a value for the diameter of the gold nucleus.

Answer

a) α-particle: $+2e = 3.20 \times 10^{-19}$ C; gold nucleus: $+79e = 1.26 \times 10^{-17}$ C

b)
- As the α-particle enters the electric field produced by the gold nucleus it experiences a repulsive force, because both particles have positive charge.
- This force gets stronger the closer the α-particle approaches to the nucleus, until all the kinetic energy of the α-particle is 'used up', i.e. stored as electric potential energy.
- The α-particle is then repelled by the nucleus, eventually regaining all its initial kinetic energy when it is well away from the gold nucleus.

c) Assuming that the distance of closest approach is approximately equal to the radius of the gold nucleus, then its diameter $\approx 2 \times 5.0 \times 10^{-14}$ m $= 1 \times 10^{-13}$ m.

Tip

Although the three bullet points could be written as a single paragraph, it is easier to follow the argument when it is laid out like this.

For a quantitative treatment of this see question 7 on page 79.

Rutherford was later able to establish conclusively that α-particles, when they gain electrons, are atoms that emit a spectrum exactly like the element helium.

Exercise

Figure 8.2 shows the key stages in an experiment first performed exactly 100 years before this book was first published. The black 'stuff' is liquid mercury, the level of which can be raised and lowered. The gas A in the glass capsule with *very* thin walls is radon. Helium gas at extremely low pressure is formed in B.

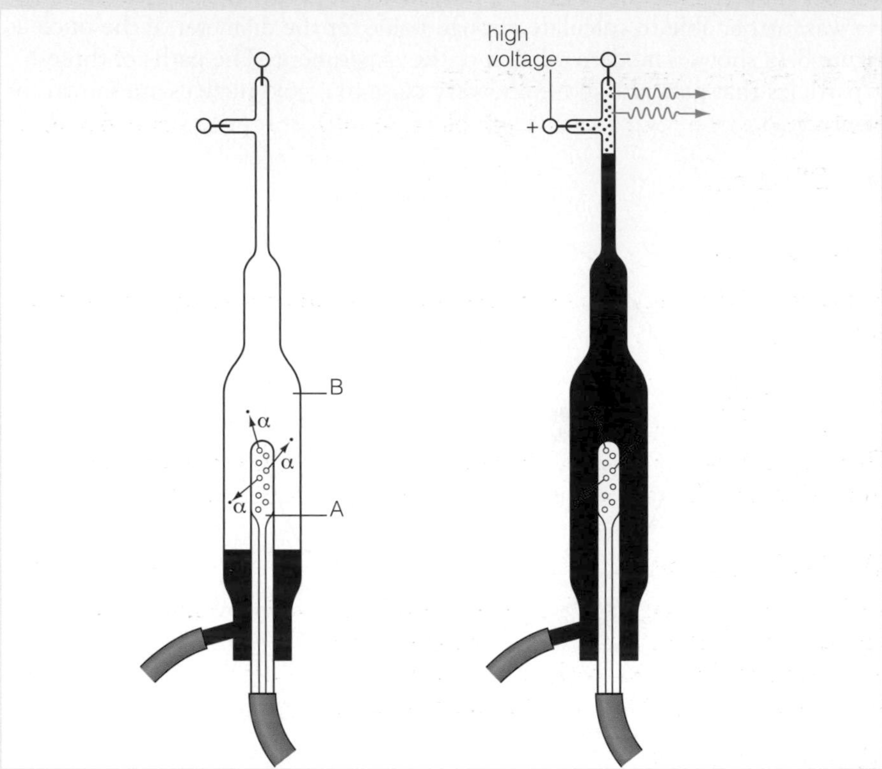

Figure 8.2 ▲

Write a bullet-point description of what is happening in this experiment and what it is designed to prove.

Tip

This is an example of a 'synoptic' exercise, where you bring an understanding of different areas of physics to focus on an unusual situation. Your examination will contain synoptic exercises.

8.3 Thermionic emission

When a piece of metal is heated to a high temperature, negatively charged electrons 'bubble' out of its surface. Of course, they will be attracted back to the surface by the positively charged protons they leave behind. But if a positively charged plate is placed near the piece of metal in a vacuum, the electrons accelerate towards it and can be made into a narrow beam. In this arrangement, called an **electron gun**, the metal is usually heated by a resistor (connected to a 6 V supply) placed behind it and the narrow beam is produced by making a small hole in the positive plate (which is at a potential of about 2000 V). Electron beams such as those found in old-style non-flat TV sets and PC monitors were described on page 58. Such beams accelerated in a vacuum through 2000 V contain electrons moving at very high speeds.

Summarising from earlier chapters:

- Force on an electron in an electric field $F_E = eE$ parallel to the field
- Force on an electron in a magnetic field $F_B = Bev$ perpendicular to the field

Figure 7.10 on page 59 shows how a beam of electrons can follow a straight path through E and B fields that are perpendicular to one another, and Figure 7.9 helps to explain why electron beams can be made to follow circular paths such as that shown in Figure 8.3.

Figure 8.3 ▲
An electron beam made visible

In Figure 8.3 the electrons are projected horizontally to the right from an electron gun in a magnetic field that is directed out of the plane of the photo. As electrons are negatively charged, the **left-hand rule** shows that the *Bev* force is centripetal and remains so as the electrons move round part of a circle. (The bright curve that enables us to 'see' where they are is caused by having a very low-pressure gas in the tube. A few electrons ionise the gas atoms, which then return to their ground state by emitting visible photons.)

8.4 Some useful algebra

Applying Newton's second law to an electron in the beam in Figure 8.3 above:

$$\frac{mv^2}{r} = Bev$$

$$\Rightarrow mv = Ber$$

As $mv = p$, the momentum of the electron, this can be rewritten as $p = Ber$ or as

$$r = \frac{p}{Be}$$

This tells us that the radius of the circle in which the electron moves is proportional to the momentum with which the electron is fired.

Worked example

In Figure 8.3 the magnetic flux density is 0.50 mT and the radius of the circle in which the electrons are moving is 20 cm.

a) Calculate the speed at which the electrons are moving.

b) Deduce through what p.d. the electrons were fired.

Answer

a) From $r = p/Be$ we get $mv = Ber \Rightarrow v = Ber/m$
Using $e = 1.6 \times 10^{-19}$ C and $m_e = 9.1 \times 10^{-31}$ kg gives

$$v = \frac{(0.50 \times 10^{-3}\,\text{N A}^{-1}\,\text{m}^{-1})(1.6 \times 10^{-19}\,\text{C})(20 \times 10^{-2}\,\text{m})}{(9.1 \times 10^{-31}\,\text{kg})}$$

$$= 1.8 \times 10^{7}\,\text{m s}^{-1}$$

b) Kinetic energy of each electron

$$= \frac{1}{2}mv^2$$

$$= \frac{1}{2}(9.1 \times 10^{-31}\,\text{kg})(1.8 \times 10^{7}\,\text{m s}^{-1})^2$$

$$= 1.5 \times 10^{-16}\,\text{J}$$

$$= \frac{1.5 \times 10^{-16}\,\text{J}}{1.6 \times 10^{-19}\,\text{J eV}^{-1}}$$

$$= 940\,\text{eV}$$

So the electrons were fired through a potential difference of 940 V.

Note

If you hold numbers in your calculator as you work through a multi-stage calculation, you sometimes get a slightly different answer.

Referring back to the above algebra:

from $mv = Ber \quad \Rightarrow \quad \frac{v}{r} = \frac{Be}{m}$

and as $\frac{v}{r} = \omega = 2\pi f$ this tells us that $2\pi f = \frac{Be}{m}$

So the frequency at which the electron circles is independent of its initial momentum, and hence of its initial kinetic energy, provided its mass is constant. This is the key to the operation of the **cyclotron**.

8.5 The cyclotron

A cyclotron is a type of **particle accelerator**. The theory above that leads to $2\pi f = Be/m$ for the frequency of charged particles circling in a magnetic field enables protons to be accelerated to high energies in a circulating beam.

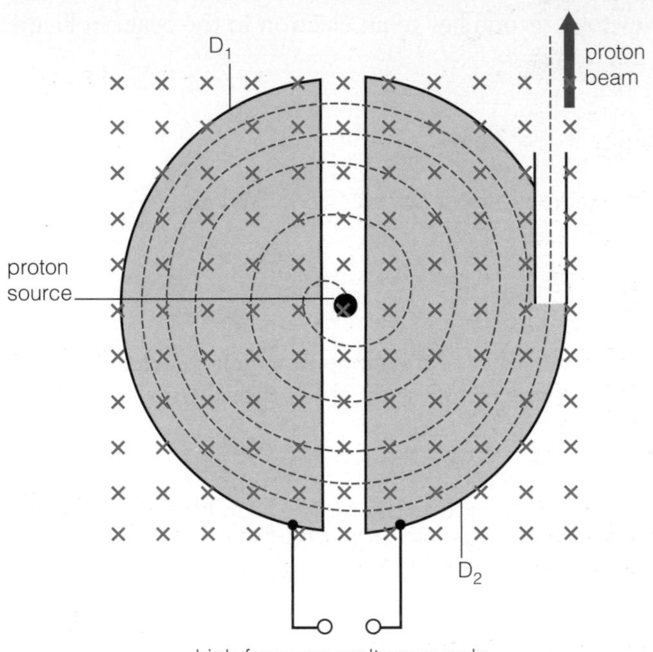

Figure 8.4 ▲
The Dees of a proton cyclotron

Figure 8.4 shows the principle of operation of a cyclotron: a source of protons – ionised hydrogen atoms – is placed at the centre of two 'Dees'. These Dees are semi-circular flat metal boxes, open at their diameter, and seen here from above. The Dees are connected to a very high-frequency alternating voltage V and are situated in a strong vertical magnetic field, indicated here by the × × ×. The whole set-up is in a good vacuum.

This is how it works:

- Some protons that emerge from the central source moving in a horizontal plane enter D_1.
- Here they follow a semi-circular path under the action of the magnetic field.
- When they next arrive at the narrow gap between the Dees they 'see' the opposite Dee to be at a negative potential and are accelerated across the gap by the electric field.
- They now enter D_2 with extra energy eV.
- They then continue in a second semi-circle at a higher speed (because r is greater) and arrive back at the gap only now to 'see' D_1 to be at a negative potential, because the voltage supply is alternating at exactly the right frequency, a frequency equal to the frequency of circulation of the protons.
- Extra energy eV is again added and this continues as the radius of the proton's path increases.

Tip

Again a detailed explanation is often best set out as a series of bullet points.

Worked example

Protons are accelerated in a cyclotron in which the magnetic flux density is 1.2 T and the voltage between the Dees when the protons cross the gap is 10 kV.

a) Show that the frequency of the voltage supply necessary for it to be synchronised with the arrival of the protons at the gap between the Dees is 18 MHz.

b) How many circles must the protons make in order to reach an energy of 10 MeV?

Answer

a) The mass and charge of a proton are $m_p = 1.7 \times 10^{-27}$ kg and $e = 1.6 \times 10^{-19}$ C.
Substituting in $2\pi f = Be/m_p$ gives

$$f = \frac{1.2\,\text{N A}^{-1}\text{m}^{-1} \times 1.6 \times 10^{-19}\,\text{C}}{2\pi \times 1.7 \times 10^{-27}\,\text{kg}}$$

$$= 1.8 \times 10^{7}\,\text{Hz} = 18\,\text{MHz}$$

b) At each crossing the protons gain 10 keV of energy. They cross twice per circle. Therefore the number of circles needed to reach 10 MeV is 10 MeV/20 keV = 500.

Note

You would not be expected to remember this formula.

The 10 MeV reached in the Worked example above is about the limit for a cyclotron accelerating protons, as the synchronism only lasts so long as m_p remains constant – see question 13 on page 80.

8.6 Linear accelerators

You have probably seen a small Van de Graaff accelerator during your physics studies. In research laboratories large ones can be used to accelerate charged particles up to energies of around 10 MeV. To accelerate protons (or other charged particles) to energies beyond this, a linear accelerator or **linac**, of a different design detail for each particle, is used. In the electron linac of Figure 8.5, the electrons are given energy as they pass between charged metal tubes. As in a cyclotron, the energy is delivered to the charged particles by the electric field in the small gap between the tubes.

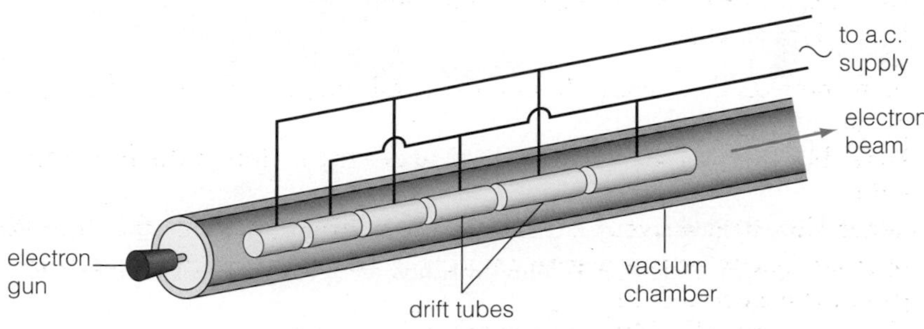

Figure 8.5 ▲
An electron linac

The tubes are connected to a high-frequency alternating voltage supply, and the lengths of the tubes are calculated so that there is always a positive charge on the 'next' tube. So a bunch of electrons fired by the electron gun is attracted to the first tube; while it is in that tube the charge on the next tube changes to positive, thus attracting the bunch leaving the first tube; and so on. The length of the tubes increases as the speed of the bunch of electrons increases.

While in any tube the electrons travel at a steady speed – they drift – there being no electric field inside the metal tubes. This is also the case for charged particles while they are inside a Dee in the cyclotron.

Electron linacs are now routinely used in hospitals to produce beams of high-energy electrons. When the beam hits a tungsten target the result is a beam of X-rays – see Figure 13.5 on page 130.

Figure 8.6 below shows the structure inside the vacuum chamber of the proton linac at Fermilab (Fermi National Accelerator Laboratory) near Chicago, USA. The drift tubes are clearly seen. This linac is 150 m long and accelerates protons to 400 MeV.

At Stanford in the USA, an electron linac that is 3 km long (see Figure 9.1 on page 81) can accelerate electrons to 50 GeV (50 000 MeV).

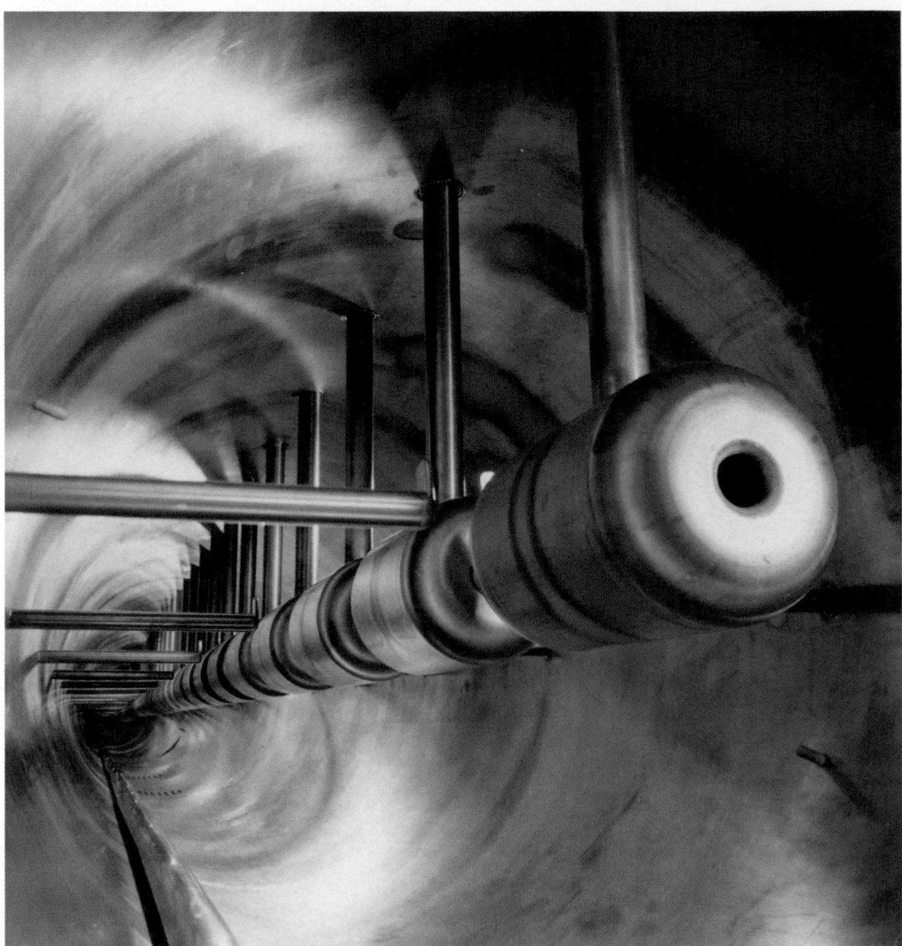

Figure 8.6 ▲
Inside the proton linac at Fermilab

Worked example

A modern proton linac has 420 metal tubes. It operates at a frequency of 390 MHz.

Calculate how long it takes a proton to travel along this linac.

Answer

The protons drift for half a cycle in each tube (because every second tube is connected to the same side of the alternating supply).

$$\therefore \text{ time in each tube} = \frac{1}{2} \times \frac{1}{f} = \frac{1}{2}(390 \times 10^6\,\text{Hz})^{-1} = 1.28 \times 10^{-9}\,\text{s}$$

$$\Rightarrow \text{time to travel down 420 tubes} = 420 \times 1.28 \times 10^{-9}\,\text{s} = 5.4 \times 10^{-7}\,\text{s}$$

8.7 Particle detectors

The history of nuclear physics in the 20th century closely followed improvements in the experimental methods available for detecting nuclear particles. Four Nobel Prizes were awarded in this field: Table 8.2 lists these, including the year of the first use of the detection method in brackets.

Date	Prize winner	Detection device
1927	C. R. T. Wilson (Scottish)	Cloud chamber (1911)
1950	C. F. Powell (English)	Photographic emulsions (1934)
1960	D. A. Glaser (American)	Bubble chamber (1960)
1992	G. Charpak (French)	Drift chamber (1992)

Table 8.2 ▲
Nobel Prizes in Physics in the field of particle detection

There may be others added to this list in the present century, if the designers of the huge detectors at CERN and other high-energy laboratories are honoured. Figure 8.7 shows a massive detector being assembled. Note the size of the physicists in their hard hats on the gantry!

Figure 8.7 ▲
One of the huge detectors for the LHC (Large Hadron Collider) at CERN

The basic principle behind the detection of charged particles has not changed, however: energetic charged particles cause **ionisation** in any material through which they pass.

The **cloud chamber** made use of ionisation in super-saturated air and the **bubble chamber** of ionisation in super-saturated liquid hydrogen. The ionised molecules along the particle's path form centres for the formation of tiny liquid water drops and tiny hydrogen gas bubbles respectively. These can be illuminated and photographed: in both cases the 'track' of the ionising particle is thus made visible.

Figure 8.8 is a photograph of a spiralling curved track in a bubble chamber. This was caused by an energetic electron that entered the chamber at the bottom right. You may guess (correctly) that there was a magnetic field perpendicular to the electron's spiral, and the left-hand rule will tell you that

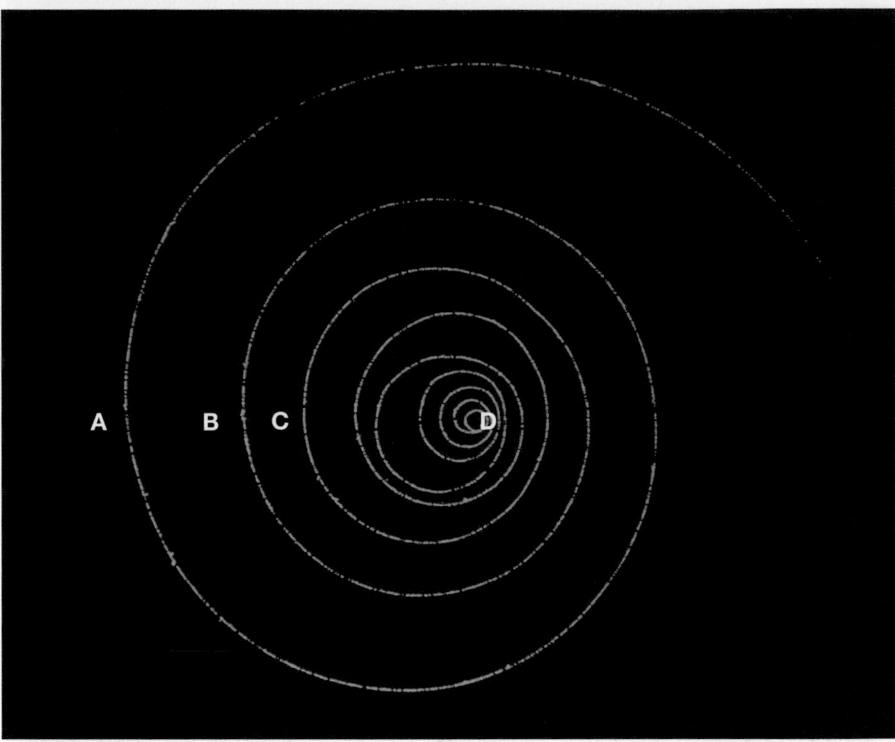

Figure 8.8 ▲
A bubble chamber photograph

the field was directed out of the plane of the paper (because an electron is negatively charged). You may also realise that the electron is gradually losing energy. There are two ways of deducing this:

1 Each ionisation will take a few electronvolts from the kinetic energy of the electron, and there are many hundreds if not thousands of ionisations in the spiral.
2 The radius of the spiral is getting smaller, and on page 71 we proved that $r = p/Be$, that is the radius of the electron's path is proportional to its momentum, and thus depends on its kinetic energy ($p^2 = 2m_e E_k$).

(The labels A, B, C and D on this photograph are used in question 14 on page 80.)

Cloud chamber photographs can be seen in Figure 3.8 (page 19) and Figure 13.12 (page 134).

High-energy charged particles can also be detected by the sparks they produce between a series of thin sheets of charged metal foil. Such a detector is called a **spark chamber** and was the forerunner of modern drift chambers.

Figure 8.9 shows how charged particles can be 'tracked' by the sparks they produce. Clearly there was no charged particle in the centre of the photograph – perhaps the incoming particle stopped there, after it had knocked a neutron out of an atom's nucleus.

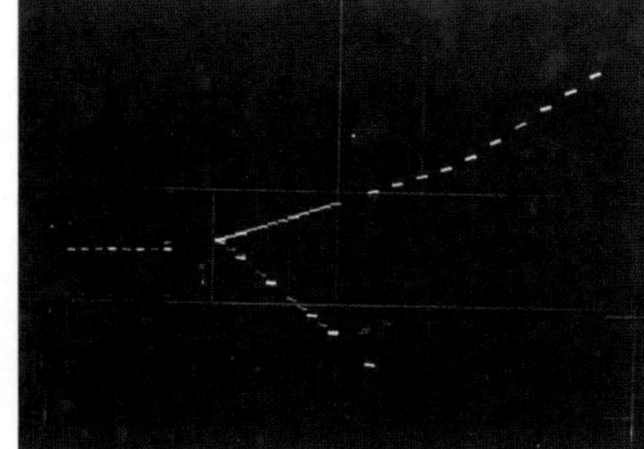

Figure 8.9 ▲
A spark chamber photograph

8.8 Einstein's equation

The famous 'ee equals em sea squared' is best written as

$$\Delta E = c^2 \Delta m$$

where $c = 3.00 \times 10^8\,\mathrm{m\,s^{-1}}$ is the speed of light in a vacuum. The Δm tells us that particles have a mass when they are at rest – their **rest mass** m_0 – but a greater mass $m_0 + \Delta m$ when they have extra energy ΔE. It further tells us what the 'rate of exchange' is between mass and energy. An extra energy (perhaps kinetic or internal or elastic) ΔE is equivalent to an extra mass $\Delta E/c^2$.

Worked example

Calculate how much 'heavier' a 12 V, 50 A h car battery is when fully charged than when totally discharged. State any assumption you make.

Answer

Assume that the battery is at the same temperature in both circumstances.

A battery with a capacity of 50 A h will discharge

$$50\,\mathrm{C\,s^{-1}} \times 3600\,\mathrm{s} = 180\,000\,\mathrm{C}$$

The total electrical energy stored in a fully charged 12 V battery is therefore

$$\Delta E = 180\,000\,\mathrm{C} \times 12\,\mathrm{J\,C^{-1}} = 2\,160\,000\,\mathrm{J}$$

This is equivalent to a mass increase of

$$\Delta m = \frac{2\,160\,000\,\mathrm{J}}{(3.00 \times 10^{8}\,\mathrm{m\,s^{-1}})^2} = 2.4 \times 10^{-11}\,\mathrm{kg}$$

The battery is therefore 2.4×10^{-10} N heavier when fully charged. (Negligible!)

Tip

Be sure to square *c* when using Einstein's equation.

The example above, like many others in our everyday world, shows us that we can normally treat the conservation of mass and the conservation of energy as two separate principles. But in the sub-atomic world, the world of high-energy electrons and nuclear particles, we need to apply the equivalence of mass and energy – Einstein's principle of the **conservation of mass–energy**.

Figure 8.10 ▲
Matter from energy!

A dramatic demonstration that kinetic energy and mass are interchangeable is shown in Figure 8.10. Here an iron nucleus approaching the Earth from deep space strikes a silver nucleus in a photographic emulsion raised high into the atmosphere by a balloon. The huge kinetic energy of the iron nucleus is used to create about 750 new ionising particles, the total mass of which multiplied by c^2 equals the loss of kinetic energy in the collision.

Worked example

Calculate the potential difference through which an electron, of rest mass $m_0 = 9.11 \times 10^{-31}$ kg, must be accelerated in order to double its effective mass.

Answer

Suppose the p.d. is V, then the energy ΔE given to the electron = eV.

In this case the equation $\Delta E = c^2\Delta m$ becomes $eV = c^2\Delta m$, and Δm becomes m_0 because the mass is doubled.

$$1.6 \times 10^{-19}\,\mathrm{C} \times V = (3.00 \times 10^{8}\,\mathrm{m\,s^{-1}})^2 \times 9.11 \times 10^{-31}\,\mathrm{kg}$$

$$\Rightarrow V = 5.1 \times 10^{5}\,\mathrm{V} \text{ or } 510\,\mathrm{kV}$$

Let's try to calculate the speed of an electron accelerated through 5000 kV in a linac producing X-rays.

Starting with $\frac{1}{2}m_e v^2 = eV$

and substituting $m_e = 9.1 \times 10^{-31}$ kg, $e = 1.6 \times 10^{-19}$ C and $V = 5\,000\,000$ V

$\Rightarrow v = 1.3 \times 10^9\,\text{m s}^{-1}$ to 2 sig. fig.

This is not possible (and we haven't made a mistake on the calculator!). Nothing can travel faster than the speed of light, $3.0 \times 10^8\,\text{m s}^{-1}$. So what has 'gone wrong'? We used Newton's formula for kinetic energy, and this assumes that objects have the same mass at all speeds. The theory of **special relativity** proposed by Einstein predicted that the faster things go, the heavier they become – as the Worked example above shows. (The graph in Figure 9.5 on page 83 shows this quantitatively.)

The electrons in the Stanford linac can acquire energies of 50 GeV, which is a factor of 10^5 greater than the 5000 keV. The rest mass of the Stanford electrons becomes negligible compared with their total effective mass – quite the opposite result from the example of the charged car battery!

8.9 Particle interactions

Figure 3.8 (page 19) shows an alpha particle striking a helium nucleus in a cloud chamber. As part of a Worked example (page 18), the *total* vector momentums before and after the collision, in the direction of the incoming α-particle, were calculated and found to be equal. In collisions momentum is always conserved; in nuclear collisions charge and mass–energy are also conserved.

In Figure 3.8 kinetic energy, a scalar quantity, is conserved as (try it):

$$\frac{1}{2}m_\alpha(1.50 \times 10^7\,\text{m s}^{-1})^2 = \frac{1}{2}m_\alpha(1.23 \times 10^7\,\text{m s}^{-1})^2 + \frac{1}{2}m_{He}(0.86 \times 10^7\,\text{m s}^{-1})^2$$

After cancelling the halves and the equal masses of m_α and m_{He}, each side is numerically 2.25×10^{14}. Here the speeds are not high enough (less that one tenth the speed of light) for the kinetic energies to represent a noticeable extra mass.

Sometimes collisions occur in which not all the particles are ionising because they are uncharged. Such particles leave no ionisation tracks in a bubble chamber.

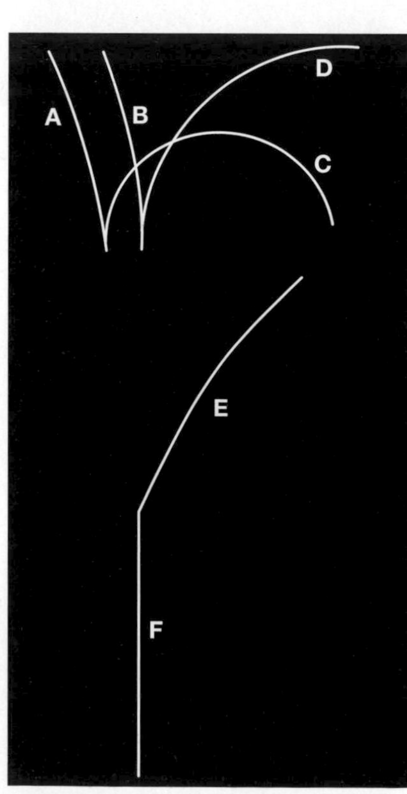

Figure 8.11 ▲

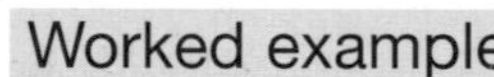

Worked example

There is a magnetic field into the plane of the bubble chamber tracks represented in Figure 8.11.

a) State and explain any deductions that can be made about the nature of the particles A, B, C, D, E and F.

b) Deduce the existence and nature of any other particles in these interactions.

Answer

a) Each of the particles A, B, C, D, E and F has an electric charge as each provides a track showing ionisation in the bubble chamber.
A and B are positively charged and the curvature of their tracks is similar. If they have equal charge then their momentums are equal.
C and D are negatively charged (from the left-hand rule). As $r_D = 2r_C$ then perhaps $Q_C = Q_D$ and their momentums would then be $p_D = 2p_C$ (since $p = BQr$).
E is a negatively charged particle. F's track is almost straight so it has a very high momentum; it must be negatively charged to conserve charge where F knocks E forward.

b) There must be two neutral particles that do not ionise, moving from the point where F makes a collision producing E. Perhaps they are γ-photons. Each then produces a pair of oppositely charged particles A/C and B/D to conserve charge where they are produced.

There is some guesswork in the answer above, but it illustrates how photographs of particle tracks and interactions can yield information. Nowadays powerful computing techniques are employed to make deductions from the information gathered by vast detecting systems at, for example, the LHC at CERN. Further examples are considered in Chapter 9, including deductions about the energies of the particles involved.

REVIEW QUESTIONS

1 The mass of a proton is N times the mass of an electron. A sensibly rounded value for N would be about:

A 200 B 550 C 2000 D 5500

2 The proton number of the element uranium is 92. The number of neutrons in an atom of the isotope ^{235}U is:

A 92 B 143 C 146 D 235

3 An energy of 12 pJ is equivalent to:

A 75 keV B 75 MeV C 7.5 GeV D 75 GeV

4 Take the density of nuclear matter as $1.0 \times 10^{16}\,\text{kg}\,\text{m}^{-3}$. It can be deduced that

a) the volume of a gold nucleus $^{197}_{79}$Au is:

A $0.2 \times 10^{-41}\,\text{m}^{-3}$ B $1.3 \times 10^{-41}\,\text{m}^{-3}$
C $2.0 \times 10^{-41}\,\text{m}^{-3}$ D $3.3 \times 10^{-41}\,\text{m}^{-3}$

b) the radius of a gold nucleus is about:

A $2.0 \times 10^{-14}\,\text{m}$ B $3.0 \times 10^{-14}\,\text{m}$
C $5.0 \times 10^{-14}\,\text{m}$ D $6.0 \times 10^{-14}\,\text{m}$

5 The curved path in Figure 8.3 (page 70) is visible because:

A electrons glow when they travel at high speeds

B electrons leave a trail of tiny water drops along their path

C gas atoms in the tube are attracted to one another and form particles

D gas atoms in the tube are ionised by the electrons.

6 Figure 8.12 shows the path of an α-particle scattered by a gold nucleus. Copy the diagram.

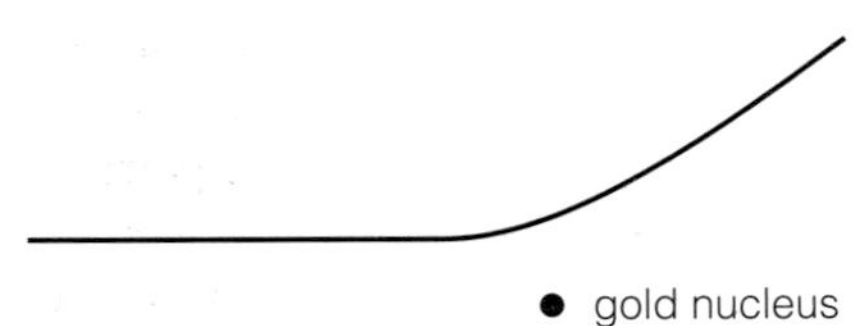

Figure 8.12 ▲

a) Label with the letter M the point on the α-particle's path where it feels the maximum force from the gold nucleus.

b) Add the path followed by an α-particle moving initially along the same line but having a smaller energy than that in the diagram.

c) Add the path followed by the original α-particle but which approaches a nucleus carrying a smaller charge than gold.

In b) and c) identify the two new tracks with suitable labels.

7 The electric potential energy (EPE) of a particle of charge Q_1 a distance r from a nucleus of charge Q_2 is given by the relationship

$$EPE = \frac{kQ_1Q_2}{r} \text{ where } k = 9.0 \times 10^9\,\text{N}\,\text{m}^2\,\text{C}^{-2}$$

a) Calculate the EPE of an α-particle $5.0 \times 10^{-14}\,\text{m}$ from the centre of a gold nucleus. (The charge on an α-particle and a gold nucleus is $2e$ and $79e$ respectively.)

b) Express this energy in MeV and explain how an α-particle with this kinetic energy will slow to zero and then reverse as it 'hits' a gold nucleus head on.

8 Draw a labelled sketch of an electron gun.

9 Protons each of momentum p move in a circle of radius r when projected perpendicular to a magnetic field of flux density B. It is shown on page 71 that $r = p/Be$.

For speeds that are not too high, derive a relationship between the kinetic energy E_k of protons moving in a circle and the radius r.

10 Prove that the synchronous frequency in a proton cyclotron is given by $f = Be/2\pi m_p$.

11 The gap between two 'tubes' in a proton linac is 25 mm. The alternating voltage has a maximum value of 200 kV. What is the maximum value of the accelerating electric field between the tubes?

12 The track of an α-particle ionising air molecules in its path is a straight line. Each ionisation requires 30 eV. The graph in Figure 8.13 shows the number of ionisations an α-particle makes as it moves from its source to the end of its track.

Use the graph to estimate the initial energy of an α-particle that travels 60 mm in air.

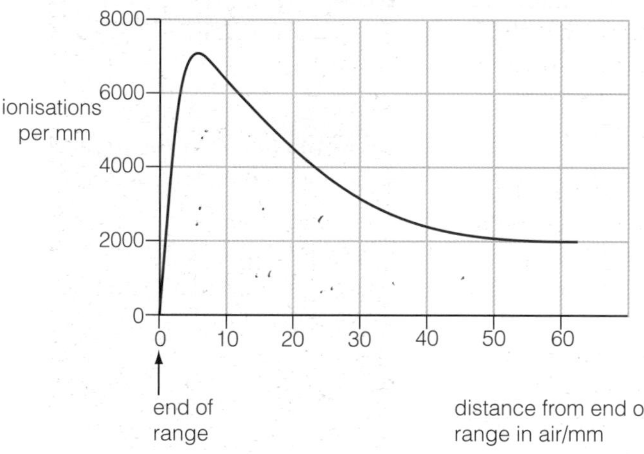

Figure 8.13 ▲

13 Protons are accelerated to an energy of 10 MeV in a cyclotron. Calculate their relativistic mass increase and express this as a percentage of their rest mass 1.67×10^{-27} kg.

14 Refer to Figure 8.8 on page 76. The momentum of an electron when moving perpendicular to a magnetic field of flux density B is given by $p = Ber$, where r is the radius of its path.

a) Measure the radius of the electron's path at A, B and C. Take the centre of its path as D. (The photograph is ⅔ real size.)

b) Calculate the momentum of the electron at A, B and C, given that $B = 1.2$ T.

c) This high-energy electron is moving at a speed $v = 3.0 \times 10^8\,\text{m}\,\text{s}^{-1}$ to 2 sig. fig. at each point. Calculate values for the mass of the electron at A, B and C.

d) How do your answers compare with the rest mass 9.1×10^{-31} kg of an electron?

15 Refer to Figure 3.8 on page 19. Show that the momentum at right angles to the initial direction of the α-particle is zero. What assumption do you make?

9 Particle physics

Some of the particles we are learning about here are the building blocks of matter that we see around us. These are protons, neutrons and electrons. We'll look at how they behave and how they make up atoms.

At a deeper sub-atomic level, high-energy experiments have revealed a mysterious world of quarks and leptons. To learn about this world you will need to abandon some Newtonian mechanics because of the very high speeds of the particles.

9.1 The discovery of quarks

The scattering of α-particles by a thin sheet of gold was described on page 69. This experiment was performed in Manchester in 1909. Sixty years later, a similar experiment was performed at Stanford in the USA, but this time the bombarding particles were high-energy electrons and the target was liquid hydrogen in a bubble chamber. At low energies the negatively charged electrons were deflected by the protons forming the nuclei of hydrogen as if the protons' positive charge was confined in a tiny volume (as in the α-particle scattering experiment). There was no net loss of kinetic energy, i.e. the scattering was elastic (Figure 9.2a).

At high electron energies – above about 6 GeV (6×10^9 electron-volts) – funny things began to happen. Now the electrons lost a lot of energy in collision and the proton fragmented into a shower of particles rather than recoiling (Figure 9.2b). This energy-to-matter transformation meant that the collision was inelastic. The conclusion was that protons were not tiny 'balls' of positive charge but contained localised charge centres. The electrons were interacting with these charge centres via the electrostatic force, which is an inverse-square law force.

The charge centres, of which there are three in the proton and also in the neutron, are called **quarks**. Figure 9.3 shows the quarks 'inside' a proton and a neutron: they are of two varieties: 'up' quarks (u, with a charge $+\frac{2}{3}e$) and 'down' quarks (d, with a charge $-\frac{1}{3}e$).

The quarks are bound together tightly (the springs in the model shown in Figure 9.3 represent this), and in order to break the quarks apart very high-energy bombarding electrons are needed. However, the experimenters did not find any individual quarks in the showers of particles emanating from the collisions, and no-one has yet found a 'free' quark. The shower of particles were mainly 'mesons' consisting of quark–antiquark pairs – see the next page and page 88.

Figure 9.1 ▲
The Stanford linear accelerator

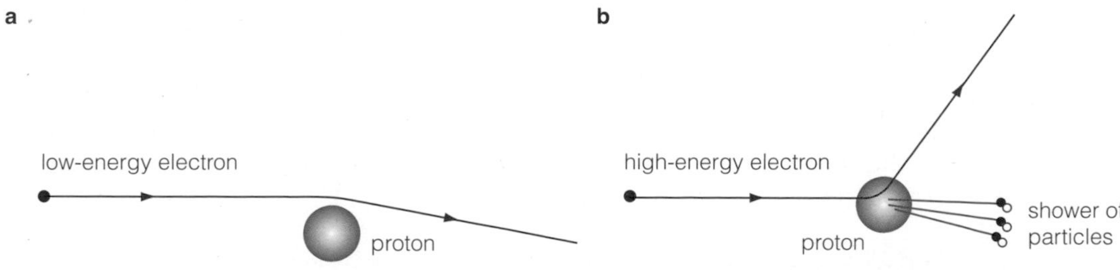

Figure 9.2 ▲
a) Elastic and b) deep inelastic scattering

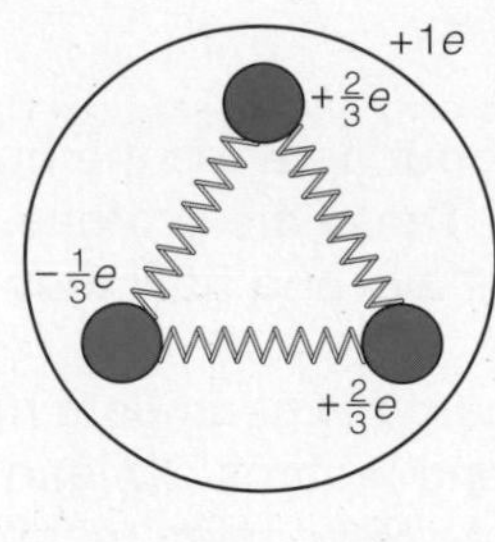

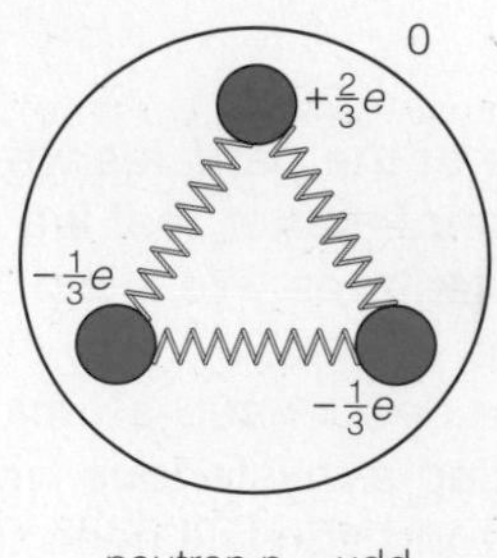

Figure 9.3 ▲
The quarks that make up a proton and a neutron

9.2 Matter and antimatter

We now believe that all fundamental particles like up and down quarks and electrons have antiparticles with exactly the same mass as their corresponding particles but with opposite charges. The antiparticle to the electron is called a **positron** and has the symbol e^+. The photograph in Figure 9.4a shows a cloud chamber in which a charged particle entering from the right is slowed down as it travels through a 6 mm lead plate across the middle of the chamber. Figure 9.4b shows a diagram of the particle track.

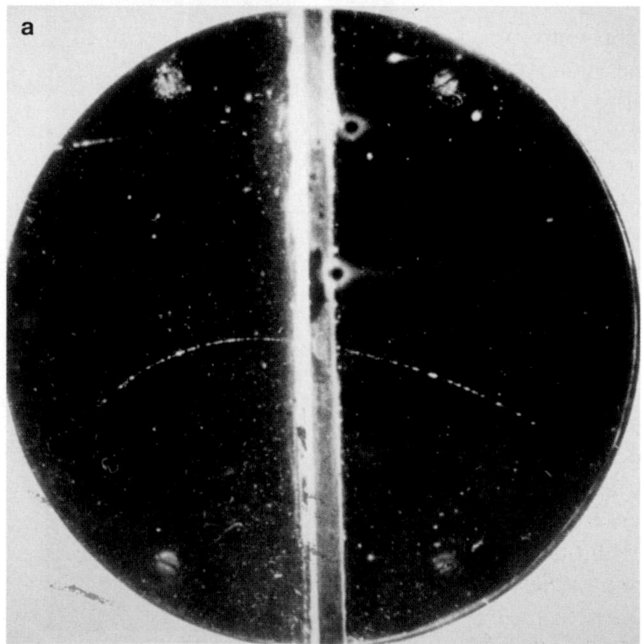

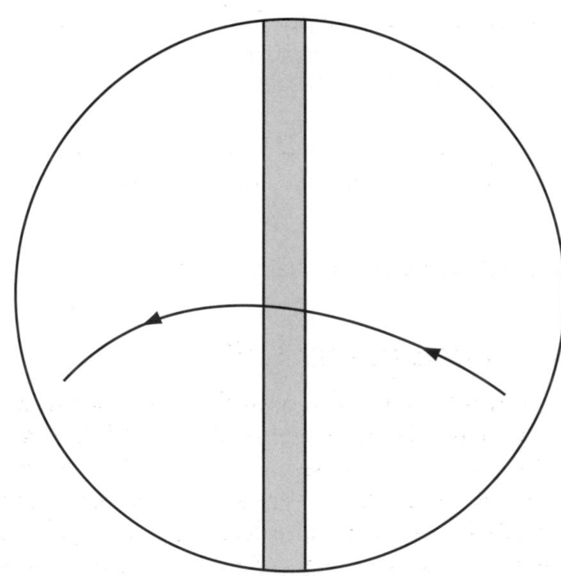

Figure 9.4 ▲
A cloud chamber photograph from 1933

On page 71 it was shown that the radius r of such a track in a magnetic field of flux density B was related to the particle's momentum p by the relationship $r = p/BQ$, where Q is the charge on the particle. The remarkable thing about this 1933 photograph is that the magnetic field was directed downwards, and so the charge on the particle was, by the left-hand rule, positive. The charge was found to be $+e$; it was a positron.

Worked example

In Figure 9.4, the magnetic flux density $B \approx 1.5$ T. Assuming the photograph is 'life-size', calculate the momentum of the positron before and after it penetrates the lead sheet.

Answer
The measured radii on the photograph are about 60 mm and 30 mm.

Using $p = BQr$ with $Q = e$ gives
$p_{\text{before}} = 1.5\,\text{T} \times 1.6 \times 10^{-19}\,\text{C} \times 0.06\,\text{m} = 1.4 \times 10^{-20}\,\text{N s}$
$p_{\text{after}} = 1.5\,\text{T} \times 1.6 \times 10^{-19}\,\text{C} \times 0.03\,\text{m} = 0.7 \times 10^{-20}\,\text{N s}$

The rest mass m_0 of both an electron and a positron is 9.1×10^{-31} kg, so a simple substitution into $p = m_0 v$ would suggest that the positron was travelling *much* faster than the speed of light $c = 3.0 \times 10^8\,\text{m s}^{-1}$. This is not possible. Einstein's theory of special relativity predicted that the faster things go, the heavier they become. The graph in Figure 9.5 puts this quantitatively and describes the mass m of an object increasing as a fraction of m_0 as its speed increases.

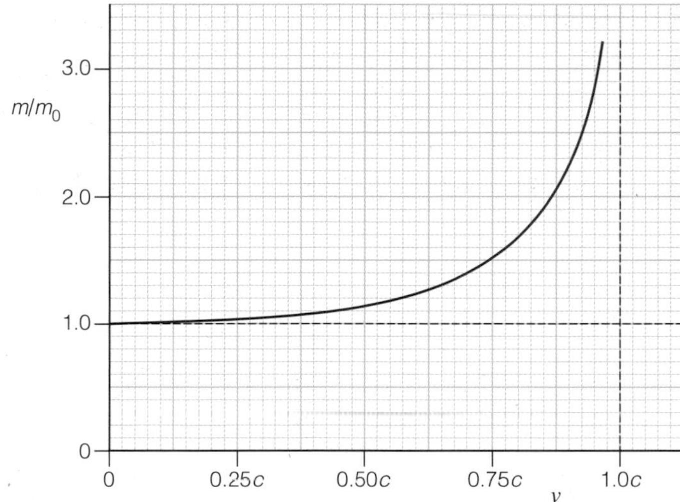

Figure 9.5 ▲
The relativistic increase of mass with speed

Note

No object can break the 'light barrier', i.e. travel faster than the speed of light, c.

9.3 Some useful units

So far in this book we have used the SI unit kilogram for mass and the SI unit joule for energy. We have also used a non-SI unit for energy, the electron-volt (eV). Of course all of these – kg, J and eV – can be used with multiples of ten, e.g. kilo-, milli-, etc. Another unit for mass, the **unified atomic mass unit**, symbol **u**, is useful in nuclear and particle physics. It measures masses on a scale where the isotope of carbon $^{12}_{6}\text{C}$ is given a mass of exactly 12 u.

As a result of Einstein's equation $\Delta E = c^2\,\Delta m$ (see page 76), another non-SI unit for mass or extra mass used in high-energy particle physics is **MeV/c^2** or **GeV/c^2**. So the rest mass of a particle might be given as 250 MeV/c^2 which, after multiplying by $1.6 \times 10^{-13}\,\text{J MeV}^{-1}$ and dividing by $(3.0 \times 10^8\,\text{m s}^{-1})^2$, equates to 4.4×10^{-28} kg.

Table 9.1 summarises these different mass units.

Definition

$m_{\text{carbon-12}}$ = 12.000 000 .... (for ever) u
This means
1 u = $1.660\,566 \times 10^{-27}$ kg or 1.66×10^{-27} kg to 3 sig. fig.

	kg	u	MeV/c^2
1 kg =	1	6.02×10^{26}	5.62×10^{29}
1 u =	1.66×10^{-27}	1	1.07×10^{-3}
1 MeV/c^2 =	1.78×10^{-30}	934	1

Table 9.1 ▲
Mass units used in particle physics

Tip

Familiarity with these non-SI units can only be achieved by practising changing from one to another. A common error is to forget to square c. It may be worth remembering that $c^2 = 9.0 \times 10^{16}\,\text{m}^2\,\text{s}^{-2}$.

9.4 Creation and annihilation of matter

The conservation of mass–energy explains two simple but astonishing events:

pair production photon → electron + positron

positron annihilation electron + positron → photon(s)

> **Note**
>
> Remember that a photon is a quantum (bundle) of electromagnetic energy.

Figure 9.6a shows pair production in a bubble chamber across which there is a magnetic field. Here a γ-ray photon creates an electron–positron pair of charged particles. Figure 9.6b is a corresponding diagram to which a dotted line has been added to show the track of the incoming photon.

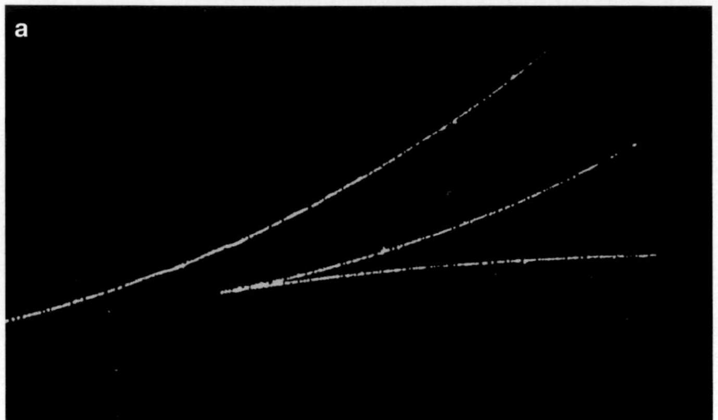

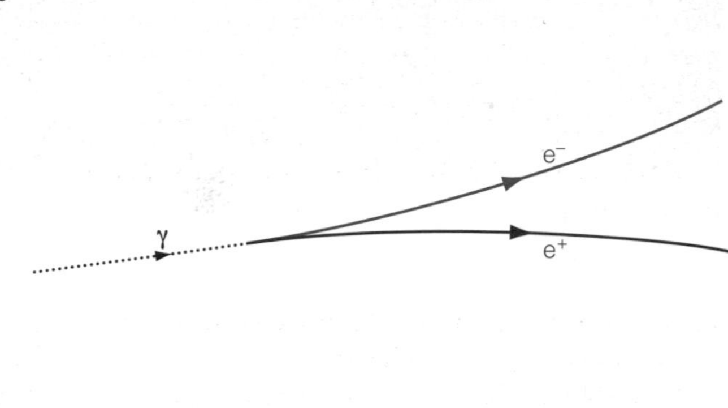

Figure 9.6 ▲
Pair production in a bubble chamber

In this event, charge is conserved: zero before and $(-e) + (+e) = 0$ after. Mass–energy is also conserved – see the Worked example below. (Momentum is conserved, as is partly shown by the equal but opposite curvature of the two particles' paths.)

Worked example

Show that the minimum γ-ray energy necessary for electron–positron pair production is 1.02 MeV.

Answer

The rest mass of an electron plus a positron is
$9.1 \times 10^{-31}\,\text{kg} + 9.1 \times 10^{-31}\,\text{kg}$.
This total mass has an energy equivalence $\Delta E = c^2\,\Delta m$, where
$\Delta m = 18.2 \times 10^{-31}\,\text{kg}$. So the minimum energy E for the photon must be

$$E = (3.0 \times 10^{8}\,\text{m s}^{-1})^2 \times (18.22 \times 10^{-31}\,\text{kg}) = 1.64 \times 10^{-13}\,\text{J or } 1.02\,\text{MeV}$$

The annihilation of an electron and a positron obviously conserves charge, but it is more difficult to demonstrate that the resulting electromagnetic radiation conserves energy, unless the resulting photons (often two, for example when the annihilating particles have equal and opposite momentums) subsequently produce detectable charged particles. What we can say is that a pair of photons from such an annihilation will each have an energy of more than 0.5×1.02 MeV and thus a minimum wavelength ($\lambda = hc/E$) of 0.6×10^{-12} m. This is a γ-photon.

PET scans

Antimatter particles are nowadays in use routinely in hospitals, whenever a PET (positron emission tomography) scan is used to image activity in a person's brain. Suppose you are asked to help with experiments to see which part of your brain is working when you look at a screen, listen to music, talk to your friends or think about your homework. The brain scans for such activities in a normal brain are shown in Figure 9.7.

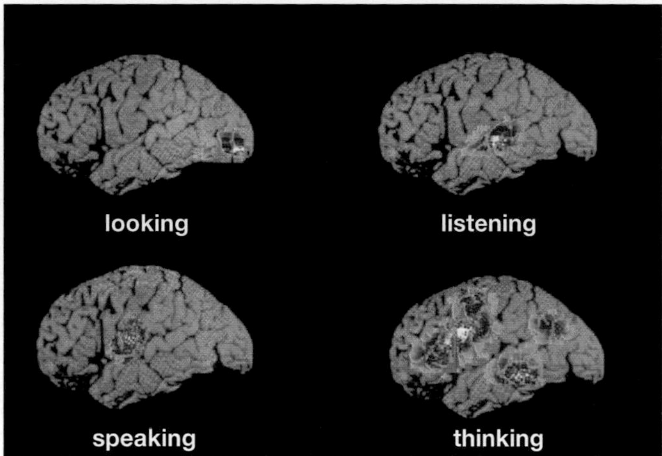

Figure 9.7 ▲

How is it done? Your blood is injected with a small amount of a liquid containing oxygen-15. This is radioactive with a half life (see page 46) of 122 seconds. In decaying in your brain, the oxygen-15 nucleus emits a positron that immediately annihilates with a local electron to produce two γ-rays, each of energy 511 keV. As the positron and electron are effectively stationary, the γ-rays move apart in opposite directions (conserving momentum) and are detected by surrounding devices called scintillators. The part of your brain they come from is deduced from the difference in time they take to reach the scintillators, and a computer produces the resulting image.

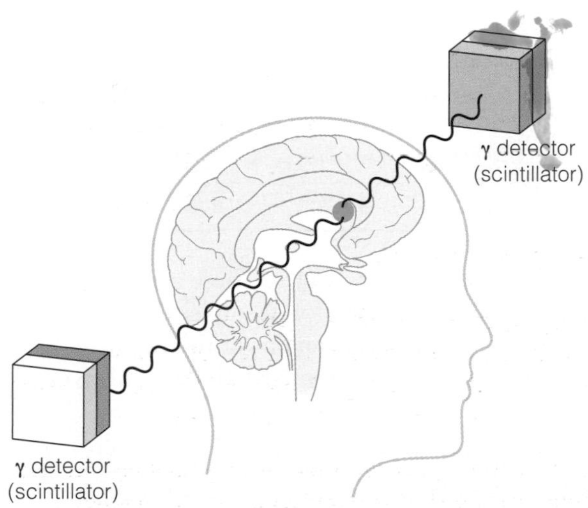

Figure 9.8 ▲
Photons from matter–antimatter annihilation enable brain activity to be imaged

Other particle–antiparticle annihilations are possible, for example that resulting from a proton meeting an antiproton. Here the energy is much higher, because the mass loss is much greater. Mass–energy is conserved by the emission of several photons or of other fundamental particles, as described below. Perhaps in a few years we will see experiments to annihilate a hydrogen atom with an anti-hydrogen atom (these are produced by their thousands in physics laboratories around the world).

9.5 The standard model

In Section 9.1 you read that the up and down quarks were discovered using high-energy electrons. All these particles – the u quark, the d quark and the electron – are 'fundamental' particles. The electron and its antiparticle the positron are members of a group of fundamental particles called **leptons**.

When electrons and positrons of very high energy annihilate each other they produce two different leptons that we call **muons** (μ), in this case a μ^- and its antiparticle a μ^+. At even higher energies the annihilation produces two more massive leptons that we call taus (τ), here τ^- and τ^+. The leptons μ^- and τ^- each have a charge of -1.6×10^{-19} C, and their antiparticles μ^+ and τ^+ have a balancing positive charge.

These fundamental leptons form the three 'generations' shown in Table 9.2.

Note

Neutrinos, which you will learn about in Chapter 13 in relation to radioactive decay, are also leptons.

	I	II	III
	electron, e^-	**muon, μ^-**	**tau, τ^-**
charge	$-e = -1.6 \times 10^{-19}$ C	$-e = -1.6 \times 10^{-19}$ C	$-e = -1.6 \times 10^{-19}$ C
rest mass	0.511 MeV/c^2	106 MeV/c^2	1780 MeV/c^2

Table 9.2 ▶ The three generations of leptons

Interactions such as radioactive decay have shown that electrons (and positrons) are associated with the up and down quarks that make up the protons and neutrons of everyday matter. High-energy experiments have shown similarly that the muon and tau leptons are associated with other quarks. These are shown in Table 9.3.

	I	II	III
	up quark, u	**charmed quark, c**	**top quark, t**
charge	$+\frac{2}{3}e$	$+\frac{2}{3}e$	$+\frac{2}{3}e$
mass	a few MeV/c^2	about 1 GeV/c^2	over 100 GeV/c^2
	down quark, d	**strange quark, s**	**bottom quark, b**
charge	$-\frac{1}{3}e$	$-\frac{1}{3}e$	$-\frac{1}{3}e$
mass	a few MeV/c^2	about 0.1 GeV/c^2	a few GeV/c^2

Table 9.3 ▶ The three generations of quarks

All quarks also have their corresponding antiparticles. These are written using the same symbols but with a bar above them: so, for example, u and $\bar{u}$, d and $\bar{d}$.

Historically, it was the symmetry of the arrangement of the three generations that led physicists to establish the **standard model**. This is summarised in Table 9.4.

e	μ	τ	leptons, charge $-e$
u	c	t	quarks, charge $+\frac{2}{3}e$
d	s	b	quarks, charge $-\frac{1}{3}e$

Table 9.4 ▶ The standard model of fundamental particles

Ordinary matter is made entirely from the 'first generation' of particles: the electron plus the up and down quarks. Figure 9.9 opposite illustrates this.

Quarks have never been isolated, which is why their masses are not precisely known. They join together in threes (baryons) and twos (mesons).

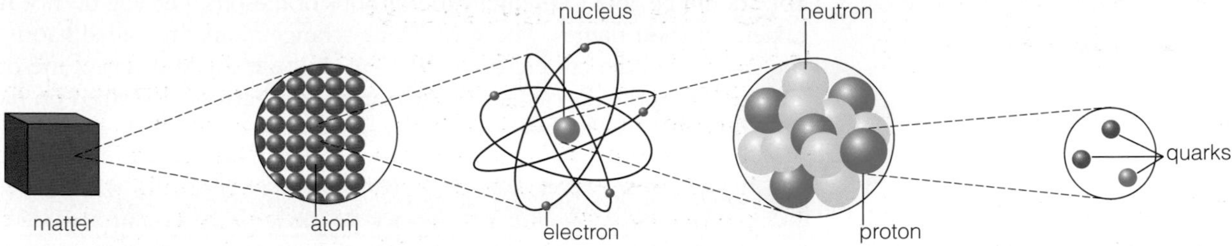

Figure 9.9 ▲
The structure of ordinary matter

9.6 Baryons and mesons

Even before quarks had been discovered, lots of previously unseen, massive particles carrying a charge of zero or $+e$ or $-e$ had been found in experiments using the increasing energies available from various accelerators. Assuming these particles are made up of two or three quarks, can you predict, using only u and d quarks plus their antiparticles ū and d̄, what some of these might be?

How about:

ddd, charge: $(-\frac{1}{3}e) + (-\frac{1}{3}e) + (-\frac{1}{3}e) = -e$

or dū, charge: $(-\frac{1}{3}e) + (-\frac{2}{3}e) = -e$

If you make use of the other four quarks, many many more combinations result. It is even possible to get particles with a charge of $+2e$ or $-2e$; can you see how?

Particles made from **three quarks**, **qqq** or **q̄q̄q̄**, are called **baryons**. Protons and neutrons are baryons. The uuu has been found with charge $+2e$, it is designated the delta particle, Δ^{++}. All baryons have their antiparticles, for example the antiproton is ūūd̄.

Tip

Try to remember:

uud is a proton

udd is a neutron.

Other than uud and udd, you do not need to remember the names or structure of any baryons.

Worked example

What is the charge on a uds baryon?

Answer

Charge $= (+\frac{2}{3}e) + (-\frac{1}{3}e) + (-\frac{1}{3}e) = 0$, i.e. it is a neutral particle.

Tip

When listing the charges for a baryon it is safest to bracket the charge on each quark so as not to get confused with $+-$ or $-+$ situations.

The uds is called a sigma zero, Σ^0, and is one of a group of Σ particles containing one strange quark. Particles with two strange quarks, e.g. dss, are called xi (Ξ) particles; the dss is xi minus, Ξ^-. The sss baryon is called omega minus, Ω^-. Its discovery was an important step in confirming the correctness of the quark model.

Particles made of **two quarks** are called **mesons**. They are all combinations of a quark and an antiquark: **qq̄**. One group of mesons, known as pions, are ud̄, uū or dd̄, and dū, as shown in Figure 9.10. These have charges of $+e$, 0, and $-e$, and are designated as π^+, π^0 and π^-.

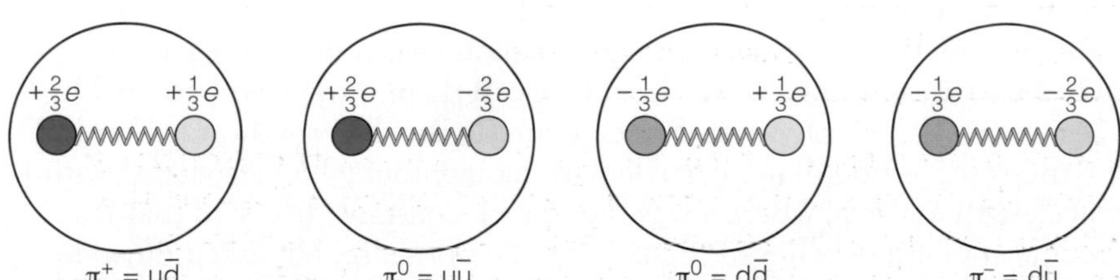

Figure 9.10 ▲
The pion group of mesons

You should be able to predict other groups of mesons, but you do not need to remember their names. Those with one strange quark are called kaons (K); the $s\bar{s}$ is called the eta meson, symbol η^0. Note it was showers of mesons that emerged from the deep inelastic scattering of electrons described on page 81 (Figure 9.2b).

Exercise

Use the standard model to predict *all* possible mesons. You should get more than 20!

It is important to remember that in all reactions the **total number of baryons is conserved**. Thus $p + p \rightarrow p + \pi^+$ is not possible (although charge is conserved), but $p + p \rightarrow p + p + \pi^0$ is possible.

Tip

In reactions involving baryons and mesons any extra quarks that materialise are always mesons of the form $q\bar{q}$.

Worked example

One possible outcome when two protons interact is: $p + p \rightarrow p + n + \pi^+$. Analyse this reaction in terms of the charge and deduce the quark structure of the π^+.

Answer

Left-hand side p + p has charge $(+e) + (+e) = +2e$.

Right-hand side $p + n + \pi^+$ has no charge from n but $(+e)$ from p and $(+e)$ from π^+, a total of $+2e$.

So charge is conserved.

Writing the known quarks in brackets:

$$p(uud) + p(uud) \rightarrow p(uud) + n(udd) + \pi^+(q\bar{q})$$

as π^+ is a meson and must consist of a quark and an antiquark. An up quark is needed on the right and an extra down quark seems to have appeared. So the meson must be a $u\bar{d}$ meson, of charge $(+\frac{2}{3}e) + (+\frac{1}{3}e) = +e$ as needed.

The quarks are thus:

$$uud + uud \rightarrow uud + udd + u\bar{d}$$

What has actually happened in the interaction is that some energy has gone to producing a $d\bar{d}$ quark–antiquark pair of zero net charge. The d then goes to form the neutron and the $\bar{d}$ to form the π meson.

Figure 9.11 ▲ Prince Louis de Broglie

9.7 Wave–particle duality

Wave–particle duality is a physicist's way of saying that waves have particle-like properties and particles behave in a wave-like manner. You have already met photons – 'particles' of electromagnetic radiation. The answer, then, to the question: 'What is light?' is that sometimes it behaves like a wave of wavelength λ and sometimes as a 'particle' or photon of energy hc/λ, where h is the **Planck constant**. ($h = 6.63 \times 10^{-34}$ J s)

The reverse, that particles with mass, such as electrons, can behave like waves seems astonishing. What should we call them – wavicles perhaps? In 1924 a young French physicist, Prince Louis de Broglie, was awarded his PhD for a thesis suggesting that a particle with momentum p had associated with it a wavelength $\lambda = h/p$, where h was the Planck constant. It is said that the awarding committee believed the thesis to be nonsense, but asked Einstein (who happened to be in Paris at the time) to look at it. Einstein said that he agreed it was without foundation, but added: 'You will feel very foolish if de Broglie turns out to be right'. And he *was* right!

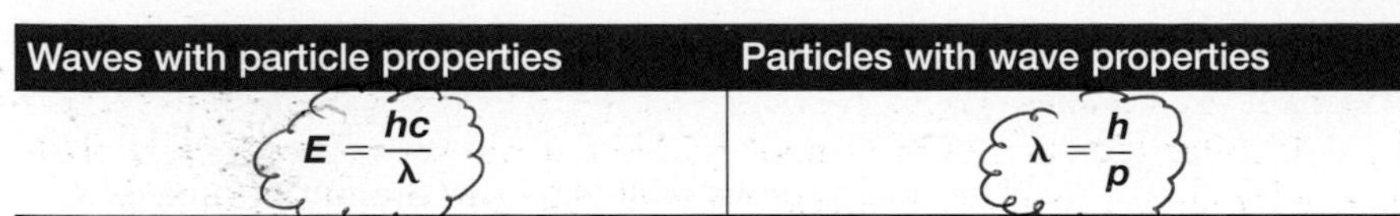

Waves with particle properties	Particles with wave properties
$E = \frac{hc}{\lambda}$	$\lambda = \frac{h}{p}$

Table 9.5 ◀
Wave–particle duality summarised

Don't you think it is amazing that the Planck constant h links energy with wavelength for electromagnetic waves *and* wavelength with momentum for particles?

Worked example

Calculate the wavelength associated with a beam of electrons that has been accelerated through 1200 V.

Answer

A voltage of 1200 V gives the electrons an energy of

$$1200\,\mathrm{J\,C^{-1}} \times 1.6 \times 10^{-19}\,\mathrm{C} = 1.92 \times 10^{-16}\,\mathrm{J}$$

For this voltage we can use the non-relativistic relationship (see page 14) between kinetic energy and momentum:

$$p^2 = 2mE \qquad \Rightarrow \qquad p = 1.87 \times 10^{-23}\,\mathrm{kg\,m\,s^{-1}}$$

using $m = 9.1 \times 10^{-31}$ kg. Hence:

$$\lambda = \frac{h}{p} = \frac{6.64 \times 10^{-34}\,\mathrm{J\,s}}{1.87 \times 10^{-23}\,\mathrm{kg\,m\,s^{-1}}} = 3.6 \times 10^{-11}\,\mathrm{m}$$

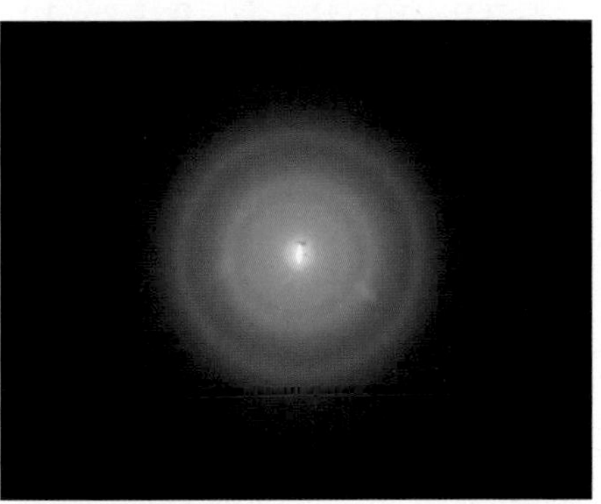

Figure 9.12 ▲
Electron diffraction rings

In your AS course you came across the fact that electrons can produce diffraction patterns. Diffraction is a key property of waves, because it is the result of wave superposition. The photograph in Figure 9.12 shows the outcome for electrons diffracted by graphite crystals in a thin sheet of graphite.

Exercise

The diameters d of electron diffraction rings vary with the accelerating voltage V. For energies up to about 5 kV it can be shown that $d \propto 1/\sqrt{V}$. Results from an experiment in which the voltage is varied provide the results given in Table 9.6.

V/kV	2.5	3.0	4.0	5.0
d/mm	37	34	29	26

Table 9.6 ▲

Show that these are consistent with the above relationship linking d to V.

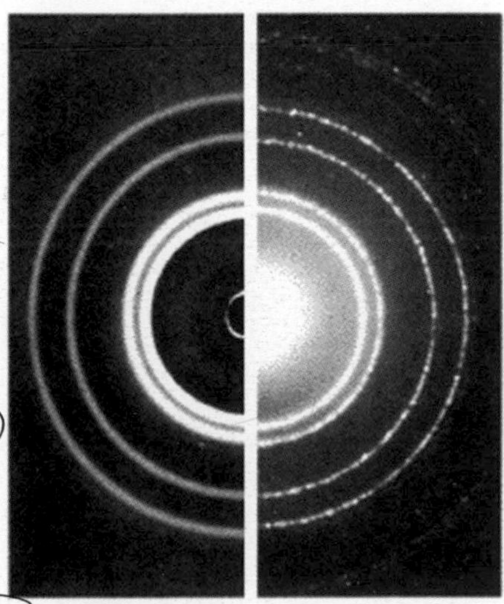

Figure 9.13 ▲
Demonstrating wave–particle duality: diffraction patterns for *left* X-rays and *right* electrons

Other diffracting crystals can be used. Figure 9.13 shows, side by side, the diffraction patterns produced by passing X-rays and electrons through the *same* very thin aluminium foil. It was arranged that the X-rays and the electrons would have the same wavelength. Do you need any further proof of the wave-like behaviour of particles?

REVIEW QUESTIONS

1 An elastic collision is one:

A between low-energy particles

B in which electrostatic forces act

C in which mass is conserved

D where no kinetic energy is lost.

2 Which of the following is **not** a unit of mass?

A kg B N C u D MeV/c^2

3 An energy of 6.4 MeV is equivalent to:

A 6.4×10^{-30} J B 1.0×10^{-15} J
C 6.4×10^{-12} J D 1.0×10^{-12} J

4 220 MeV/c^2 is equivalent, to 2 sig. fig., to a mass of:

A 3.9×10^{-28} kg B 3.7×10^{-25} kg
C 3.1×10^{-8} kg D 2.1×10^{8} kg

5 The minimum photon energy needed to create a proton–antiproton pair is:

A 3.0×10^{-10} u B 3.0×10^{-10} J
C 1.0×10^{-19} u D 1.0×10^{-19} J

6 Which of the following is an impossible baryon?

A ddu B $\bar{u}\bar{d}\bar{d}$ C $ud\bar{s}$ D $\bar{u}\bar{d}\bar{s}$

7 Which of the following is a possible meson?

A ds B $u\bar{d}$ C $\bar{d}\bar{s}$ D us

8 How many baryons are there in an atom of $^{7}_{3}$Li?

A zero B 3 C 4 D 7

9 This drawing of a bubble chamber event showing electron–positron pair production at A is incorrect.

Figure 9.14 ▲

The tracks shown in Figure 9.14 are not possible because:

A the particle tracks curve in different senses

B charge is not conserved

C momentum is not conserved

D the magnetic field was upwards.

10 The graph in Figure 9.5 on page 83 can be described by the equation

$$m = \frac{m_0}{\sqrt{\left(1 - \frac{v^2}{c^2}\right)}}$$

By taking at least three readings from the graph, show that it has been correctly drawn.

11 The Stanford accelerator can accelerate electrons beyond 30 GeV.

a) What is the mass in kg of a particle of mass 30 GeV/c^2?

b) Express this energy as a multiple of the rest mass of an electron.

12 The initial (lower) interaction in Figure 8.11 on page 78 is the decay of a K^- meson into two pi mesons, π^0 and π^+. Express this decay in terms of the quarks involved, indicating how the charge is conserved by your quark equation.

13 A K^+ meson can decay into three pions as shown at D in Figure 9.15.

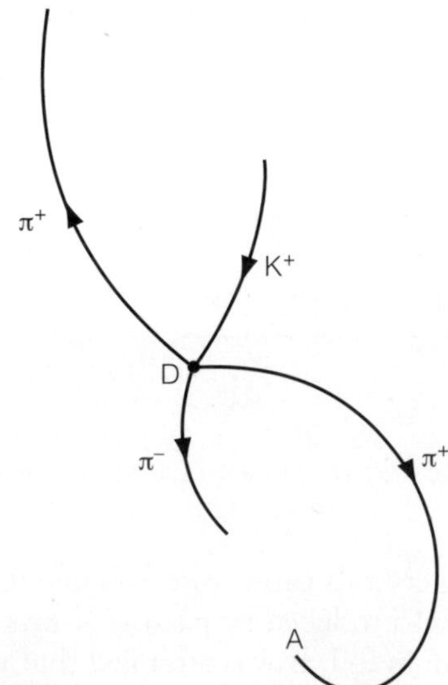

Figure 9.15 ▲

a) State the direction of the magnetic field affecting the paths shown in Figure 9.15.

b) Why does the 'downward' π^+ meson follow a path DA that reduces gradually in radius?

14 The famous Ω^- particle, with quark structure sss, was first found as a result of a K^- meson interacting with a proton to give a K^0 and a K^+ plus the Ω^-. Given that the K^- meson is $s\bar{u}$, discuss the quarks involved in this reaction.

15 Calculate the wavelength associated with a tennis ball of mass 57.5 g that is served at $220\,\text{km}\,\text{h}^{-1}$, and comment on the result.

16 Figure 9.16 shows how an electron orbiting a proton in a hydrogen atom might be thought of as forming a standing wave. Suggest how this model might predict discrete or quantised energy levels for hydrogen.

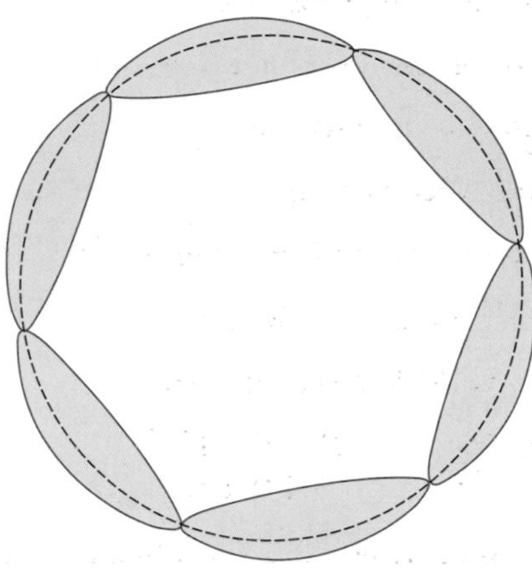

Figure 9.16 ▲

Unit 4 test

Time allowed: 1 hour 35 minutes
Answer **all of** the questions.
There is a data sheet on page 225.

For Questions 1–10, select one answer from A–D.

1 Which of the following is **not** a possible unit for momentum?

A Ns B $\mathrm{kg\,m\,s^{-2}}$ C $\mathrm{J\,m^{-1}\,s}$ D $\mathrm{kg\,m\,s^{-1}}$

[Total: 1 mark]

2 The basic principle on which a firework rocket is propelled is:

A by chemicals in the rocket exploding

B by pushing exhaust gases against the air

C by throwing out exhaust gases into the air

D by Isaac Newton's first law of motion.

[Total: 1 mark]

3 A singly ionised helium atom and an ionised hydrogen atom are each accelerated in a vacuum through a potential difference of 200 V. The kinetic energy gained by the helium is:

A a quarter of the energy gained by the hydrogen

B half the energy gained by the hydrogen

C the same as the energy gained by the hydrogen

D twice the energy gained by the hydrogen.

[Total: 1 mark]

4 When in a car speeding round a corner at $13\,\mathrm{m\,s^{-1}}$ (just under 30 m.p.h.), a passenger with a mass of 65 kg experiences a sideways centripetal force of 550 N. The road is part of a corner of radius (to 2 sig. fig.):

A 20 m B 15 m C 11 m D 10 m

[Total: 1 mark]

5 The effect of the forces on an electric dipole placed across a uniform electric field is:

A to accelerate the dipole against the direction of the field

B to accelerate the dipole in the direction of the field

C to twist the dipole to lie perpendicular to the field

D to twist the dipole to lie parallel to the field.

[Total: 1 mark]

6 Which of the following statements about electric fields and magnetic fields is **incorrect**?

A They are both produced by electric currents.

B They both occur naturally around the Earth.

C They both exert forces on moving charges.

D They can both be described by field lines.

[Total: 1 mark]

7 A proton enters a uniform magnetic field of magnetic flux density 0.50 T. The proton is moving at $1.2 \times 10^{8}\,\mathrm{m\,s^{-1}}$ along the magnetic field lines. The magnetic force on the proton:

A is about 10^{-11} N

B is about 10^{-10} N

C is zero

D cannot be calculated because the proton is moving close to the speed of light.

[Total: 1 mark]

8 Protons being accelerated in a cyclotron move at a constant speed while inside the Dees because:

A they are shielded from the magnetic field by the Dees

B there is no force acting on them along their path

C only while in the Dees are they travelling in a vacuum

D the alternating voltage supply is synchronised to their movement.

[Total: 1 mark]

9 The mass of a quark thought to be $1.6\,\mathrm{GeV}/c^2$ would be:

A 1.1×10^{-8} kg B 3.3×10^{-19} kg

C 2.8×10^{-27} kg D 2.8×10^{-30} kg

[Total: 1 mark]

10 Which of the following is a possible reaction?

A $p + \pi^{+} \rightarrow p + p$

B $p + n \rightarrow \pi^{+} + \pi^{-}$

C $p + p \rightarrow p + p + n$

D $p + p \rightarrow p + p + \pi^{0} + \pi^{0}$

[Total: 1 mark]

11 Figure 1 shows how the horizontal push of the ground on a sprinter varies with time during one of his first few strides.

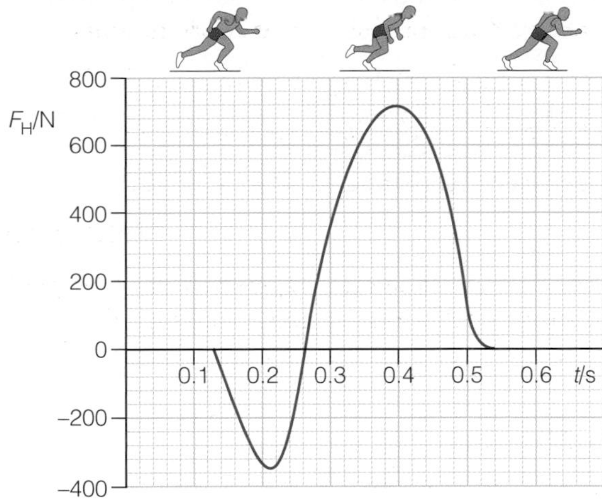

Figure 1 ▲

a) Explain what information can be deduced from such a graph. [3]

b) Describe the general shape of a similar graph for a stride just after the sprinter has crossed the winning line. [2]

[Total: 5 marks]

12 Figure 2 shows a small charged polystyrene sphere weighing only 1.1×10^{-3} N on the end of a thread 440 mm long. It is being pushed aside from the vertical by a charged, fixed sphere. The distance between the centres of the spheres is 37 mm.

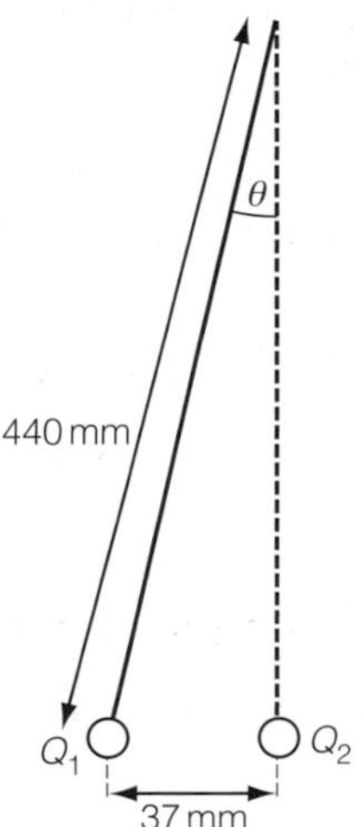

Figure 2 ▲

a) Draw a free-body force diagram for the sphere with charge Q_1.

Use the diagram to prove that the electric force F between the spheres is given by $F = W \tan\theta$. [3]

b) Hence show that F is approximately 0.1 mN. [2]

c) If $Q_1 = 5.0 \times 10^{-9}$ C, calculate Q_2. [3]

[Total: 8 marks]

13 Figure 3 shows a bicycle dynamo with the dynamo housing and lamp in a cut-away style. Describe and explain the workings of the dynamo and lamp unit.

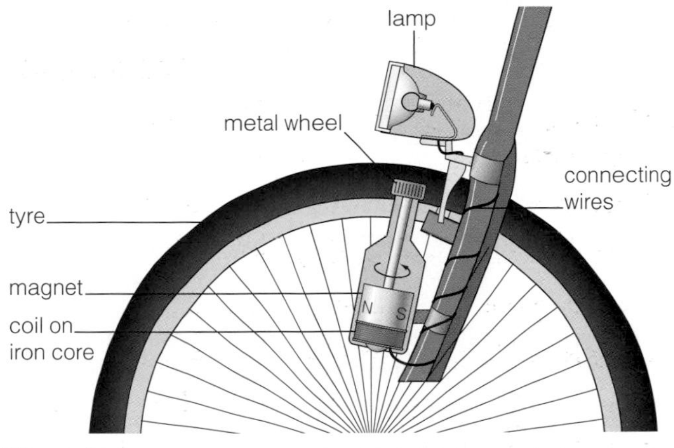

Figure 3 ▲

[Total: 5 marks]

14 The charge on a capacitor is found as it charges through a fixed resistor from a 12 V supply. The graph of Figure 4 shows the results.

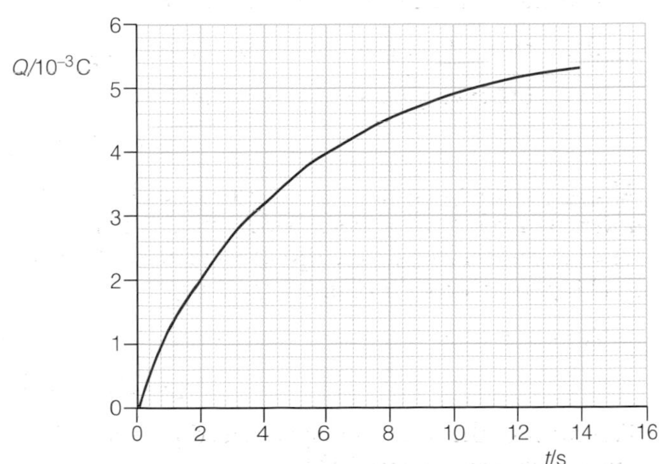

Figure 4

a) Estimate a value for the maximum charge Q_0 stored. [1]

b) Explain how the current I in the circuit at any time t can be found from the graph. [2]

c) Explain how the time constant RC of the circuit can be deduced from the graph and give a value for RC. [2]

d) The resistor used had a resistance of 8.5 kΩ. Calculate, using two different methods, two values for the capacitance C of the capacitor. [4]

[Total: 9 marks]

15 The Moon circles the Earth every 27.3 days at an average distance from Earth of 383×10^3 km. The radius of the Earth is 6.38×10^3 km.

a) Calculate the Moon's centripetal acceleration as it circles the Earth. [3]

b) Compare your value with the free-fall acceleration g_0 at the Earth's surface and suggest how this is related to the ratio 383/6.38. [2]

[Total: 5 marks]

16 In a hydrogen atom the average distance apart of the proton and the electron is 5.3×10^{-11} m. Assuming the electron to be a particle orbiting the proton, explain how the frequency *f* with which the electron is orbiting can be calculated and deduce a value for *f*.

[Total: 6 marks]

17 Figure 5 shows three drift tubes that form part of an electron linac – a linear accelerator. The charges on the tubes at one instant are shown.

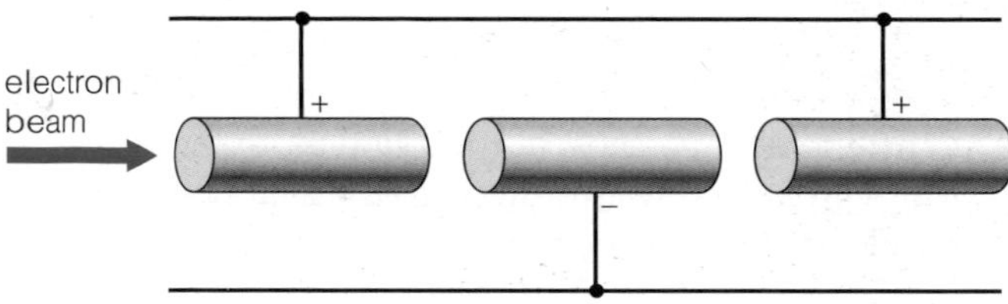

Figure 5 ▲

a) Copy the diagram. Add electric field lines to your diagram and explain the function of these electric fields in the workings of this electron linac. [4]

b) i) Why are the tubes referred to as 'drift tubes'?

ii) Suggest why the tubes appear to be of equal lengths in the diagram. [3]

c) Calculate the increase in mass of an electron accelerated through a p.d. of 8.4 GV and comment on the result. [4]

[Total: 11 marks]

18 Figure 6 shows two vertical metal plates supported in a vacuum. The left plate is positively charged and the right plate is earthed.

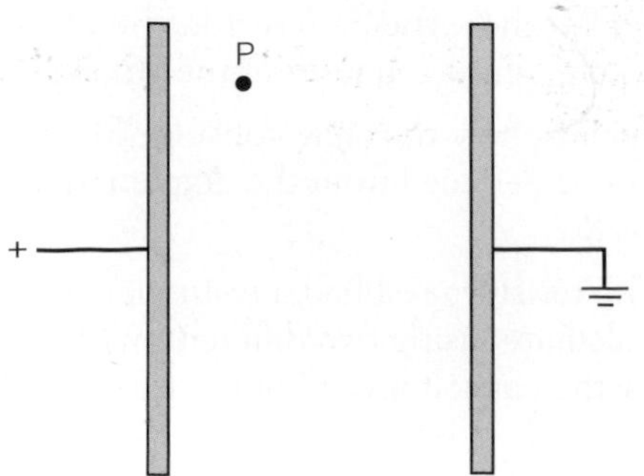

Figure 6 ▲

a) Describe the shapes of the *g*-field and the *E*-field between the plates. [3]

b) Draw a free-body force diagram for a positively charged dust particle P that has a mass *m* and carries a charge *Q*. [2]

c) Describe and explain the path that the dust particle will follow as it moves in the vacuum between the plates. [3]

[Total 8 marks]

19 Figure 7 represents tracks in a bubble chamber before and after a particle interaction.

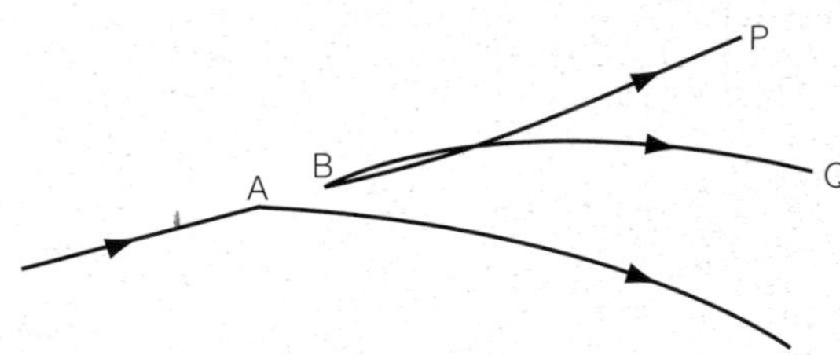

Figure 7 ▲

a) The quark components of the particles involved in the reaction at A where a xi particle decays are:

dss → uds + ūd

i) State the nature of the particles involved and determine the charge on each.

ii) Hence deduce the sense of the magnetic field across the cloud chamber. [6]

b) One of the particles appearing as a result of the decay at B is a proton.

i) State which, P or Q, is the proton.

ii) What is the charge on the other particle?

iii) Give a quantitative explanation of which particle has the greater momentum, using the usual symbols for the quantities involved. [7]

[Total: 13 marks]

[TOTAL: 80 marks]

Unit 5

Topic 5 Thermal energy

10 Specific heat capacity
11 Internal energy and absolute zero
12 Gas laws and kinetic theory

Topic 6 Nuclear decay

13 Nuclear decay

Topic 7 Oscillations

14 Oscillations

Topic 8 Astrophysics and cosmology

15 Universal gravitation
16 Astrophysics
17 Cosmology

Topic 9 Practical work

18 A guide to practical work

10 Specific heat capacity

Specific heat capacity is a property of a material. It is a measure of how much energy we need to supply to a given mass of that material in order to raise its temperature. Knowledge of the specific heat capacity of different substances is important in the choice of materials for particular applications, for example storage heaters. It is quite easy to determine reasonable values for the specific heat capacity of solids and liquids in a school or college laboratory, so you will be expected to be familiar with such experiments and their limitations. You also need to be clear what is meant by the terms *heat* and *temperature* so that you can use these terms correctly and in the appropriate context.

10.1 Heat and temperature

'Heat' and 'temperature' are two words that we often use in everyday life. They are such common words that you would think it would be easy to explain what we understand by each of them – but it isn't! We know what we mean by 'hot' and 'cold' – we can feel the difference – but ask your friends to tell you what the words 'heat' and 'temperature' mean and see what responses you get!

They might say that temperature is 'how hot something is', but then what does 'hot' mean – it's just another, less scientific, word for 'high temperature'! What we do know from simple observation is that, if two bodies are at different temperatures, *energy* flows from the 'hot' body to the 'cold' body. In physics this flow of energy, from a higher temperature to a lower temperature, due to conduction, convection or radiation, is what we call 'heat'. As heat is a form of energy it is measured in the units of energy, i.e. joules (J).

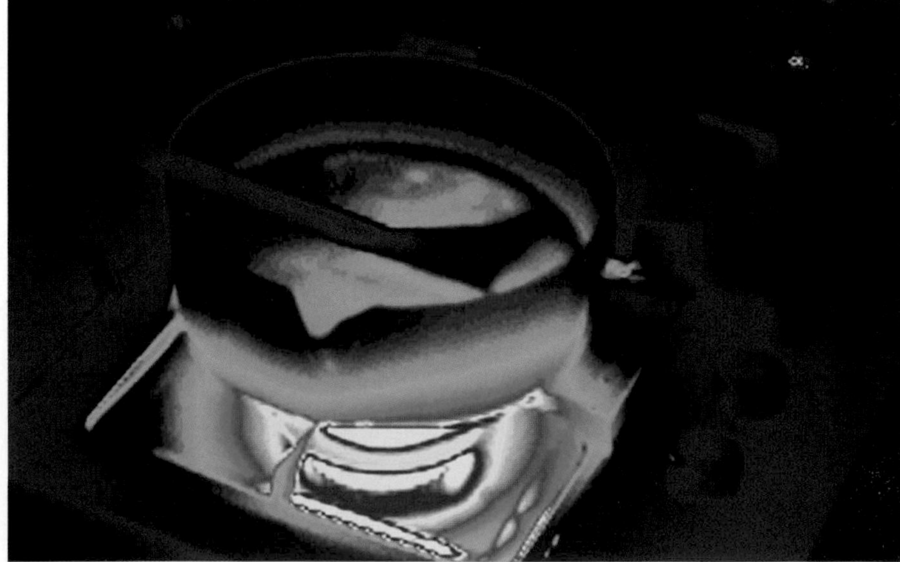

Figure 10.1 ▲
A thermogram

Figure 10.1 shows a thermogram of soup being heated in a pan on a gas hob. Thermography records the temperature of surfaces by detecting long-wavelength infrared **radiation**. The colours show variations in temperature, with a scale that runs from white (hottest), through red, yellow, green and blue to pink (coldest). The blue of both the pan and the soup show good heat **conduction** by the metal pan. The plastic handle of the spoon is a poor conductor and stays pink. The intense temperature of the gas flames is seen under the pan. Heat from the flames to the pan is mainly transferred by **convection**.

You should understand that heat is a *flow* of energy. The idea that heat is something contained *in* a body is wrong. What a body *does* have is **internal energy**. This is made up of the **potential energy** contained within the inter-atomic bonds and the **kinetic energy** of the vibrations of the atoms. The potential component arises from the energy stored in the inter-atomic bonds that are being continuously stretched and compressed as the atoms vibrate – rather like springs. The kinetic energy depends on the temperature – the hotter a body is, the more rapidly its atoms vibrate and so the greater their kinetic energy becomes. So:

the **temperature** of a body is a measure of the mean, random, kinetic energy of its vibrating atoms.

We will look at this kinetic theory of matter again in Chapter 11, where we will see that the internal energy of a body (and hence its temperature) can also be increased by doing '**work**' on it – for example, rubbing your hands together makes them hot!

Definition

Heat is the flow of energy from a region of higher temperature to a region of lower temperature.

10.2 Units of temperature

Temperature is one of the fundamental (base) SI quantities. The base SI unit of temperature is the **kelvin**, symbol K (not °K). This is defined in terms of what is called the absolute thermodynamic scale of temperature, which has **absolute zero** as its zero and *defines* the melting point of ice as 273 K (to 3 sig. fig.). The meaning of 'absolute zero' will be explained in Chapter 11.

In practical everyday situations, we measure temperature on the Celsius scale, which defines the melting point of ice as 0 °C and the boiling point of water as 100 °C under certain specified conditions.

However, a temperature *interval* of 1 K is exactly equivalent to a temperature interval of 1 °C – for example, a temperature rise of 1 °C from 20 °C to 21 °C is precisely the same as a temperature rise of 1 K from 293 K to 294 K.

We usually give the symbol θ to a temperature recorded in °C and the symbol T if the temperature is in K. From the above, it follows that a *change* in temperature of $\Delta\theta$ measured in °C is numerically the same as the corresponding *change* of temperature ΔT in K. As the kelvin is the SI unit of temperature we should, strictly speaking, always use 'K' when we are quoting a temperature change.

Absolute zero (0 K) corresponds to a temperature of approximately −273 °C on the Celsius scale, therefore temperatures measured in °C can be converted to temperatures in K by using

$$T/\text{K} = \theta/°\text{C} + 273$$

Figure 10.2 ▲ Lord Kelvin, who introduced the absolute scale of temperature

Tip

Remember:

Temperature	°C	K
absolute zero	−273	0
ice point of water	0	273
boiling point of water	100	373

Table 10.1 ▲

Worked example

Calculate the missing temperatures in Table 10.2, which shows the boiling points of some common elements.

Element	°C	K
aluminium	2350	a)
argon	−186	b)
copper	c)	2853
helium	d)	4

Table 10.2 ▲

Answer

a) 2350 °C = 2350 + 273 = 2623 K
b) −186 °C = −186 + 273 = 87 K
c) 2853 K = 2853 − 273 = 2580 °C
d) 4 K = 4 − 273 = −269 °C

10.3 Specific heat capacity

Imagine that you were to heat 1 kg of copper, 1 kg of aluminium and 1 kg of water by means of an electric heater, connected to a joule-meter, and measure how much energy was needed to raise the temperature of each by, say, 10 °C – whoops, we should say 10 K!

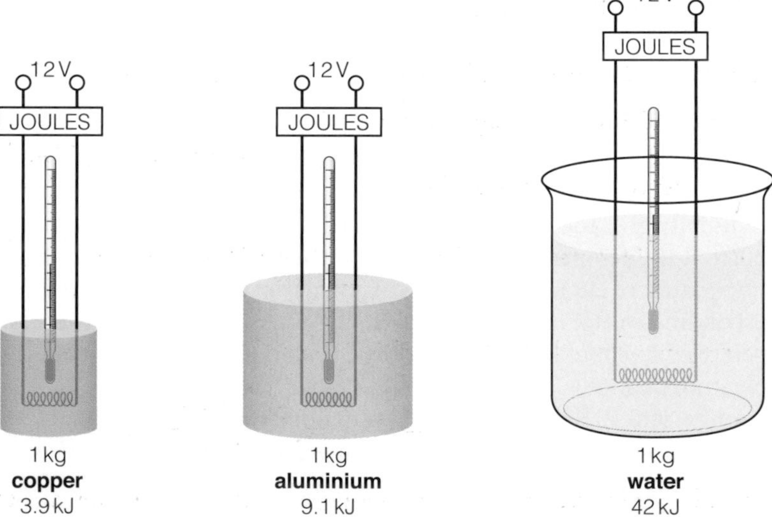

Figure 10.3 ▲

Definition

Specific heat capacity is the quantity of energy needed to raise the temperature of a material, per kilogram per degree rise in temperature.

You would find that it took about 3.9 kJ for the copper, 9.1 kJ for the aluminium and 42 kJ for the water. It is clear from this that a significantly different quantity of energy is needed to raise the temperature of equal masses of different materials by the same amount.

The property of a material that quantifies this is called its **specific heat capacity**, which is given the symbol *c*.

If we had *m* kg of the material and we raised its temperature by ΔT K, the energy needed would be

$$\Delta E = mc\,\Delta T$$

Rearranging:

$$c = \frac{\Delta E}{m\,\Delta T}$$

giving the units of c as $\mathbf{J\,kg^{-1}\,K^{-1}}$.

In Figure 10.3 we had 1 kg of each material and the temperature rise was 10 K in each case. This tells us that the specific heat capacities of copper, aluminium and water are as given in Table 10.3.

Material	Specific heat capacity/J kg^{-1} K^{-1}
copper	390
aluminium	910
water	4200

Table 10.3 ▲

Worked example

a) An aluminium saucepan of mass 400 g, containing 750 g of water, is heated on a gas hob. How much energy would be required to bring the water from a room temperature of 18 °C to the boil?

b) An electric kettle rated at 2.1 kW contains 1.2 kg of water at 25 °C. Calculate how long it will take for the water to come to the boil. Explain why it will actually take longer than you have calculated.

Answer

a) You have to remember that the saucepan will have to be heated to 100 °C, as well as the water, so:

$$\Delta E = m_a c_a \Delta T + m_w c_w \Delta T \text{ where } \Delta T = (100-18)\,°\text{C} = 82\,\text{K}$$

$$\Delta E = 0.400\,\text{kg} \times 910\,\text{J kg}^{-1}\text{K}^{-1} \times 82\,\text{K} + 0.750\,\text{kg} \times 4200\,\text{J kg}^{-1}\text{K}^{-1} \times 82\,\text{K}$$

$$= 30\,\text{kJ} + 260\,\text{kJ} = 290\,\text{kJ}$$

b) Energy to heat water $\Delta E = mc\,\Delta T = 1.2\,\text{kg} \times 4200\,\text{J kg}^{-1}\text{K}^{-1} \times 75\,\text{K} = 378\,\text{kJ}$

Energy supplied electrically $= P\,\Delta t = 378\,\text{kJ}$

$$\Rightarrow \Delta t = \frac{378 \times 10^3\,\text{J}}{2.1 \times 10^3\,\text{J s}^{-1}} = 180\,\text{s}$$

It will take longer than this, because some energy will be needed to heat the element of the kettle and the kettle itself, and some energy will be lost to the surroundings.

10.4 Measuring specific heat capacity

The specific heat capacity, of both solids and liquids, can be found by simple electrical methods using the principle:

electrical energy transferred by heater = increase in internal energy of material

$$\Delta E = IV\,\Delta t = mc\,\Delta T$$

Experiment

Measuring the specific heat capacity of aluminium

Weigh the block of aluminium to find its mass m and then place it in the lagging. Insert a little cooking oil into the two holes to ensure good thermal contact when the heater and the thermometer are inserted.

Take the initial temperature θ_i of the block. Switch on the power supply and start the stopclock. Record the current I and the potential difference V.

After a time Δt of 3.00 minutes, switch off the power and record the highest steady temperature θ_f reached by the block. Then, from:

$$IV\Delta t = mc(\theta_f - \theta_i)$$

you can calculate the specific heat capacity c:

$$c = \frac{IV\Delta t}{m(\theta_f - \theta_i)}$$

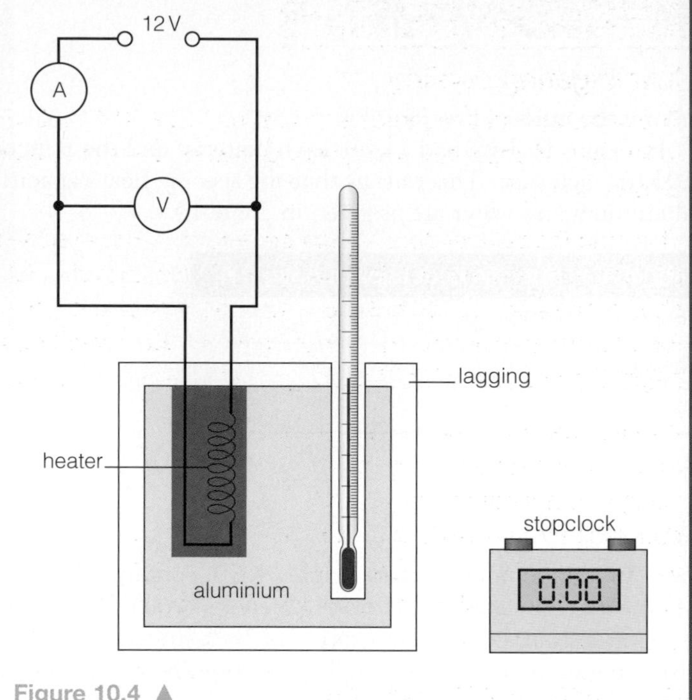

Figure 10.4 ▲

Worked example

The following data were recorded in an electrical method to find the specific heat capacity of aluminium:

$m = 993\,g$	$V = 10.3\,V$	$\theta_i = 21.4\,°C$
$I = 3.1\,A$	$\Delta t = 3.00$ minutes	$\theta_f = 27.3\,°C$

Use these data to calculate the specific heat capacity of aluminium.

Answer

Using the formula above, these data give:

$$c = \frac{3.1\,A \times 10.3\,V \times 3.00 \times 60\,s}{0.993\,kg \times (27.3 - 21.4)\,K} = 980\,J\,kg^{-1}\,K^{-1}$$

The experiment described above is quick and easy to perform, and gives quite a good value for the specific heat capacity. There are, however, several sources of error:

- energy is absorbed by the heater itself; this is the main source of error and will make the value of c too large because it means that not all the energy supplied is used to raise the temperature of the block of aluminium;
- energy is lost to the surroundings, despite the lagging (also making c too large);
- a little energy will be taken by the lagging (again making c too large);
- inaccuracy of the thermometer, especially as $\Delta\theta$ is fairly small (this could make c too large if $\Delta\theta$ were too small or too small if $\Delta\theta$ were too large);
- inaccuracy of the meters (again, this could affect the value of c either way).

This method, nevertheless, is perfectly adequate for examination purposes, providing you are aware of its limitations and understand why the value of c obtained is likely to be too large.

Experiment

Investigating cooling

The same apparatus can be used as in Figure 10.4, but *without* the lagging, to investigate more fully what happens when the block is heated and then allowed to cool.

This time the temperature should be recorded during heating at regular intervals (at least every half minute) for 12 minutes, after which the heater should be switched off, but the temperature should continue to be recorded for a further 8 minutes. The current and voltage should be recorded at the beginning and end of the heating period in order to get average values.

A better way of doing this would be to record the data electronically using a temperature sensor and data logger; the results can then be recorded and displayed on a computer. A typical such printout is shown in Figure 10.5.

In the experiment that gave rise to the curve in Figure 10.5, the block had a mass of 1.00 kg; the heater was on for 12.0 minutes, with the average current and voltage being 1.65 A and 9.59 V respectively.

A value for the specific heat capacity of the aluminium can be found as before by reading off θ_i as 25.5 °C and θ_f as 37.0 °C from the graph. Check for yourself that this gives a value for c of 990 J kg^{-1} K^{-1} (don't forget to convert g to kg and minutes to seconds!).

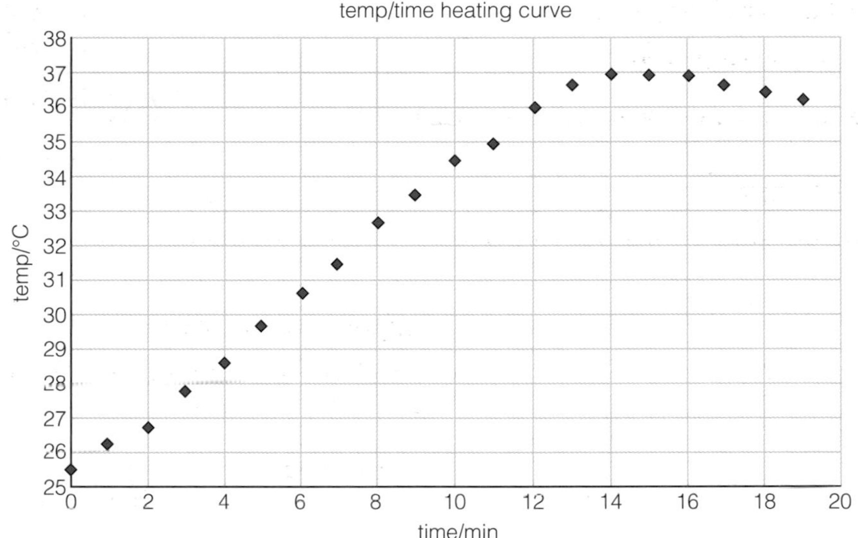

Figure 10.5 ▲

Taking account of energy loss to the surroundings

The graph in Figure 10.5 provides us with a lot of information about the transfer of energy that takes place when the block is heated and then allowed to cool. It is therefore worth studying in some detail.

Initially, the rate at which the temperature rises is fairly slow, because it takes time for the heat to conduct from the heater, through the block to the thermometer, even though aluminium is a good conductor.

Then the line becomes almost linear when, to a good approximation, the rate of transfer of electrical energy in the heater is equal to the rate at which energy is absorbed by the block. After a few minutes, the rate at which the temperature rises gradually gets less. This is because the energy transferred to the surroundings gets greater as the temperature of the block increases.

After the power has been switched off (at 12 minutes), the temperature of the block continues to rise for about a further 2 minutes, because it takes a finite time for all the heat to conduct from the heater through the block to the thermometer.

Tip

Always work in base units when doing specific heat calculations and remember that $\Delta\theta$/°C = ΔT/K (there is no need to add or subtract 273).

Once all the energy from the heater has been transferred to the block, the transfer of heat to the surroundings becomes clearly apparent as we can now see the temperature of the block starting to fall. The block, of course, has been transferring energy to the surroundings all the time. The value of 990 J kg^{-1} K^{-1} that we obtained for the specific heat capacity is likely to be too large because of this heat transferred to the surroundings and the energy absorbed by the heater itself.

One way of making allowance for the heat transferred to the surroundings is to determine the gradient $\Delta\theta/\Delta t$ of the graph line at the point where it first becomes almost linear – in this case after about 2 minutes. At this point, the temperature of the block is less than 2 K above room temperature and so the rate of loss of energy to the surroundings is negligible. We can therefore reasonably assume that:

rate of electrical energy transferred by heater = rate of energy gained by block

$$IV = mc\frac{\Delta\theta}{\Delta t}$$

Exercise

Determine the gradient of the graph in Figure 10.5 where it is linear. You should find that $\Delta\theta/\Delta t$ is about 0.0167 K s^{-1}. Check that this gives a value for c of 950 J kg^{-1} K^{-1} (to 2 sig. fig.). You should remember that it is essential to express $\Delta\theta/\Delta t$ in units of K s^{-1} and not K min^{-1}.

This is much closer to the accepted value of c for aluminium, but does not take into account the energy taken by the heater itself. This energy could be determined from a separate experiment and subtracted from the electrical energy supplied. Alternatively, it could be assumed that the specific heat capacity of the heater material is not significantly different from that of aluminium and so the mass of the heater could be added to that of the block. In practice, this is quite a reasonable approximation to make.

Finding the specific heat capacity of a *liquid* poses a slight problem, as the liquid has to be in some form of container, such as a glass beaker. To get an accurate value for the specific heat capacity of the liquid, we must take into account the energy taken by the container as well as by the liquid.

This apparent difficulty can be largely overcome by using an *expanded* polystyrene cup to contain the liquid. This has the two-fold advantage of having very little mass and also being a very good insulator. This means that it both absorbs very little energy itself and at the same time minimises heat transfer from the liquid to the surroundings.

Experiment

Measuring the specific heat capacity of water

The heater is a 15 Ω, 11 W ceramic body, wire-wound resistor. A 12 V power supply should be used, with digital meters set on the 20 V and 2 A ranges respectively.

Use a measuring cylinder to pour 120 ml (m = 120 g) of water into the expanded polystyrene cup. The set-up is shown in Figure 10.6. Take the initial temperature θ_i of the water and then switch on the power supply and start the stopclock. Record the current I and the voltage V.

After a time Δt of 5.0 minutes has elapsed, switch off, stir well and record the highest steady temperature θ_f reached by the water.

Calculate the specific heat capacity c of the water from:

$$c = \frac{IV\Delta t}{m(\theta_f - \theta_i)}$$

The same method can be used to determine the specific heat capacity of cooking oil, but this can be very messy!

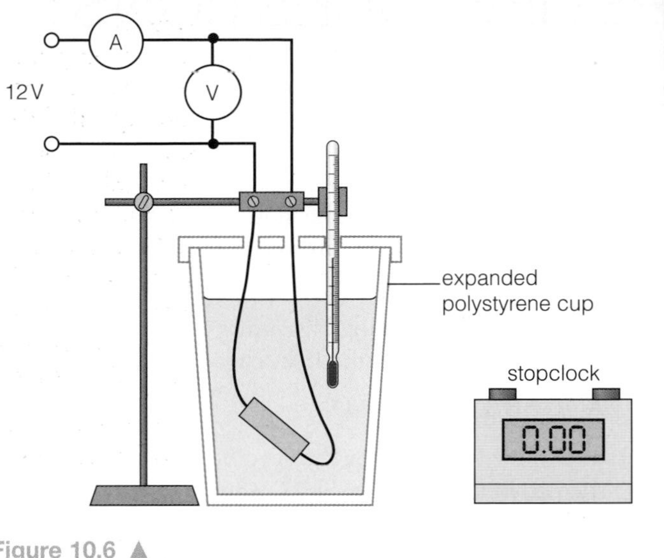

Figure 10.6 ▲

Worked example

In an experiment similar to that described above, 120 g of water was used. The water was heated for 5.0 minutes, during which time the average current was 0.77 A and the potential difference was 11.4 V. The temperature of the water rose from 21.2 °C to 26.3 °C.

a) What value do these data give for the specific heat capacity of water?

b) What are the main sources of error in such an experiment and how would each of them affect this value?

Answer

a) $IV\Delta t = mc\Delta\theta$

$$c = \frac{IV\Delta t}{m(\theta_f - \theta_i)} = \frac{0.77\,\text{A} \times 11.4\,\text{V} \times 5.00 \times 60\,\text{s}}{0.120\,\text{kg} \times (26.3 - 21.2)\,\text{K}}$$

$$= 4300\,\text{J}\,\text{kg}^{-1}\,\text{K}^{-1}$$

b) The main sources of error are:

- energy taken by the heater itself (making c too large);
- heat lost to the surroundings, mainly from the top of the cup (also making c too large);
- thermometer error, especially as $\Delta\theta$ is only about 5 K (could make c too large or too small, depending on whether $\Delta\theta$ were too small or too large respectively);
- meter errors (which could affect the value of c either way);
- small amount of energy taken by the thermometer and cup (making c too large).

Tip

Learn the two basic experiments to find the specific heat capacity for a solid and for a liquid and understand the main sources of error in each.

Remember to use SI units when doing calculations – particularly kg and seconds.

REVIEW QUESTIONS

1 In an experiment to find the specific heat capacity of cooking oil, 139.3 g of the oil is poured into a thin plastic cup. The temperature of the oil is taken and found to be 21.4 °C. The oil is then heated for 3.0 minutes by means of an immersion heater, which is rated at 11 W. The maximum steady temperature reached by the oil is 29.3 °C.

a) Which of the following base units does **not** occur in the base units of specific heat capacity?

A K B kg C m D s

b) The value obtained for the specific heat capacity of the oil, in units of $\text{J kg}^{-1}\text{K}^{-1}$, is approximately:

A 0.03 B 1.8 C 30 D 1800

c) The experimental value obtained for the specific heat capacity of the oil should be quoted to:

A 1 significant figure

B 2 significant figures

C 3 significant figures

D 4 significant figures

d) Which of the following could be a reason for the value obtained being too small?

A Energy has been lost to the surroundings.

B The mass of the cup has been ignored.

C The thermometer is reading systematically low.

D The heater is operating at less than its stated value.

2 Copy and complete the table, which shows the temperature of the melting points of some common elements.

Element	°C	K
hydrogen	−259	a)
iron	1540	b)
nitrogen	c)	63
sulphur	d)	388

Table 10.4 ▲

3 The hot water for a bath is supplied from a hot water tank in which there is a 3.0 kW immersion heater.

a) How much energy would be needed to heat 0.25 m^3 of water for a bath from 15 °C to 35 °C? The specific heat capacity for water is $4200\,\text{J kg}^{-1}\text{K}^{-1}$.

b) Show that it would take nearly 2 hours to heat this amount of water.

c) Why, in practice, would it take longer than this?

d) Suggest why it would save energy by taking a shower instead of a bath.

4 A freezer has an internal volume of 0.23 m^3 and operates at an internal temperature of −18 °C. How much energy must be removed to cool the air inside from 20 °C to the operating temperature? You may assume that the density of air is $1.3\,\text{kg m}^{-3}$ and that the specific heat capacity of air is $1.0 \times 10^3\,\text{J kg}^{-1}\text{K}^{-1}$.

5 A student pours 117 cm^3 of water (specific heat capacity $=4200\,\text{J kg}^{-1}\text{K}^{-1}$) into a thin plastic coffee cup and heats it with a 100 W immersion heater. He records the temperature as the water heats up and then plots his data as shown in Figure 10.7.

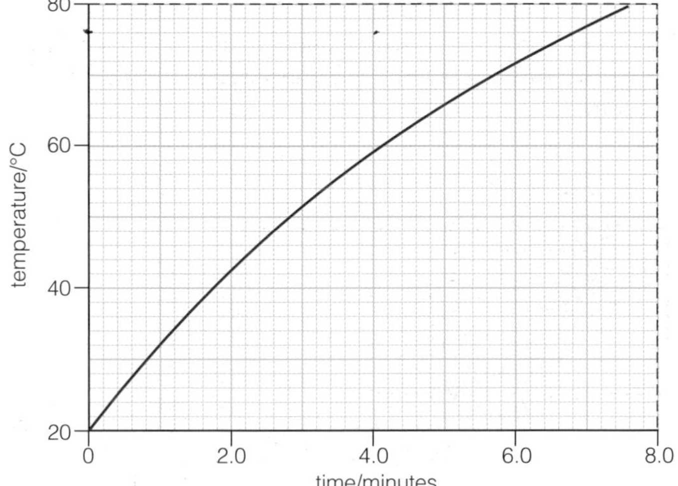

Figure 10.7 ▲

a) Use the graph to show that the rate of rise of the temperature of the water at the beginning of the heating process is about $0.2\,\text{K s}^{-1}$.

b) Hence determine the rate at which the water is absorbing energy. Comment on your answer.

c) Explain why the rate at which the temperature rises slows down as the heating process continues.

6 The following is taken from a manufacturer's advertisement for an electric shower:

The beauty of electric showers is that they draw water directly from a cold water supply and heat it as it is used, so you don't need to have a stored hot water supply. Because they are easy to install, electric showers are extremely versatile. In fact, virtually every home – new and old – can have one.

a) Write a word equation to describe the energy changes that take place in an electric shower.

Rewrite the equation using the appropriate formulae.

b) The technical data supplied by the manufacturer states that a particular shower has a 10.8 kW heater and can deliver hot water at a rate of 14 litres per minute.

Calculate the temperature of the hot water delivered by this shower when the temperature of the cold water supply is 16 °C.

(1 litre of water has a mass of 1 kg.)

7 A student uses the arrangement shown in Figure 10.8 to determine a value for the specific heat capacity of water.

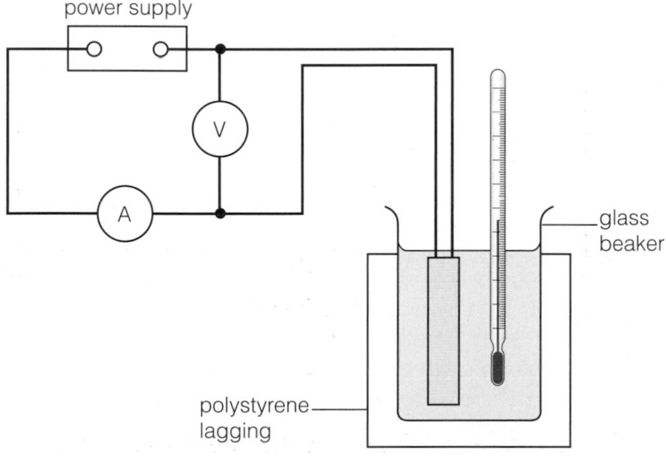

Figure 10.8 ▲

He recorded the following data:

mass of water	$m = 250\,\text{g}$
potential difference	$V = 11.9\,\text{V}$
current	$I = 4.12\,\text{A}$
time	$\Delta t = 4.00$ minutes
temperature rise	$\Delta\theta = 10.2\,\text{K}$

a) Show that the student would obtain a value for the specific heat capacity of water of about $4600\,\text{J kg}^{-1}\,\text{K}^{-1}$ from these data.

b) His teacher suggests that this value is significantly higher than the accepted value of $4200\,\text{J kg}^{-1}\,\text{K}^{-1}$ because he has ignored the energy taken by the glass beaker. The student finds out from a data book that the specific heat capacity of glass is $780\,\text{J kg}^{-1}\,\text{K}^{-1}$; he then weighs the beaker and finds its mass to be 134 g.

i) What is the percentage difference between the student's value and the accepted value for the specific heat capacity of water?

ii) How much energy is needed to raise the temperature of the water by 10.2 K?

iii) How much energy is needed to raise the temperature of the glass beaker by 10.2 K?

iv) Comment on your answers in relation to the teacher's suggestion.

c) Explain **two** advantages of using an expanded polystyrene cup instead of a glass beaker in such an experiment.

8 A storage heater consists of a concrete block of mass 45 kg and specific heat capacity $800\,\text{J kg}^{-1}\,\text{K}^{-1}$. It is warmed up overnight when electricity is cheaper.

a) Show that it needs about 2 MJ of energy to raise its temperature from 10 °C to 70 °C.

b) How much will it cost to heat it up if 1 unit of electricity (3.6 MJ) costs 10p?

c) What will be the average power given out if it takes 5 hours to cool from 70 °C to 10 °C?

d) Explain how the power emitted will change as it cools down.

11 Internal energy and absolute zero

We saw in Chapter 10 that all bodies have internal energy and that this is made up of the potential energy and kinetic energy of the atoms or molecules of the body – potential energy due to the inter-atomic or inter-molecular bonds and kinetic energy due to the vibration of the atoms or molecules. We will look at this kinetic theory again in this chapter and see how the internal energy of a body can be increased by supplying heat or doing work on the body. We'll also see how the idea of internal energy leads to the concept of an absolute zero of temperature. Experiments at temperatures close to absolute zero have led to the discovery of superconductors and superfluids. Superconductors have far-reaching applications – from MRI scanners to high-speed trains.

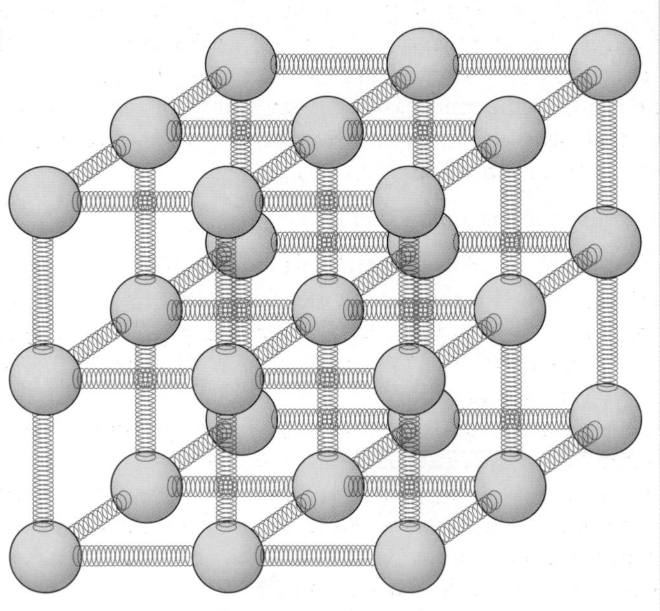

Figure 11.1 ▲
Atomic model of a solid. Polystyrene spheres represent the atoms and springs represent the bonds. If the model is held by one sphere, which is then shaken, the springs (bonds) transmit this vibration to the other spheres (atoms) and the whole model vibrates. The model then has potential energy stored in the springs (bonds) and kinetic energy of the vibrating spheres (atoms).

11.1 Internal energy of an ideal gas

In a gas the molecules have kinetic energy due to their random motion. In addition, molecules that are made up of two or more atoms (e.g. O_2 or CO_2) can have kinetic energy due to rotation and vibration.

In an **ideal gas** we assume that the inter-molecular forces are negligible, except during collision (see Chapter 12, page 124). Furthermore, we assume that the collisions of the gas molecules with one another and with the walls of any container are *elastic*. Imagine two molecules approaching each other head-on as in Figure 11.2a.

Just before collision each molecule has kinetic energy. As the molecules get closer, there are repulsive forces between them which slow them down until momentarily both are stationary (Figure 11.2b). At this point all the kinetic energy has been converted into potential energy as a result of the work done against the repulsive forces. These repulsive forces become extremely large

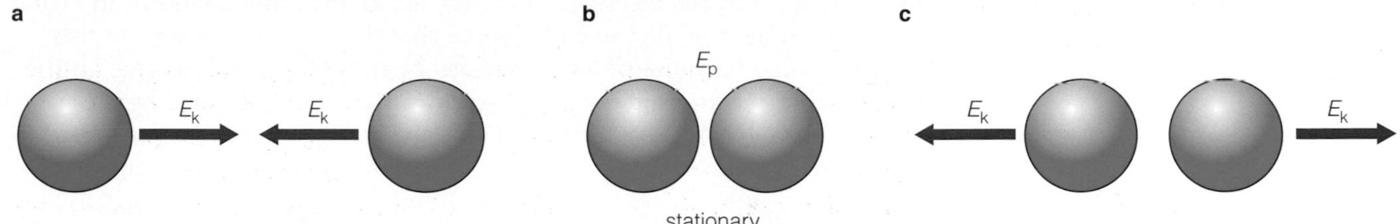

Figure 11.2 ▲

when the molecules become very close and so do work pushing the molecules apart again. As the collision is **elastic**, all the potential energy is converted back to kinetic energy once more (Figure 11.2c). As collisions between gas molecules take place randomly, there is a continuous interchange of kinetic and potential energy in this way.

In the case of an ideal gas, another assumption is that the duration of collisions is negligible compared with the time spent in between collisions, and so we consider the internal energy to be entirely kinetic. Since temperature is a measure of the kinetic energy of the molecules, this means that for an ideal gas the temperature is a measure of the total internal energy.

We'll look at this in more detail in Chapter 12.

Tip

Remember, the internal energy of an ideal gas is entirely the random kinetic energy of its molecules.

11.2 Heating and working

By considering internal energy, we can get a better understanding of what is meant by 'hot' and 'cold'. A hot body has a much greater *concentration of internal energy* compared with a cold body, so when we 'heat' a body we are increasing its internal energy.

We define **heat** as the *random* interchange of energy between two bodies in thermal contact, resulting in *energy flowing from hot to cold.*

This transfer of energy may be by means of conduction, convection or radiation. The bodies may not necessarily be in *physical* contact; for example the Earth receives energy from the Sun by means of radiation travelling through nearly 150 million kilometres of space!

For heat to flow there must be a *temperature difference*. For example, let's imagine that you hold a metal screw, which is at room temperature, say 20 °C, between your fingers, which are at your body temperature of 37 °C (Figure 11.3). Energy will flow from your fingers to the screw.

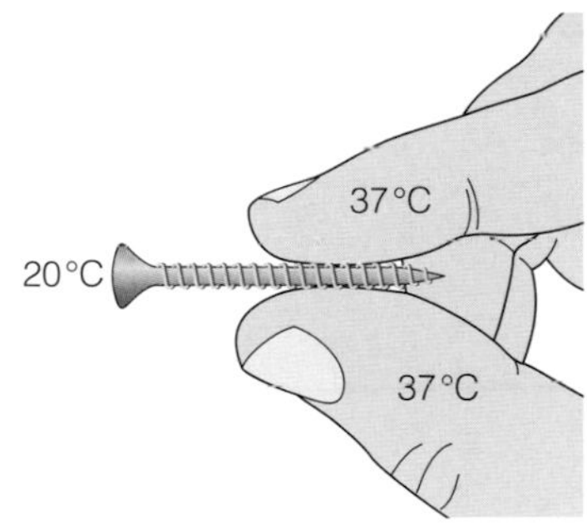

Figure 11.3 ▲
A temperature difference causes a flow of energy

This flow of **heat** increases the internal energy of the molecules of the screw until its temperature is also 37 °C. Energy has been taken from your finger, which gives the sensation of the screw feeling 'cold'. The amount of energy ΔE taken from your finger will be given by:

$$\Delta E = mc\,\Delta\theta$$

where m is the mass of the screw, c is its specific heat capacity and $\Delta\theta$ is the temperature difference between your fingers and the initial temperature of the screw.

Energy can also be transferred between two bodies in the form of **work**, irrespective of any temperature difference. This work can either be *mechanical* or *electrical.*

Consider screwing the same screw into a piece of hard wood (Figure 11.4). After a time you will notice that the screw gets hot and you get tired! Energy has been transferred *mechanically* from you to the screw by means of the work done against the frictional force between the screw and the wood. The work done, or energy transferred, is

$$\Delta E = F\,\Delta x$$

where Δx is the distance moved by the screw against the frictional force F.

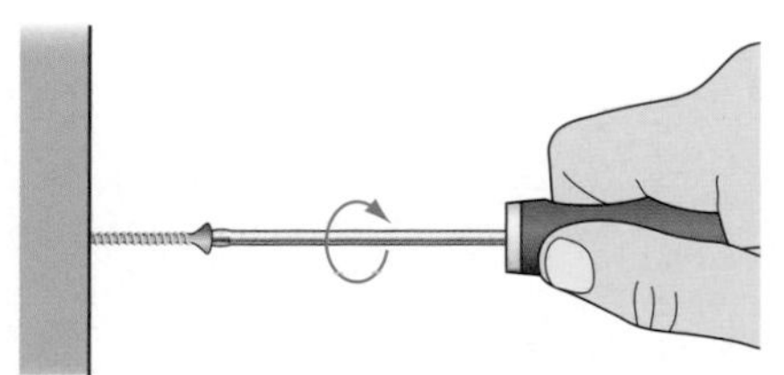

Figure 11.4 ▲
Doing mechanical work

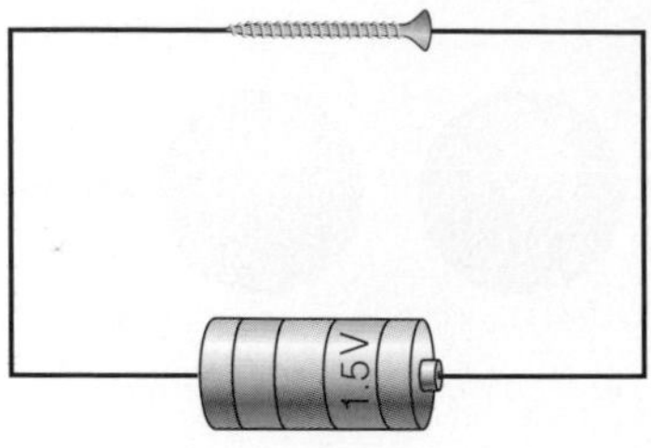

Figure 11.5 ▲
Doing electrical work

Alternatively, you could connect a battery across the screw as shown in Figure 11.5. After a short while you will observe that the screw gets warm. In this case energy has been transferred *electrically* by the cell exerting a force on the electron charge carriers in the metal screw. This force, multiplied by the distance moved by the electrons, will give the amount of energy transferred to the internal energy of the screw. From Unit 2 we have that this energy is:

$$\Delta E = IV\,\Delta t$$

Energy transfer by working is an **ordered** process and is independent of any temperature difference. By 'ordered' we mean that the force, and therefore the energy transfer, is in a defined direction, not random as in the case of heat.

Worked example

A car of mass 1600 kg has four brake discs, each of mass 1.3 kg. The discs are made from an iron alloy of specific heat capacity 480 J kg^{-1} K^{-1}.

a) Calculate the rise in temperature of the discs when the car brakes from a speed of 90 kph (25 m s^{-1}) to rest.

b) State any assumptions that you make.

c) Explain the energy changes that have taken place.

d) Suggest why manufacturers supply discs with holes drilled in them for high-performance cars.

e) In order to reduce the weight of racing cars, and therefore increase performance, aluminium brakes are being developed. The specific heat capacity of aluminium is 910 J kg^{-1} K^{-1}. Suggest why aluminium is a suitable material for this purpose.

Answer

a) Using $\Delta E = mc\,\Delta T$, where $\Delta E = \frac{1}{2}mv^2$ is the kinetic energy of the car, we have

$$\Delta E = \frac{1}{2} \times 1600\,\text{kg} \times (25\,\text{m s}^{-1})^2 = 5.0 \times 10^5\,\text{J}$$

$$\Delta T = \frac{\Delta E}{mc} = \frac{5.0 \times 10^5\,\text{J}}{4 \times 1.3\,\text{kg} \times 480\,\text{J kg}^{-1}\,\text{K}^{-1}} = 200\,\text{K}$$

b) We have to make a number of assumptions:

- the energy is shared equally between the four discs;
- all the energy is given to the discs – in practice some will be given to the pads that press against the discs;
- the discs do not dissipate any of the energy into the air or conduct any energy away during the braking time.

c) The ordered kinetic energy of the car has been converted into the disordered, random internal energy of the brake discs.

d) The holes enable air to pass through the discs. This ventilation dissipates the energy from the discs to the surrounding air more quickly and so prevents the discs from over-heating.

e) The crucial factor is that the specific heat capacity of aluminium is almost twice that of the iron alloy. This means that aluminium discs having half the mass of iron discs can be used without them getting any hotter than iron discs.

11.3 Change of state

We have just seen that heating or doing work on a body increases the internal energy of its molecules and usually this raises the temperature of the body. But what happens when a substance melts or vaporises? This is what we call a **change of state**.

We usually think of there being three 'states of matter' – solid, liquid and gas – although there is in fact a fourth state, called a **plasma**. Plasma, consisting of ionised gas, usually at a very high temperature, makes up over 99% of the visible universe and perhaps most of that which is not visible. In a 'plasma TV', with which you may be familiar, collisions excite (energise) the xenon and neon atoms in the plasma, causing them to release **photons** of light energy.

The physics of plasmas has come to be very important – and not just because of TV screens! **Nuclear fusion** is the mechanism that powers our main source of energy, the Sun (see page 198). If we can develop a technique for controlled nuclear fusion (as opposed to the uncontrolled energy of a thermonuclear bomb), we will create an infinite supply of 'clean' energy and go a long way to solving the world's energy crisis and environmental problems. At the temperature required for fusion, the reacting material is in the plasma state. Because a plasma is ionised, i.e. made up of charged particles, it can be controlled by magnetic fields. In modern fusion experiments, the plasma is confined in a doughnut-shaped vessel with magnetic coils called a '*tokamak*'.

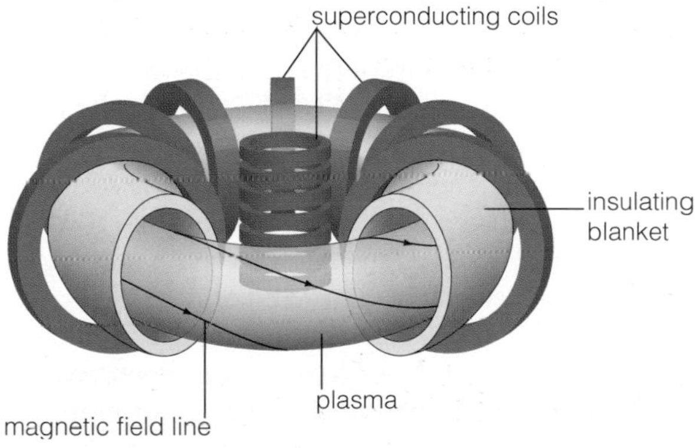

Figure 11.6 ▲
A tokamak

Figure 11.6 shows the principle of a tokamak. The plasma is contained in the doughnut-shaped vessel, called a 'torus'. Using superconducting coils (blue) a magnetic field is generated, which causes the extremely high-temperature plasma particles to run around in circles, without touching the vessel wall. In reality, a number of other coils are present, which produce subtle changes to the magnetic field.

Let's get back to what happens when a substance changes state. Imagine taking a cube of ice out of the freezer and leaving it on the laboratory bench to melt (preferably in a beaker!). Its initial temperature will be about −18 °C, the normal temperature inside a freezer. The laboratory is probably about 22 °C, so we can say that the ice will heat up as energy is transferred from the room to the ice because of a temperature difference. The molecules of the ice will gain internal energy, both potential and kinetic. Because the molecules gain kinetic energy, the temperature of the ice will increase. This continues until the ice reaches a temperature of 0 °C and starts to melt. At this point *all* the energy it receives is used to do work, increasing the potential energy and overcoming the bonds that keep the ice as a solid. The solid changes into a liquid. During this process there is *no* increase in the kinetic energy of the molecules and so the ice remains at 0 °C until it has all melted. Although the ice is still receiving heat from the warmer surroundings, the effect of this heat is not observed as a rise in temperature of the ice. It is called **latent heat**, 'latent' meaning 'hidden'.

Once the ice has melted, the water thus formed continues to receive heat from the room, but the kinetic energy of the molecules now increases, as well as the potential energy, and the temperature of this water rises until it reaches room temperature and **thermal equilibrium** is established.

If this water were now to be poured into an electric kettle and heated, electrical energy would be converted into thermal energy – internal energy of the water molecules – and the temperature of the water would increase until it reached 100 °C. All the electrical energy supplied by the element of the kettle is now used to do work pulling the molecular bonds apart to change the liquid water into a vapour. All the energy goes into increasing the potential energy of the molecules and there is *no* increase in their kinetic energy. Once again, the temperature remains constant while the water is boiling and changing state from a liquid into a vapour. The energy required to change the water into vapour is again called the latent heat.

11.4 Absolute zero

We saw in Chapter 10 that the SI unit of temperature, the kelvin, was defined in terms of what is called the absolute thermodynamic scale of temperature. This has **absolute zero** as its zero. A temperature of 0 K is called absolute zero because it is the lowest temperature that can theoretically be reached. Its value on the Celsius scale is −273.15 °C, but we usually round this to −273 °C for calculations.

Although it is not possible in practice to cool any substance to absolute zero, scientists have achieved temperatures very close to 0 K, where matter exhibits quantum effects such as superconductivity and superfluidity.

A superconductor is a material that will conduct electricity *without resistance* when cooled below a certain temperature. Once set in motion, the current will flow forever in a closed loop in the superconducting material, making it the closest thing in nature to perpetual motion. There is, however, a catch! The superconductor has to be cooled to a very low temperature by, for example, liquid nitrogen and, unfortunately, energy is needed to maintain this low temperature. The Large Hadron Collider at CERN in Geneva, Switzerland, which is the world's largest and highest-energy particle accelerator, needs 96 tonnes of liquid helium to keep its electromagnets (Figure 11.8) at their operating temperature, 1.9 K (−271 °C).

Nevertheless, the uses of superconductors are diverse. For example, they are being used to improve the efficiency of motors and the transmission of electricity, in MRI scanners (see pages 63 and 172) and in magnetic levitating trains.

Magnetic levitation ('maglev') is an example of the application of electromagnetic induction (see Unit 4, page 61). When the magnet in Figure 11.7 is dropped and falls towards the superconductor, the change in magnetic flux induces an electric current in the superconductor (Faraday's law). By Lenz's law, the current flows in a direction such that its magnetic field opposes that of the magnet and so the magnet is repelled. As the superconductor has no resistance, the current in it continues to flow, even though the magnet is no longer moving. The magnet is permanently repelled and hovers above the semiconductor.

This principle is used in maglev trains. The Japanese train shown in Figure 11.9 reached a world record speed of 581 km/h in 2003.

Figure 11.7 ▲
A magnet suspended above a superconducting coil

Figure 11.8 ▲
Superconducting electromagnet for the Large Hadron Collider

Figure 11.9 ▲
Maglev train

The lowest temperature recorded in a laboratory is about 100 pK (pico = 10^{-12}). The coldest known region of the universe is in the Boomerang Nebula, 5000 light years away in the constellation of Centaurus (Figure 11.10). Its temperature is about 1 K above absolute zero.

In terms of the kinetic theory, absolute zero is the temperature at which the molecules of matter have their lowest possible average kinetic energy. In a simplified model, the molecules are considered to have *no* average kinetic energy at absolute zero, in other words they have no random movement. In practice, quantum mechanics requires that they have a *minimum* kinetic energy, called the **zero-point energy**.

Figure 11.10 ▲
Boomerang Nebula

Worked example

The following passage is taken from an article on zero-point energy.

> In conventional quantum physics, the origin of zero-point energy is the Heisenberg uncertainty principle, which states that, for a moving particle such as an electron, the more precisely one measures the position, the less exact is the best possible measurement of its momentum. The least possible uncertainty of position times momentum is specified by Planck's constant, *h*. A parallel uncertainty exists between measurements involving time and energy. This leads to the concept of zero-point energy.

a) The Heisenberg uncertainty principle can be expressed by the equation

$$\Delta p\, \Delta x = h$$

where p represents momentum in the direction of displacement x. Show that this is consistent with Planck's constant h having units of J s.

b) An electron has a speed of $7.2 \times 10^7\,\mathrm{m\,s^{-1}}$. What is the uncertainty in its position? Comment on your answer.

c) Express the statement 'A parallel [i.e. equivalent] uncertainty exists between measurements involving time and energy' as an equation of similar form to that given in part a). Write down this equation and suggest how this leads to the concept of zero-point energy.

Answer

a) Units of $h = \Delta p\, \Delta x$ are $\mathrm{kg\,m\,s^{-1}} \times \mathrm{m} = \mathrm{kg\,m^2\,s^{-1}}$
$\mathrm{J\,s} = \mathrm{N\,m\,s} = \mathrm{kg\,m\,s^{-2}} \times \mathrm{m} \times \mathrm{s} = \mathrm{kg\,m^2\,s^{-1}}$
So units of h are J s.

b) $$\Delta x = \frac{h}{\Delta p} = \frac{h}{\Delta(mv)} = \frac{6.63 \times 10^{-34}\,\mathrm{J\,s}}{9.11 \times 10^{-31}\,\mathrm{kg} \times 7.2 \times 10^7\,\mathrm{m\,s^{-1}}}$$
$$= 1.0 \times 10^{-11}\,\mathrm{m}$$

This is of the order of the size of an atom, which means that we can only tell where the electron is to within about an atomic diameter.

c) $\Delta E\, \Delta t$ = a constant value. In fact $\Delta p\, \Delta x = h$ is equivalent to $\Delta E\, \Delta t = h$. ΔE is the minimum uncertainty in, or fluctuation from, a particular measured value of energy, which means that zero energy can never be achieved.

REVIEW QUESTIONS

1 The molecules of an ideal gas do **not** have:

A internal energy

B kinetic energy

C potential energy

D random energy

2 When ice is melting, which of the following does **not** change?

A volume

B internal energy of the molecules

C potential energy of the molecules

D kinetic energy of the molecules

3 Which of the following quantities could **not** be measured in joules?

A energy

B heat

C power

D work

4 The table below lists the melting points and boiling points for various common substances. Calculate the missing temperatures.

Substance	Melting point		Boiling point	
	/°C	/K	/°C	/K
water	0	273	100	373
mercury	−39	a)	b)	630
alcohol	c)	156	79	d)
oxygen	e)	54	−183	f)
copper	1083	g)	h)	2853

Table 11.1 ▲

5 A physics textbook states that 'The *internal energy* of a body can be increased by *heating* it or by *doing work* on it.'

How would you explain what this means to a friend? Illustrate your answer with reference to the terms in *italics* and examples from everyday life.

6 Explain the molecular energy changes that take place when a tray of water at room temperature is put into a freezer and ice cubes are formed.

7 In 1913, Heike Kamerlingh Onnes was awarded the Nobel Prize in Physics 'for his investigations on the properties of matter at low temperatures, which led, amongst other things, to the production of liquid helium'.

The Nobel Prize in Physics in 2003 was awarded jointly to Alexei Abrikosov, Vitaly Ginzburg and Anthony Leggett 'for pioneering contributions to the theory of superconductors and superfluids'.

a) Helium liquefies at about 4 °C above absolute zero. Explain what is meant by the term 'absolute zero'.

b) Discuss the significance of Kamerlingh Onnes' achievement of 'the preparation of liquid helium'.

c) Explain what is meant by 'superconductors'.

d) Discuss the practical uses now being made of superconductors.

8 In an article entitled 'Coal Power Station Aims for 50% Efficiency', the manufacturer states that this 'is possible due to a special steam turbine with a steam temperature of 700 instead of 600 degrees Celsius'.

The maximum thermal efficiency of a turbine is given by

$$\text{maximum efficiency} = \frac{T_1 - T_2}{T_1} \times 100\%$$

where T_1 and T_2 are the respective temperatures, in kelvin, of the steam entering and leaving the turbine.

a) The efficiency of a turbine with a steam temperature of 600 °C is 46%. Show that the steam leaves the turbine at a temperature of approximately 200 °C.

b) Assuming that the temperature at which the steam leaves the turbine remains the same, is the manufacturer's claim that an efficiency of 50% will be achieved with a steam temperature of 700 °C valid?

c) Suggest how the remaining 50% of energy could be usefully used.

12 Gas laws and kinetic theory

The development of the steam engine in the 18th century revolutionised industry and transport, transforming people's lives. Progress was accelerated in the 19th century by the work of physicists in the field of thermodynamics – the study of how gases behave under various conditions of pressure, volume and temperature. By making some very simple assumptions about the properties of the molecules of a gas, we can explain much about the behaviour of gases, both qualitatively and quantitatively. More recently this has led to practical applications such as the development of more efficient car engines, power stations and refrigerators.

12.1 Pressure

In Unit 1 we defined pressure as the force per unit area acting at right angles to a surface:

$$p = \frac{F}{A}$$

From the above equation, we can see that the unit of pressure is $\mathrm{N\,m^{-2}}$. As pressure is a very commonly used quantity, it is given its own unit, the **pascal** (Pa); a pressure of $1\,\mathrm{Pa} \equiv 1\,\mathrm{N\,m^{-2}}$.

Worked example

A brick has dimensions of 200 mm × 100 mm × 50 mm and a mass of 1.80 kg. Calculate:

a) the density of the brick,

b) the pressure it exerts when it i) rests on its flat face, and ii) stands on one of its ends.

Answer

a) $\text{Density} = \frac{\text{mass}}{\text{volume}} = \frac{1.80\,\mathrm{kg}}{0.200\,\mathrm{m} \times 0.100\,\mathrm{m} \times 0.050\,\mathrm{m}} = 1800\,\mathrm{kg\,m^{-3}}$

b) i) On flat face: $p = \frac{F}{A} = \frac{1.80\,\mathrm{kg} \times 9.8\,\mathrm{N\,kg^{-1}}}{0.200\,\mathrm{m} \times 0.100\,\mathrm{m}} = 880\,\mathrm{Pa}$

ii) On one end: $p = \frac{F}{A} = \frac{1.80\,\mathrm{kg} \times 9.8\,\mathrm{N\,kg^{-1}}}{0.100\,\mathrm{m} \times 0.050\,\mathrm{m}} = 3.5\,\mathrm{kPa}$

Tip

Make sure you always use SI units throughout when doing pressure and density calculations, by having

- masses in kg
- *all* dimensions in m, areas in $\mathrm{m^2}$ and volumes in $\mathrm{m^3}$.

12.2 Fluid pressure

We say that a substance is a 'fluid' if it has the ability to flow. Therefore both liquids and gases are fluids. We will look at the particular behaviour of gases in more detail later on.

The pressure at a point in a fluid is defined as the force per unit area acting on a very small area round the point. We saw in Unit 1 that the pressure p at a point of depth h in a liquid of density ρ is given by:

$$p = h\rho g$$

Note that this is the pressure due to the weight of liquid above the point. There will be *additional* pressure acting at the point due to the pressure p_A of the atmosphere. The total pressure will therefore be $p = p_A + h\rho g$.

Note

You do not need to memorise this equation but you may be given it to use in the examination.

Atmospheric pressure is not a constant quantity – for example we use its changing value to forecast weather – but we take a 'standard atmosphere' to be 1.01×10^5 Pa, or 101 kPa.

If a liquid is stationary, it follows that the pressure at any point in the liquid must act equally *in all directions*. If not, there would be a resultant force, which would cause the liquid to flow.

Figure 12.1 ▲
A free diver

Worked example

In 2008, a world record for free diving (diving without breathing apparatus) was set at a depth of 122 m. Assuming that the density of sea water has an average value of $1.03 \times 10^3\,\text{kg}\,\text{m}^{-3}$ over this depth and that atmospheric pressure is 101 kPa, calculate:

a) the pressure exerted by the sea water at a depth of 122 m, and

b) the total pressure, in 'atmospheres', acting on a diver at this depth.

The pressure p due to a depth h of liquid of density ρ is $p = h\rho g$.

Answer

a) Pressure due to water, $p_w = h\rho g = 122\,\text{m} \times 1.03 \times 10^3\,\text{kg}\,\text{m}^{-3} \times 9.81\,\text{N}\,\text{kg}^{-1}$

$$= 1.23 \times 10^6\,\text{Pa}$$

b) The total pressure acting on the diver will be atmospheric pressure (1.01×10^5 Pa) plus the water pressure of 1.23×10^6 Pa, so

$$p_{tot} = 1.01 \times 10^5\,\text{Pa} + 1.23 \times 10^6\,\text{Pa} = 1.33 \times 10^6\,\text{Pa}$$

In atmospheres,

$$p_{tot} = \frac{1.33 \times 10^6\,\text{Pa}}{1.01 \times 10^5\,\text{Pa}} = 13 \text{ atmospheres}$$

Tip

Remember that atmospheric pressure must be added to the pressure due to a depth of liquid in order to determine the total pressure acting.

12.3 The gas laws

When we are looking at the behaviour of gases we have four variables to consider – the mass, pressure, volume and temperature of the gas. In order to investigate how these quantities are related, we need to keep two constant while we see how the other two vary with one another.

Boyle's law

The relationship between pressure and volume for a fixed mass of gas at constant temperature was discovered by Robert Boyle in the 17th century. We can investigate this for ourselves by means of a simple experiment as described below.

Experiment

Investigating the relationship between pressure and volume for a fixed mass of gas at constant temperature

The simple apparatus shown in Figure 12.2a can be used. The volume V of the fixed mass of air under test is given by the length l of the column of air multiplied by the area of cross section of the glass tube. If the tube is uniform, this area will be constant, so we can assume that $V \propto l$. (Some tubes may actually be calibrated to give the volume V directly.) The total pressure of the air is read straight off the pressure gauge.

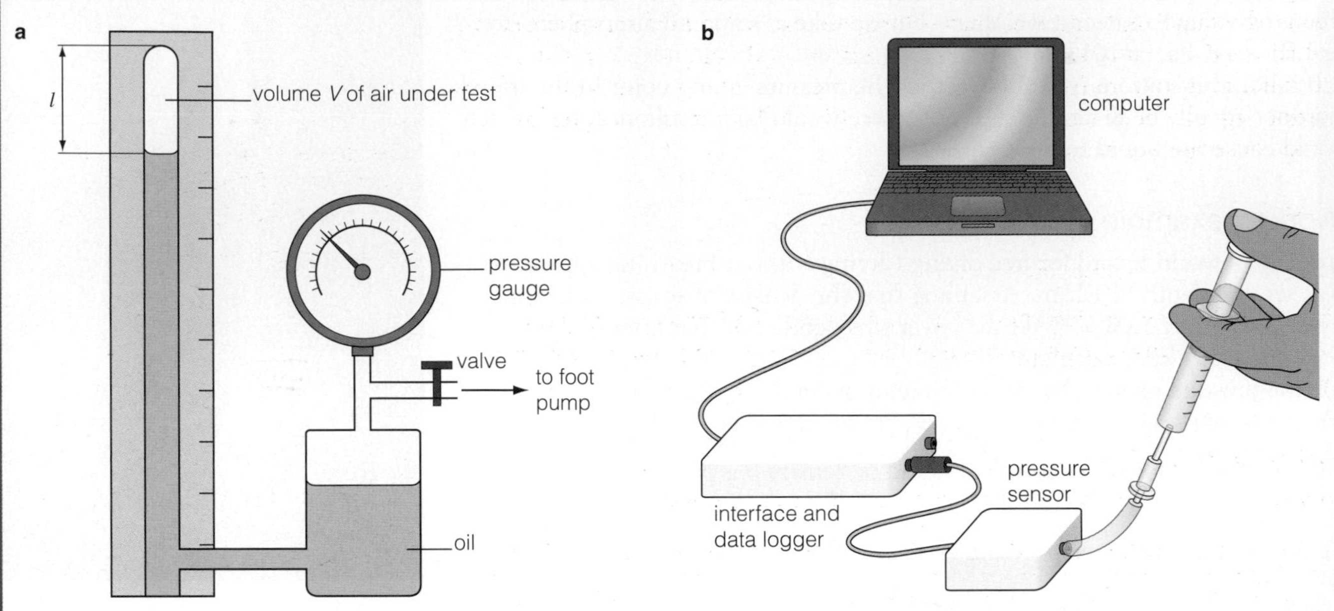

Figure 12.2 ▲

Open the valve so that the air starts at atmospheric pressure and then increase the pressure by means of the foot pump. This pressure is transmitted through the oil and compresses the air. The pressure p and the corresponding length l (or volume V) of the air column should be recorded for as wide a range of values as possible. It is important to leave sufficient time after each change in pressure for the air to reach thermal equilibrium with its surroundings so that its temperature remains constant. Put your results in a table, together with values of $1/l$ (or $1/V$), like Table 12.1.

Safety note

Avoid excessive pressure and make sure connections are secure, otherwise oil may be sprayed at high pressure over observers.

p/kPa							
l/mm (or V/cm^3)							
$(1/l)$/m^{-1} (or $(1/V)$/m^{-3})							

Table 12.1 ▲

Plot graphs of p against l (or V) and of p against $1/l$ (or $1/V$). You should get graphs like those shown in Figure 12.3a and b.

This experiment can also be done using a syringe to contain the air, with a pressure sensor and data logger (Figure 12.2b).

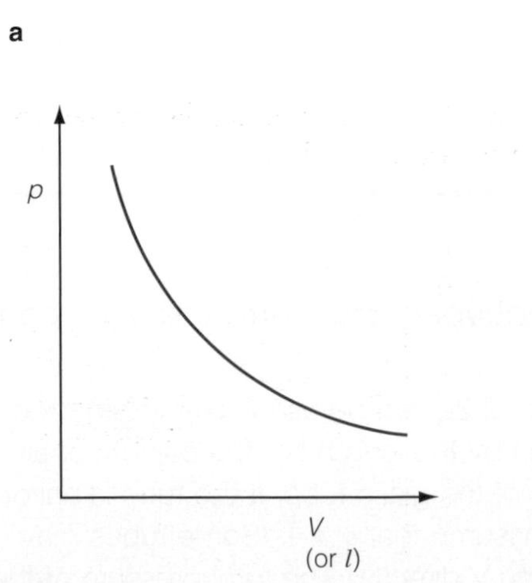

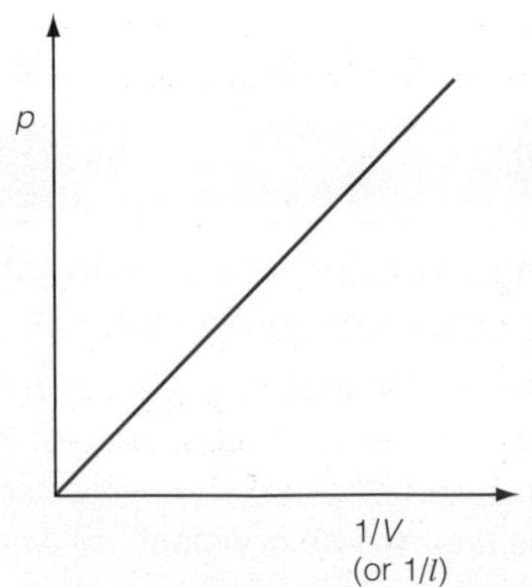

Figure 12.3 ▲

The graph of p against V shown in Figure 12.3a is called an **isothermal** ('iso' means 'the same' and 'thermal' means 'temperature'). An isothermal is a curve that shows the relationship between the pressure and volume of a gas at a particular temperature (see also Figure 12.4a below). The precise nature of this relationship can be determined from Figure 12.3b. As this is a straight line *through the origin* we can deduce that:

$$p \propto \frac{1}{V} \qquad \text{or} \qquad p = \text{constant} \times \frac{1}{V}$$

giving $\quad pV = \textbf{constant}$

This is **Boyle's law**, which states that the pressure of a fixed mass of gas is inversely proportional to its volume provided that the temperature is kept constant.

Tip

Remember to give the conditions when stating Boyle's law, namely

- a fixed mass of gas
- at constant temperature.

Worked example

In a Boyle's law experiment, the following data were recorded:

p/kPa	100	120	140	160	180	200	220	240
V/cm³	20.1	16.6	14.4	12.4	11.2	9.9	9.1	8.4

Table 12.2 ▲

Plot a suitable graph to investigate the extent to which Boyle's law is obeyed and comment on your result.

Answer

You need to plot a graph of p against $1/V$. You should therefore calculate values of $1/V$ (preferably expressed in m^{-3}).

p/kPa	100	120	140	160	180	200	220	240
V/cm³	20.1	16.6	14.4	12.4	11.2	9.9	9.1	8.4
(1/V)/cm⁻³	0.050	0.060	0.070	0.081	0.089	0.101	0.110	0.119
(1/V)/m⁻³	50 000	60 000	70 000	81 000	89 000	101 000	110 000	119 000

Table 12.3 ▲

The proportionality can be seen, as the resulting graph of p against $1/V$ is a straight line through the origin. This shows that Boyle's law is obeyed over the range of pressures investigated.

Tip

A convenient way of remembering Boyle's law for calculations is:

$p_1V_1 = p_2V_2$

You need to be familiar with the three possible graphs that can be plotted to illustrate Boyle's law. These are shown in Figure 12.4a, b, c.

Figure 12.4 ▼
Graphs illustrating Boyle's law

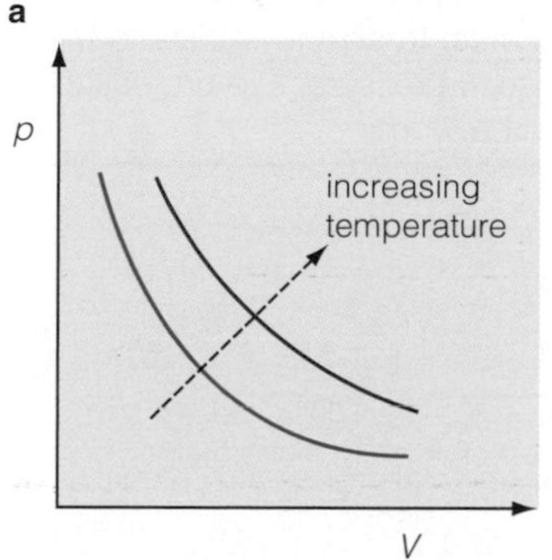

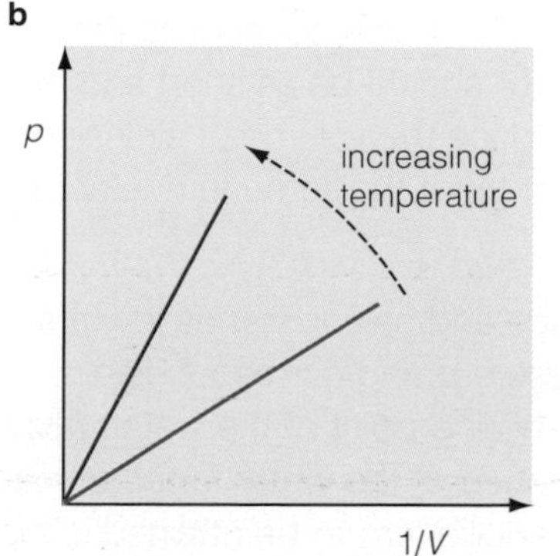

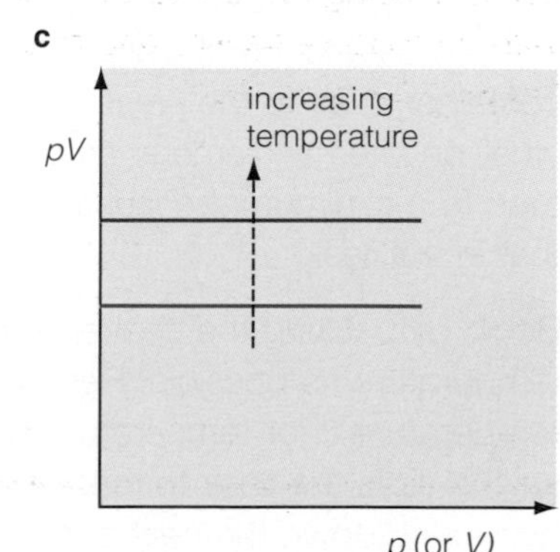

Note

The pressure p due to a depth h of liquid of density ρ is $p = h\rho g$.

Worked example

A deep-sea diver descends to a depth of 190 m in sea water to investigate a wreck. The sea water has an average density of $1.02 \times 10^3\,\text{kg m}^{-3}$ and atmospheric pressure is 1.01×10^5 Pa.

a) Show that the pressure at this depth is approximately 20 atmospheres.

b) The diver releases a bubble of air having a volume of $3.0\,\text{cm}^3$. Calculate the volume of this bubble when it reaches the surface. What assumption do you have to make in your calculation?

Answer

a) The pressure due to the depth of water is

$$p_w = h\rho g = 190\,\text{m} \times 1.02 \times 10^3\,\text{kg m}^{-3} \times 9.81\,\text{N kg}^{-1} = 1.90 \times 10^6\,\text{Pa}$$

We must now add atmospheric pressure to this, giving a total pressure of

$$p_{tot} = 1.90 \times 10^6\,\text{Pa} + 1.01 \times 10^5\,\text{Pa} = 2.00 \times 10^6\,\text{Pa}$$

In atmospheres this is

$$\frac{2.00 \times 10^6\,\text{Pa}}{1.01 \times 10^5\,\text{Pa}} = 19.8 \approx 20 \text{ atmospheres}$$

b) Using $p_1V_1 = p_2V_2$ for the air bubble, where $p_1 = 20$ atmospheres, $V_1 = 3.0\,\text{cm}^3$ and $p_2 = 1$ atmosphere (at the surface),

$$V_2 = \frac{p_1V_1}{p_2} = \frac{20 \text{ atmospheres} \times 3.0\,\text{cm}^3}{1 \text{ atmosphere}} = 60\,\text{cm}^3$$

In doing this calculation, we have to assume that the density of the sea water is uniform, which it won't quite be, and, more significantly, that the temperature of the water at the surface is the same as that at a depth of 190 m. In practice, it is likely to be much colder at this depth.

Tip

When using $p_1V_1 = p_2V_2$, it doesn't matter what the units of p and V are as long as they are the *same on both sides of the equation*. In the worked example we have used non-SI units for both quantities – atmospheres for p and cm^3 for V.

Pressure and temperature

The relationship between the pressure and temperature of a fixed mass of gas at *constant volume* can be investigated as described below.

Experiment

Investigating the relationship between pressure and temperature of a fixed mass of gas at constant volume

This experiment can be performed either using a thermometer and a pressure gauge (Figure 12.5a) or using temperature and pressure sensors with a suitable data logger (Figure 12.5b). Note that the tube connecting the container of air to the pressure gauge/sensor should be as short and of as small an internal diameter as possible to reduce the volume of air inside the tube to a minimum, because this air will not be at the same temperature as that in the flask.

Safety note

The flask chosen as the pressure vessel must to able to withstand the increase in pressure. Eye protection should be worn.

Begin by packing the beaker with ice (or ice and salt) to get as low a starting temperature as possible. Record the pressure p and corresponding temperature θ for temperatures up to the boiling point of the water. Allow time for the air in the flask to reach the temperature of the water for each reading by turning down the heat and waiting for equilibrium to be attained. (This is

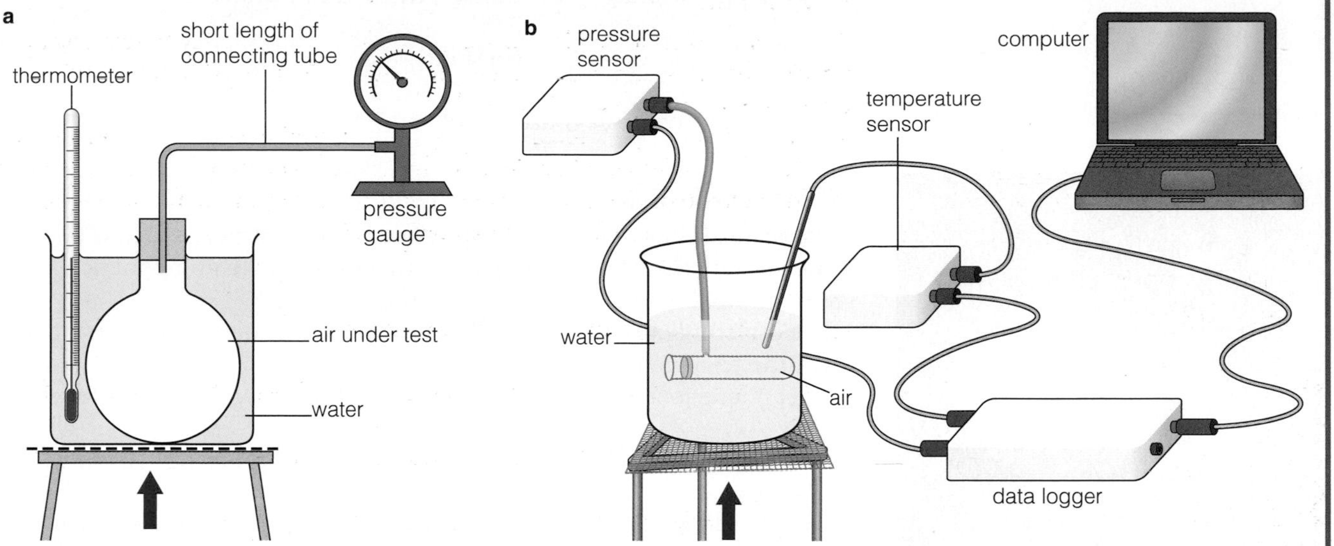

Figure 12.5 ▲

not necessary using the data logger.) The water should be continuously stirred to ensure an even temperature distribution.

Although the range of p is very small, it is instructive to plot a graph of p against θ with the pressure scale starting at zero (this gives a very small scale, which would not normally be acceptable) and the temperature scale going from −300 °C to +100 °C. You should get a graph that looks like Figure 12.6.

When the graph is extrapolated to zero pressure, the corresponding temperature is found to be −273 °C (or slightly different because of experimental error). This is, of course, absolute zero (see page 110).

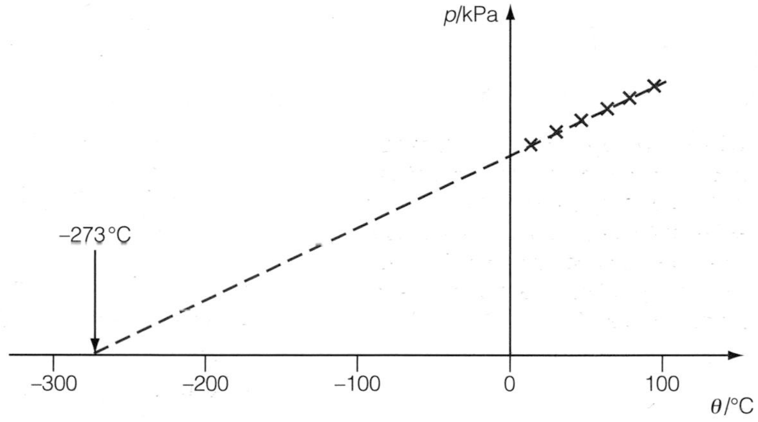

Figure 12.6 ▲

If the data from the experiment above were re-plotted as a graph of p against T in kelvin, the graph would be like that shown in Figure 12.7.

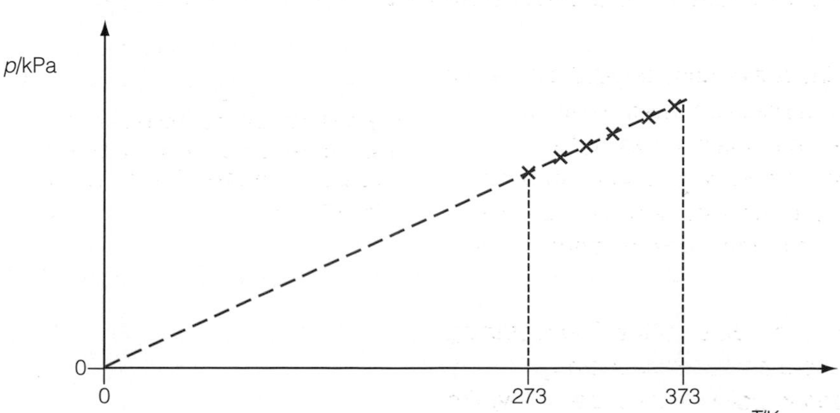

Figure 12.7 ▲
Pressure against kelvin temperature for a gas of fixed mass and volume

Tip

If you are describing a data-capture method in an examination, you must draw *all* the apparatus as in Figure 12.5b and state that you would select 'x-axis = temperature' and 'y-axis = pressure' from the 'Graph' menu.

As the graph is a straight line through the origin, it shows that

$$p \propto T \qquad \text{or} \qquad \frac{p}{T} = \text{constant}$$

for a fixed mass of gas at constant volume.

Figure 12.8 ▲
Using a tyre pressure gauge

Worked example

A car manual states that the 'tyre pressure must be 36 psi (= 250 kPa)'. (This is the pressure that the air inside the tyre must be *above* atmospheric pressure, which you may assume to be 100 kPa.)

The driver inflates the tyres to 35 psi (pounds per square inch) on a day when the temperature is 14 °C.

After a long journey, she checks the pressure again and finds it is now 39 psi.

a) Use the data from the manual to show that 100 kPa is equivalent to 14.4 psi.
b) Show that after the journey the temperature of the air in the tyre would be approximately 37 °C.
c) State what assumption you have to make in your calculation.
d) Suggest why the temperature of the air inside the tyre has increased.

Answer

a) If 250 kPa = 36 psi, then

$$100\,\text{kPa} = 36\,\text{psi} \times \frac{100\,\text{kPa}}{250\,\text{kPa}} = 14.4\,\text{psi}$$

b) Using the fact that p/T is constant, and remembering that T must be in kelvin and that p is the *total* pressure of the air in the tyre,

$$\frac{p}{T} = \text{constant} \quad \Rightarrow \quad \frac{p_1}{T_1} = \frac{p_2}{T_2} \quad \Rightarrow \quad T_2 = \frac{p_2 T_1}{p_1}$$

where $p_1 = (35.0 + 14.4)\,\text{psi} = 49.4\,\text{psi}$
$p_2 = (39.0 + 14.4)\,\text{psi} = 53.4\,\text{psi}$
$T_1 = (14 + 273)\,\text{K} = 287\,\text{K}$

$$\text{So } T_2 = \frac{p_2 T_1}{p_1} = \frac{53.4\,\text{psi} \times 287\,\text{K}}{49.4\,\text{psi}} = 310.2\,\text{K} \approx 310\,\text{K} = 37\,°\text{C}$$

c) You have to assume that the volume of the air remains constant – that is the tyre does not expand.

d) The rubber of the tyres exhibits a property called hysteresis (see page 69 of the AS book and Figure 12.9 opposite). The part of the tyre in contact with the road gets compressed, and as it moves round it relaxes again. This cycle is repeated several times per second. Some of the energy is transferred to the internal energy of the rubber each cycle and so the tyres become warm and conduct some of their internal energy to the air inside, increasing the pressure.

You might like to show for yourself that the energy per unit volume transferred to internal energy for each cycle in Figure 12.9 is about $1.5\,\text{MJ}\,\text{m}^{-3}$.

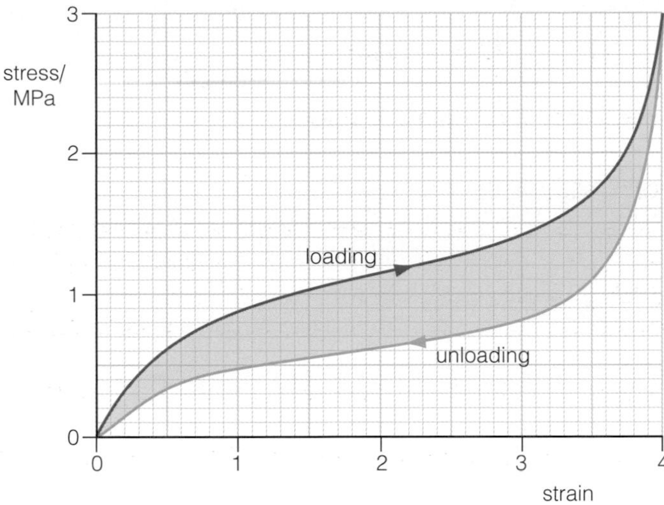

Figure 12.9 ◀
Hysteresis loop for rubber: the area of the loop represents the energy per unit volume transferred to internal energy during the cycle

12.4 Equation of state for an ideal gas

Combining p/T = constant with Boyle's law, pV = constant, for a fixed mass of gas we get:

$$pV = \text{constant} \times T$$

If we take **one mole** as the 'fixed mass' we find that the constant in the above expression is *the same for all gases*. It is called the **universal molar gas constant**, symbol R, and has a value of $8.31\,\text{J}\,\text{K}^{-1}\,\text{mol}^{-1}$.

If we have n moles of a gas we then have

$$pV = nRT$$

An alternative way of expressing this relationship is in terms of another constant, called the **Boltzmann constant**, symbol k, after Ludwig Boltzmann, who did much to advance the development of thermodynamics in the latter half of the 19th century. The equation becomes

$$\mathbf{pV = NkT}$$

where N is the number of *molecules* of the gas and $k = 1.38 \times 10^{-23}\,\text{J}\,\text{K}^{-1}$.

A gas that obeys this equation under *all* conditions is called an **ideal gas**, and hence this equation is called the **ideal gas equation** or the **equation of state for an ideal gas**. Of course, there is no such thing as an 'ideal gas', but in practice most gases obey this equation provided that the pressure is not too large and the gas is above a certain temperature, called its critical temperature. This is a very low temperature for most gases, for example $-118\,°\text{C}$ for oxygen and $-147\,°\text{C}$ for nitrogen, the two main constituents of air. This means that under normal laboratory conditions air behaves like an ideal gas.

We showed earlier that experimentally $p \propto T$ for a fixed mass of gas at constant volume. It can also be shown that $V \propto T$ for a fixed mass of gas at constant pressure. It therefore follows that at absolute zero, when $T = 0$, the pressure p and volume V both become zero, which is further confirmation that absolute zero is the lowest temperature theoretically possible. However, this argument is only true for an ideal gas – most real gases liquefy well above absolute zero, and even helium is liquid at 4 K.

Figure 12.10 ▲
Ludwig Boltzmann

Tip

Remember, in the equation $pV = NkT$ you must have:

- p in Pa
- V in m^3
- N as the number of molecules
- T in K.

Worked example

a) Show that the units of Boltzmann's constant k are $\text{J}\,\text{K}^{-1}$.

b) Show that the volume of one mole (6.02×10^{23} molecules) of an ideal gas at 'standard temperature and pressure' (i.e. $T = 273.15\,\text{K}$ and $p = 101.3\,\text{kPa}$) is 22.4 litres.

Answer

a) From $pV = NkT$, we have

$$k = \frac{pV}{NT} = \frac{\text{N m}^{-2} \times \text{m}^3}{\text{K}} = \frac{\text{N m}}{\text{K}} = \text{J K}^{-1}$$

b) From $pV = NkT$, we have

$$V = \frac{NKT}{p} = \frac{6.02 \times 10^{23} \times 1.38 \times 10^{-23}\,\text{J K}^{-1} \times 273.15\,\text{K}}{101.3 \times 10^3\,\text{Pa}}$$

$$= 0.0224\,\text{m}^3 = 22.4 \text{ litres}$$

Rearranging the ideal gas equation $pV = NkT$, we get

$$k = \frac{pV}{NT} = \text{constant}$$

So for a fixed mass of gas (i.e. N = constant) we get

$$\frac{pV}{T} = \text{constant}$$

If, therefore, a fixed mass of gas has initial values p_1, V_1 and T_1, and final values p_2, V_2 and T_2, we can say

$$\frac{p_1V_1}{T_1} = \frac{p_2V_2}{T_2}$$

This equation is a very useful one to remember for working out problems.

Tip

Remember that in the equation T must be in K.

Worked example

A fish swimming at a depth of 20.0 m in the sea, where the temperature is 7.0 °C, expels an air bubble of volume 0.40 cm³. The bubble then rises to the surface, where the temperature is 24.0 °C.

a) Assuming that atmospheric pressure is 101 kPa and that the density of sea water is $1.03 \times 10^3\,\text{kg m}^{-3}$, calculate the pressure inside the bubble at a depth of 20.0 m.

b) Give two reasons why the bubble expands as it rises to the surface.

c) What is the volume of the bubble when it just reaches the surface?

Note

The pressure p due to a depth h of liquid of density ρ is $p = h\rho g$.

Answer

a)
$$\begin{aligned} p_{\text{tot}} &= p_A + h\rho g \\ &= 1.01 \times 10^5\,\text{Pa} + (20\,\text{m} \times 1.03 \times 10^3\,\text{kg m}^{-3} \times 9.81\,\text{N kg}^{-1}) \\ &= 1.01 \times 10^5\,\text{Pa} + 2.02 \times 10^5\,\text{Pa} \\ &= 3.03 \times 10^5\,\text{Pa} \end{aligned}$$

b) As the bubble rises to the surface, h gets less and so the pressure acting on it will also get less. As the volume of air is inversely proportional to pressure, the volume will increase. Additionally, the temperature increases towards the surface and, since $V \propto T$, this will also cause the air to expand.

c) Using $\frac{p_1V_1}{T_1} = \frac{p_2V_2}{T_2}$

$$V_2 = \frac{p_1V_1T_2}{p_2T_1} = \frac{3.03 \times 10^5\,\text{Pa} \times 0.40\,\text{cm}^3 \times 297\,\text{K}}{1.01 \times 10^5\,\text{Pa} \times 280\,\text{K}} = 1.3\,\text{cm}^3$$

Tip

It is always a good idea to rearrange the equation first, before putting in the data, particularly when the numbers are complex such as in this example.

12.5 Kinetic model of temperature

The application of kinetic theory to the molecules of an ideal gas enables us to derive an expression for the pressure p of a gas in terms of the density ρ of the gas and the **mean square speed** of the molecules, namely:

$$p = \frac{1}{3}\rho\langle c^2\rangle$$

The 'mean square speed' is the average of the squares of the speeds of all the individual molecules of the gas, and is given the symbol $\langle c^2\rangle$. This means we have to square the speed of each molecule, add up all these squared speeds and divide by the total number of molecules to give the average.

Tip

For examination purposes you need to be able to recognise and use this equation but you don't need to derive it.

Worked example

Calculate the mean square speed of gas molecules having the following speeds measured in m s^{-1}: 310, 320, 330, 340 and 350. (You might like to do this using a spreadsheet to make the number crunching easier).

Answer

Speed/m s^{-1}	Speed squared/$\text{m}^2\,\text{s}^{-2}$
310	96 100
320	102 400
330	108 900
340	115 600
350	122 500
Sum of the squared speeds	545 500
Mean square speed (÷ 5)	109 100

Table 12.4 ▲

Note that, although the mean square value (109 100) is close to, it is *not* the same as, the mean value (330) squared (108 900).

By combining the equation $p = \frac{1}{3}\rho\langle c^2\rangle$ with the ideal gas equation $pV = NkT$, it can be shown that:

$$\frac{1}{2}m\langle c^2\rangle = \frac{3}{2}kT$$

The expression $\frac{1}{2}m\langle c^2\rangle$ is the **average kinetic energy** of the randomly moving molecules of the gas. As k is a constant, it follows that

> the average kinetic energy of the molecules of a gas is *proportional to the absolute temperature* of the gas.

It also follows from the above equation that, in this simple, non-quantum model, the average random kinetic energy of the molecules will be zero at absolute zero, when $T = 0$.

Tip

For examination purposes, you must be able to recognise and use the expression

$\frac{1}{2}m\langle c^2\rangle = \frac{3}{2}kT$

remembering that T must be in kelvin.

Worked example

Calculate the mean square speed of the oxygen molecules in air at a temperature of 24 °C, given that an oxygen molecule has a mass of 5.34×10^{-26} kg.

Answer

Rearranging $\frac{1}{2}m\langle c^2\rangle = \frac{3}{2}kT$ gives us

$$\langle c^2\rangle = \frac{3kT}{m} = \frac{3 \times 1.38 \times 10^{-23}\,\mathrm{J\,K^{-1}} \times 297\,\mathrm{K}}{5.34 \times 10^{-26}\,\mathrm{kg}} = 2.30 \times 10^5\ \mathrm{m^2\,s^{-2}}$$

It is worth noting that the square root of the mean square speed – the 'root mean square speed' – is approximately equal to the average speed of the molecules. If we take the square root of $2.30 \times 10^5\,\mathrm{m^2\,s^{-2}}$ we get $480\,\mathrm{m\,s^{-1}}$. In Unit 2 we saw that the speed of sound in air is about $340\,\mathrm{m\,s^{-1}}$. This is about 25% less than the average random speed of the air molecules that carry the sound energy, which would appear to be very reasonable.

12.6 Evidence for the kinetic theory

Direct experimental evidence for the kinetic theory of matter ('kinetic' means 'having motion') was provided by a discovery by a Scottish botanist, Robert Brown, in 1827. He noticed that tiny grains of pollen, when suspended in water and viewed under a microscope, continually moved backwards and forwards with small, random, jerky paths. We now attribute this to unequal bombardment of the very fine grains of pollen by the invisible water molecules, which themselves must therefore be in continuous motion.

Scientific evidence is much stronger if we can quantify it. Einstein did just this – he analysed this 'Brownian motion' mathematically and was able to determine a value for the Avogadro constant (the number of particles, 6.02×10^{23}, in a mole of a substance), which agreed very closely with the value obtained by chemical means.

Brownian motion provides strong evidence for particles of matter being in continuous motion. Kinetic theory relates the macroscopic (large-scale) behaviour of an ideal gas, in terms of its pressure, volume and temperature, to the microscopic properties of its molecules.

The assumptions we make in order to establish a kinetic theory for an *ideal gas*, with the corresponding experimental justification for these assumptions, are summarised in Table 12.5.

Assumption	Experimental evidence
A gas consists of a very large number of molecules.	Brownian motion.
These molecules are in continuous, rapid, random motion.	
Collisions between molecules and between molecules and the walls of a container are perfectly elastic (i.e. no kinetic energy is lost).	If not, the molecules would gradually slow down – this cannot be the case as Brownian motion is observed to be continuous. It would also mean that the gas would gradually cool down.
The volume occupied by the molecules themselves is negligible compared with the volume of the container.	It is easy to compress a gas by a large amount.
Intermolecular forces are negligible except during a collision.	From the above it follows that on average the molecules are very far apart relative to their size and so the intermolecular forces become very small.
The duration of collisions is negligible compared with the time spent in between collisions.	This also follows from the fact that the molecules, on average, are very much further apart than their size.

Table 12.5 ▶

12.7 Historical context

Our study of thermodynamics would not really be complete without a brief look at the historical context in which thermodynamics was developed. The impetus was provided by the Industrial Revolution, which took place from the mid-18th to the mid-19th century. The steam engine allowed improvements in mining coal and in the production of iron, evocatively portrayed in the painting 'Coalbrookdale by Night'. The use of iron and steel in machinery led to major changes in agriculture, manufacturing and transport. This all had a profound effect on the socio-economic and cultural conditions in Britain and around the world, in the same way that electronics and the computer have revolutionised our lives in more recent times.

Figure 12.11 ▲
Coalbrookdale by Night, painted in 1801 by P. J. De Loutherbourg

Following James Watt's refinements to the design of the steam engine towards the end of the 18th century, engineers sought to improve its efficiency further by the application of the laws of physics. Thus began the intensive study of the behaviour of gases and the development of kinetic theory, thermodynamics and eventually quantum mechanics, which forms the foundation for most theories in physics today.

As we saw in Unit 2, the development of quantum mechanics led to wide-ranging experiments on solid materials ('solid state physics') in the 1920s and 1930s. Following the Second World War (when the attention of physicists was diverted elsewhere) Bardeen, Brattain and Shockley invented the transistor in 1947. They were jointly awarded the Nobel Prize in Physics in 1956 'for their researches on semiconductors and their discovery of the transistor effect'. Their work on semiconductors led to the development of silicon chips, without which we would not have today's computers, nor your iPod!

Figure 12.12 ▲
James Watt's steam engine, 1785

REVIEW QUESTIONS

1 a) Which of the graphs in Figure 12.13 does **not** represent the behaviour of a gas that obeys Boyle's law?

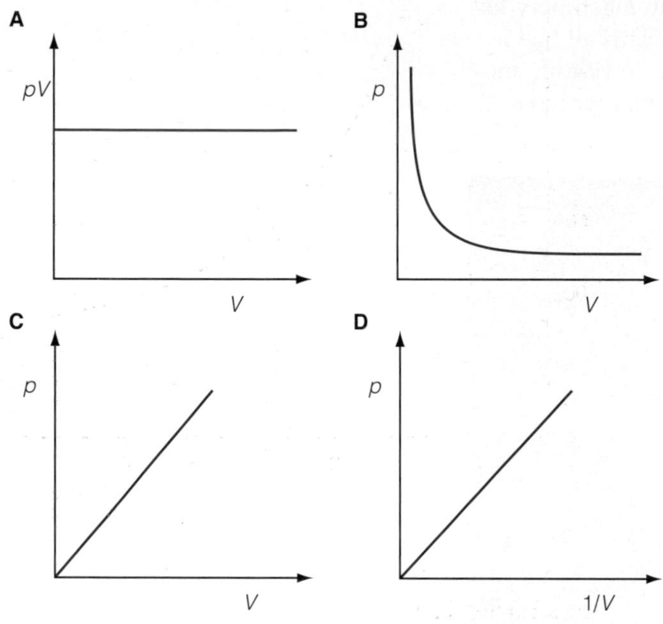

Figure 12.13 ▲

b) Which of the graphs in Figure 12.14 does **not** represent the behaviour of an ideal gas?

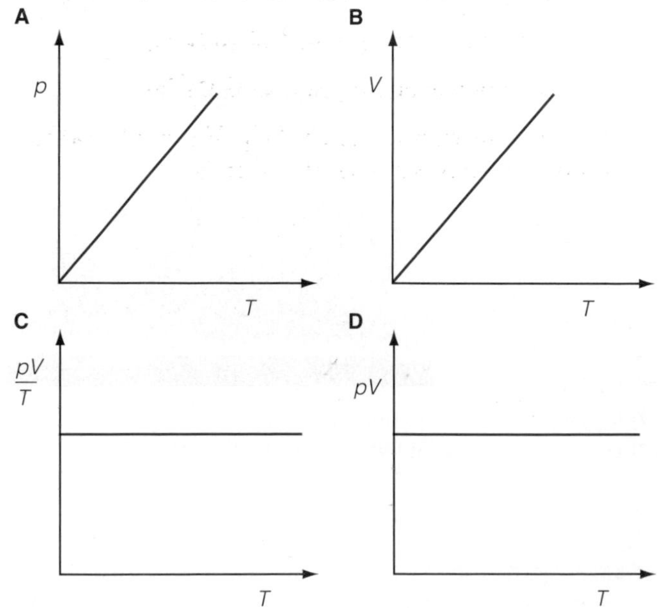

Figure 12.14 ▲

2 The pressure exerted on the ground by

a) the heel of a shoe having an area of 0.80 cm^2 when a girl of mass 57 kg puts all her weight on it is:

A 0.07 MPa B 0.70 MPa C 7.0 MPa D 70 MPa

b) the wheel of a car of mass 1600 kg, assuming that each tyre has an area of 56 cm^2 in contact with the road and that the weight of the car is evenly distributed, is:

A 0.07 MPa B 0.70 MPa C 7.0 MPa D 70 MPa

3 In an experiment to investigate how the pressure of air at constant volume depends on the temperature of the air, a flask contains air at a pressure of 101 kPa and a temperature of 17 °C. When the air is heated to a temperature of 100 °C, the pressure will become about:

A 17 kPa B 79 kPa C 130 kPa D 594 kPa

4 a) Show that the equation for pressure in a liquid, $p = h\rho g$, is homogeneous with respect to units.

b) Calculate the pressure difference between offices at the top and bottom of the Canary Wharf building in London, which is 240 m high. Assume that the density of air is 1.29 kg m^{-3}. What other assumption do you have to make?

c) Discuss whether office workers would notice such a pressure difference on a day when atmospheric pressure was 101 kPa.

d) Explain why a calculation like you performed in part b) could not be used to calculate the pressure experienced by an aeroplane flying at a height of 10 000 m.

5 The speed of sound in helium at 20 °C is about 1000 m s^{-1}, compared with 340 m s^{-1} in air. The high speed of sound in helium is responsible for the amusing 'Donald Duck' voice that occurs when someone has breathed in helium from a balloon.

a) Show that the equation $\frac{1}{2}m\langle c^2\rangle = \frac{3}{2}kT$ is homogeneous with respect to units.

b) Explain what is meant by the term $\langle c^2\rangle$.

c) What is the mean square speed of five air molecules that have speeds of 300, 400, 500, 600 and 700 m s^{-1}?

d) Calculate the mean square speed for i) air molecules and ii) helium molecules at a temperature of 20 °C. Assume that an 'air molecule' has a mass of 4.8×10^{-26} kg and a helium molecule has a mass of 6.7×10^{-27} kg.

e) Comment on your answer in relation to the speed of sound in helium compared with that in air.

6 The equation of state for an ideal gas is $pV = NkT$.

a) What is meant by an *ideal gas*?

b) What do the symbols N and k represent in this equation?

c) Show that the product pV has the same units as energy.

d) Hence show that the units of k are $\mathrm{J\,K^{-1}}$.

e) How many molecules of air would there be in a laboratory measuring 10.0 m × 8.0 m × 3.0 m on a day when the temperature is 19 °C and the pressure is 101 kPa? ($k = 1.38 \times 10^{-23}\,\mathrm{J\,K^{-1}}$.)

7 A 60 W light bulb contains 110 $\mathrm{cm^3}$ of argon at a pressure of 87 kPa when it is at a room temperature of 17 °C. When the lamp has been switched on for some time, the temperature of the argon becomes 77 °C.

a) Calculate the pressure of the argon when the lamp is on.

b) Calculate the number of molecules of argon in the bulb, assuming that $k = 1.38 \times 10^{-23}\,\mathrm{J\,K^{-1}}$.

c) If 6.0×10^{23} molecules of argon have a mass of 40 g, what is the mass of argon in the bulb?

8 The laws of Association Football state that the pressure of the ball must be 'between 0.6 and 1.1 atmospheres (600–1100 $\mathrm{g/cm^2}$) above atmospheric pressure at sea level'.

a) Show that the above data gives a value for atmospheric pressure of approximately 100 kPa.

b) A football is pumped up to a pressure of 0.75 atmospheres above atmospheric pressure in the warm dressing room, where the temperature is 21 °C. It is then taken out onto the pitch, where the temperature is only 7 °C. Determine whether or not the pressure will still be within the legal limit.

c) State what assumption you have made in arriving at your answer.

9 The graph shows a p–V curve for a gas at a temperature of 300 K. Using the plotted points A, B, C as a guide, sketch the graph on a sheet of graph paper.

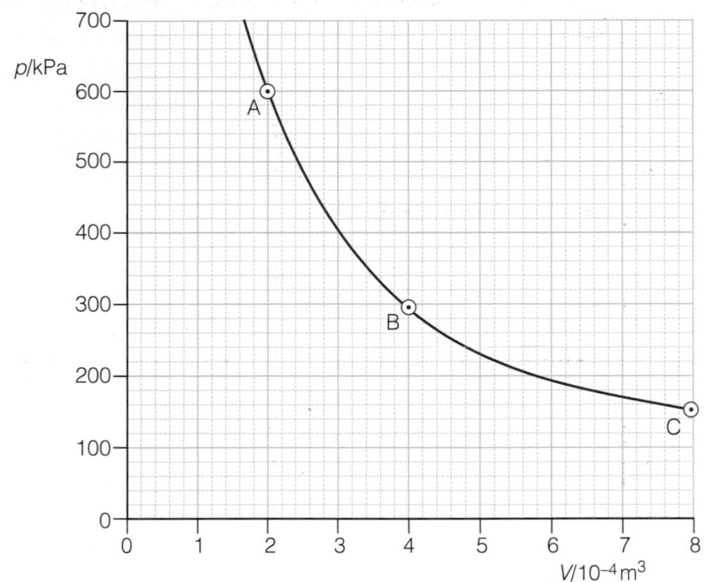

Figure 12.15 ▲

a) State Boyle's law.

b) By considering the points A, B and C, show that the gas obeys Boyle's law.

c) The temperature is now increased to 400 K. What will now be:

i) the pressure for a volume of $8.0 \times 10^{-3}\,\mathrm{m^3}$,

ii) the volume at a pressure of 400 kPa,

iii) the volume at a pressure of 640 kPa?

d) Use your answers to sketch the p–V curve for 400 K on the same axes as your first sketch.

13 Nuclear decay

The word 'nuclear' often conjures up the wrong image – even to the extent that Nuclear Magnetic Resonance (NMR) scanners are now called Magnetic Resonance Imaging (MRI) scanners to avoid the use of the word 'nuclear'. The bad image, of course, is generated by the thought of nuclear weapons and, to a lesser extent, by nuclear power stations. Whilst few would argue against the evil of nuclear weapons, the debate about nuclear energy is finely balanced and one that as a physicist you should be able to discuss in an informed and rational way. The study of radioactivity and its effects and consequences should help you to develop a better understanding of the 'nuclear debate'. Whether we like it or not, the nuclear genie has been released from its bottle and cannot be squeezed back in again.

13.1 Discovery of radioactivity

Ionising radiation, or 'radioactivity' as we now call it, was discovered by accident in 1896 by the French scientist Henri Becquerel. He found that certain uranium salts affected a photographic plate even when it was covered with black paper. The effect was similar to that of X-rays, which had been discovered by Wilhelm Röntgen a year earlier while working with high-voltage cathode ray tubes.

Becquerel showed that the rays emitted from the uranium caused gases to *ionise* and that they differed from X-rays in that they could be *deflected by a strong magnetic field*. For his discovery of spontaneous radioactivity Becquerel was awarded the Nobel Prize in Physics in 1903, which he shared with Marie and Pierre Curie, who had discovered two further radioactive elements, which they named radium and polonium.

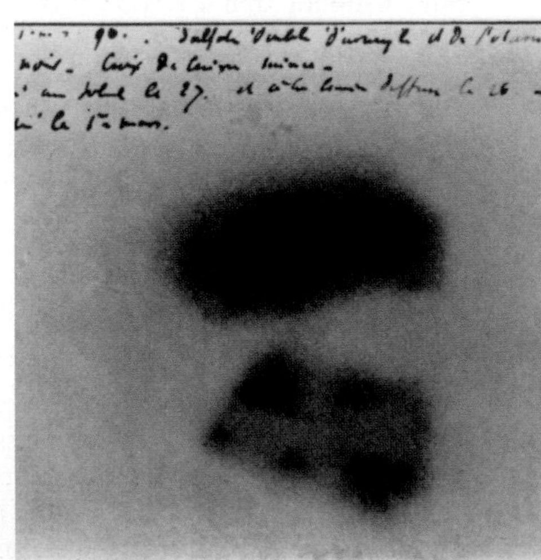

Figure 13.1 ▲
Henri Becquerel and the exposed photographic plate that led him to propose a new form of radiation

13.2 Background radiation

We live in a radioactive environment. The air you breathe, the ground you walk on, the house you live in, the food you eat and the water you drink all contain radioactive isotopes (this term is defined on page 135). This radiation, which is constantly present in our environment, is called **background radiation**.

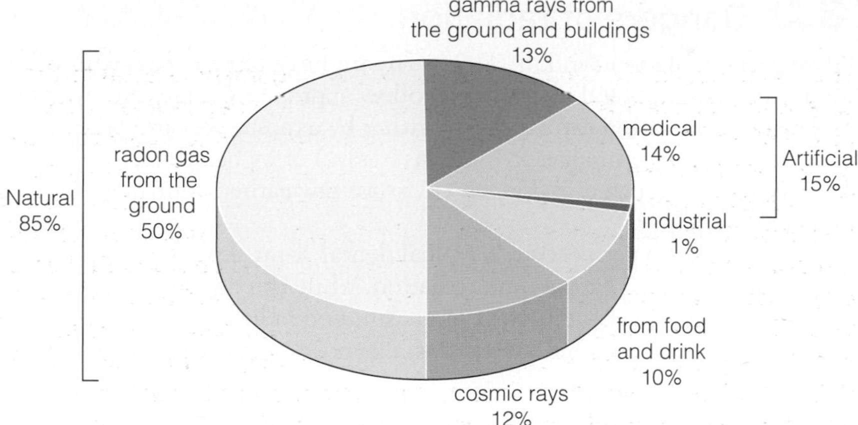

Figure 13.2 ◀ Sources of background radiation

The main natural sources of background radiation are:

- radioactive gases (mainly radon) emitted from the ground, which can be trapped in buildings and build up to potentially dangerous levels – high levels of radon can greatly increase the risk of lung cancer;
- radioactive elements in the Earth's crust – mainly uranium and the isotopes it forms when it decays – these give rise to gamma radiation, which is emitted from the ground and rocks (e.g. granite) and from building materials (e.g. bricks and plaster);
- cosmic rays from outer space which bombard the Earth's atmosphere producing showers of lower-energy particles such as muons, neutrons and electrons and also gamma rays;
- naturally occurring radioactive isotopes present in our food and drink, and in the air we breathe, including carbon-14 and potassium-40.

These natural sources make up about 85% of the radiation to which we are exposed (see Figure 13.2). The remaining 15% comes from artificial sources, mainly medical, e.g. X-rays; a tiny fraction comes from nuclear processes in industry.

Figure 13.3 summarises the natural sources of background radiation to which we are exposed.

Note

The risk of death due to radon-induced cancer at the average radon level in the UK is estimated to be 3 in 1000, or about the same as from a pedestrian traffic accident. This is increased by a factor of 10 for someone who smokes 15 cigarettes a day.

Tip

The 'background count' must be measured and then subtracted from subsequent readings when undertaking quantitative experiments on radioactivity – see Section 13.6.

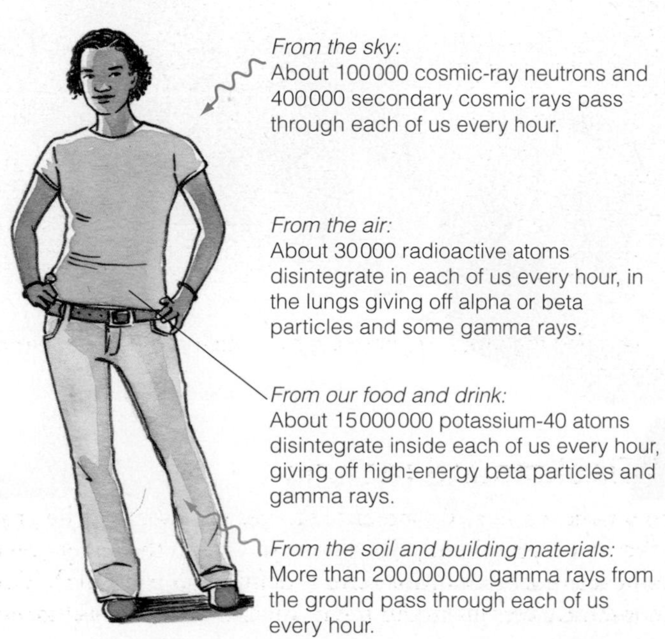

Figure 13.3 ◀ Natural background radiation

Figure 13.4 ▲
Radiation warning symbol

13.3 Dangers of radiation

WARNING – radiation can kill! That's why we have to be very careful when dealing with radiation and must always follow appropriate safety precautions. The dangers of radiation can be caused either by exposure of our bodies to some external source of radiation (such as X-rays) or by breathing or ingesting radioactive matter (such as radon which, as we saw earlier, contributes to the background radiation).

To put things into perspective, a typical dental X-ray dose is equivalent to about 3 days of natural background radiation, while the dose from a chest X-ray is equivalent to about 10 days of background radiation. Having said this, radiation exposure is accumulative and so diagnostic X-rays should always be kept to a minimum.

A detailed discussion of the biological effects of radiation is beyond the scope of this book, except to say that even a very low dose of ionising radiation, whether natural or artificial, has a chance of causing cancer because of the damage it can do to our DNA. So, it is essential to always observe the following precautions:

- keep as far away as possible (at least 30 cm) from all laboratory sources of ionising radiation;
- do not touch radioactive materials – use a handling tool;
- keep sources in their lead storage containers when not in use;
- during an investigation keep the source pointed away from the body, especially the eyes;
- limit the time of use of sources – return to secure storage as soon as possible;
- wash hands after working with a radioactive source.

Worked example

Figure 13.5 shows a radiotherapist using a remote control to operate a linear accelerator, which produces high-energy X-rays.

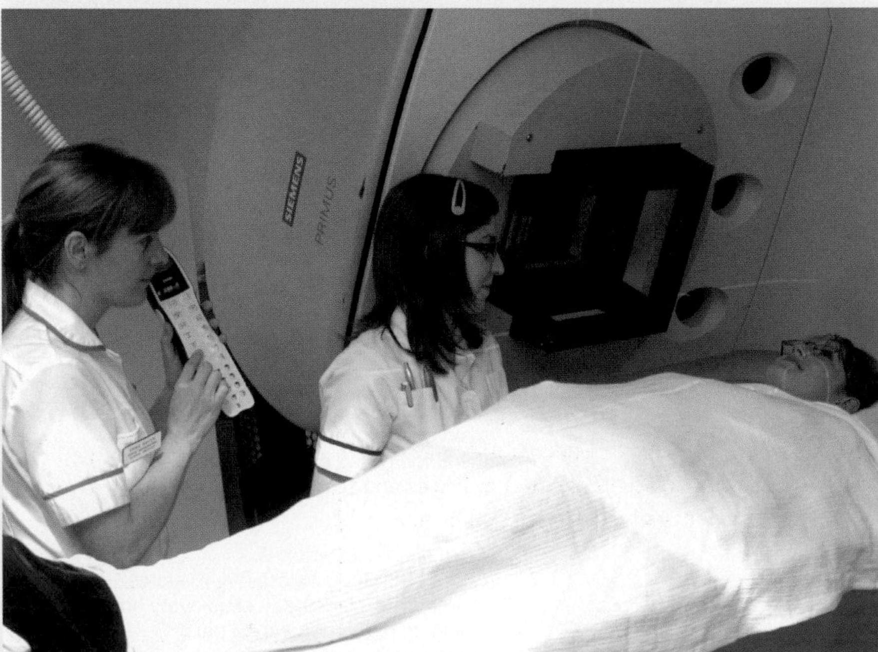

Figure 13.5 ▲

Explain why the radiotherapist
a) uses a remote control, and
b) wears a radiation badge.

Answer

a) X-rays are particularly dangerous if received in repeated doses because they can inflict biological damage to humans and, in particular, can cause cancer. The radiographer needs to be as far away as possible from the X-rays to minimise the dose she receives, and so she uses a remote control to operate the machine.

b) It is important to monitor on a regular basis the amount of radiation to which workers using X-rays or radioactive materials are exposed. This is done by wearing a 'badge', which can measure the amount of radiation to which the worker has been subjected. If the dose exceeds the safe limit, the worker has to be moved to different tasks.

13.4 Alpha, beta and gamma radiation

Within two years of Becquerel's discovery of radiation, Ernest Rutherford, working at the Cavendish Laboratory in Cambridge, showed that there were two distinct types of radiation, which he called alpha (α) rays and beta (β) rays (α and β being the first two letters of the Greek alphabet). By 1900, a third type of radiation had been identified by the French physicist Paul Villard. This was called gamma (γ) radiation. The property of all these radiations that enables them to be detected and distinguished is their ability to cause **ionisation**. They are often collectively referred to as '**ionising radiation**'.

Experiment

Demonstrating ionisation

The set-up is shown in Figure 13.6. The energy from the candle flame gives electrons in the air sufficient kinetic energy to break free from their atoms. This process is called **ionisation**. The negative electrons move towards the left-hand positive plate while the much heavier positive ions of air move towards the right-hand negative plate.

The movement of these ions creates a 'wind' which blows the candle towards the right-hand negative plate. The electrons and positive ions act as 'charge carriers', creating a small current in the circuit, which is detected by the nanoammeter.

If the candle is replaced by an α-emitter, such as americium-241, a small current is again observed. This is because the energy from the α-radiation causes the air to become ionised, just like the flame.

Safety note

This experiment involves the use of a high-voltage supply and a radioactive source. The voltage supply should be limited to give a current of no more than 5 mA. Stringent safety procedures should be followed with the radioactive source (see page 130).

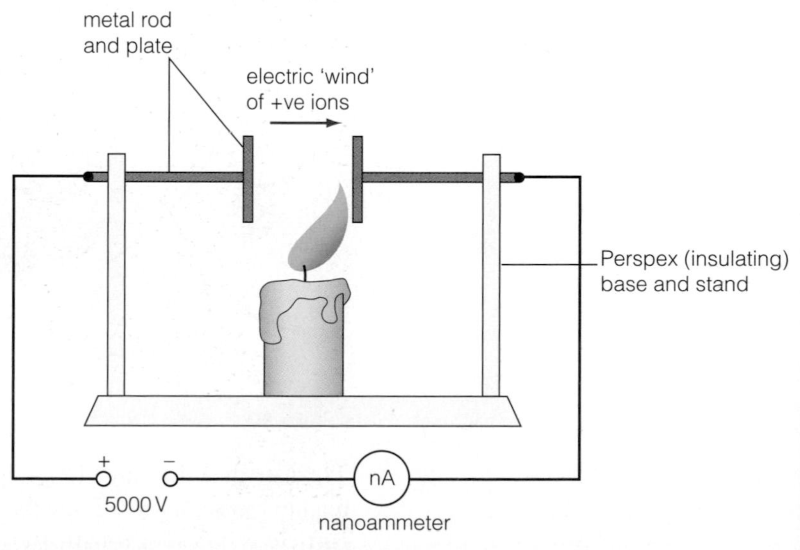

Figure 13.6 ▲

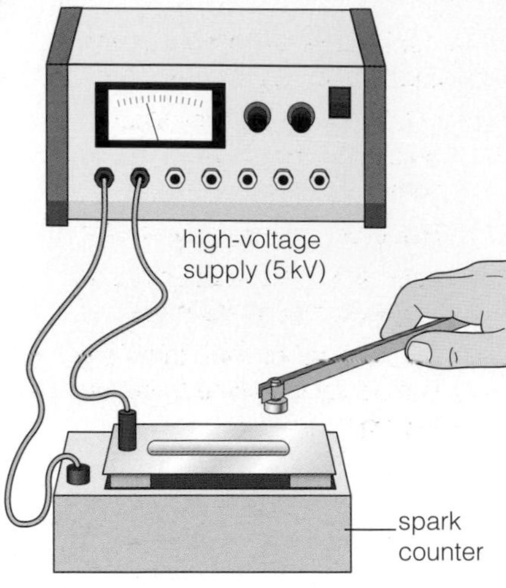

Figure 13.7 ▲
Using a spark counter

Alpha radiation

In the school laboratory, alpha radiation can be detected using either a **spark counter** or an **ionisation chamber**. Details of how these devices work are not needed, except to understand that both depend on the ionising properties of alpha radiation.

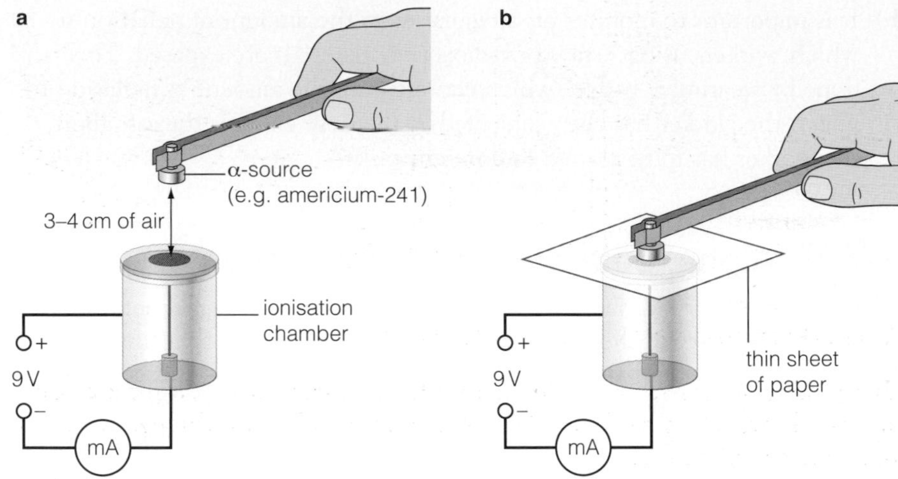

Figure 13.8 ▲
Using an ionisation chamber to find the range of alpha radiation

With a spark counter connected to a high-voltage supply (Figure 13.7) or with an ionisation chamber and sensitive current detector arranged as in Figure 13.8, it can be shown that the range of α-radiation is a few centimetres in air (Figure 13.8a) and that α-radiation is completely absorbed by a thin sheet of paper (Figure 13.8b).

Rutherford showed that α-radiation can be deflected by very strong magnetic and electric fields (see Figure 13.12a on page 134). From such deflections it can be shown that α-radiation is not actually 'radiation' but consists of positively charged particles. As we saw in Unit 4 (page 70), a classic experiment undertaken by Rutherford and Royds in 1909 showed that α-particles were **helium nuclei**, i.e. helium atoms without their electrons (Figure 13.9).

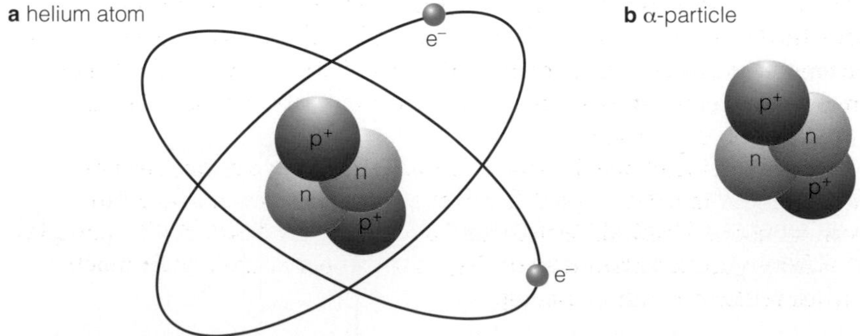

Figure 13.9 ▲

The fact that they are charged, and have a large mass, means that α-particles readily ionise matter. Thus the energy of an α-particle is rapidly lost and the particle is quickly brought to rest – in other words, the particle is easily absorbed as it passes through matter. In the ionisation process the α-particle gains two electrons and becomes an atom of helium gas.

Beta radiation

In order to detect beta radiation we use a **Geiger–Müller tube (G-M tube)** connected to a **scaler** or a **rate-meter**. Such an arrangement cannot be used to detect α-particles because these cannot penetrate the thin mica window of the G-M tube.

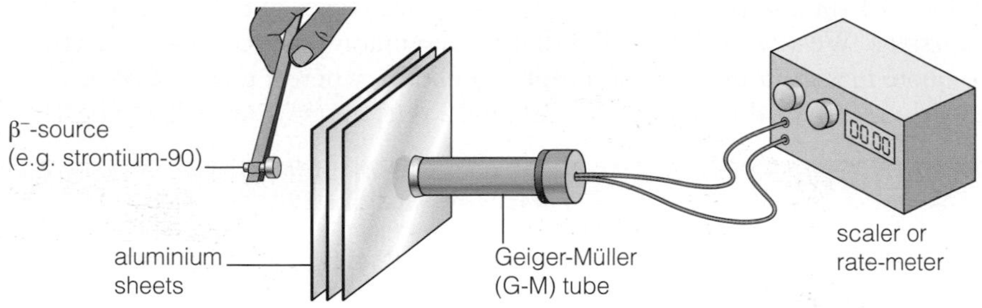

Figure 13.10 ▲
Using a G-M tube to find the range of beta radiation

Using the arrangement shown in Figure 13.10, we can show that β-radiation can travel through as much as 50 cm of air and can also pass through a few millimetres of aluminium.

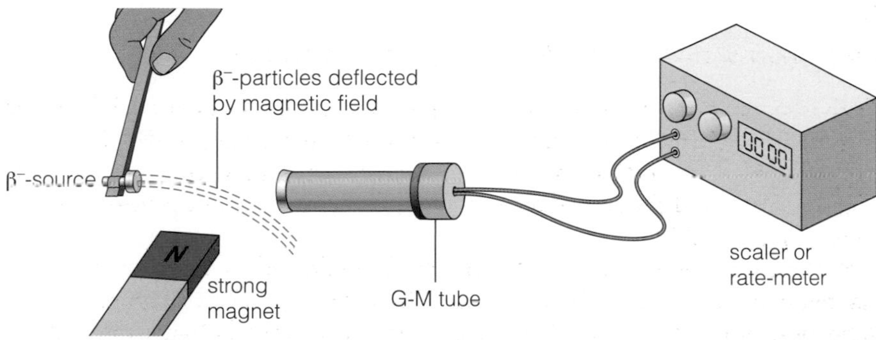

Figure 13.11 ▲
Showing the deflection of beta radiation in a magnetic field

If a very strong magnet is brought up as shown in Figure 13.11, the count-rate is observed to drop, showing that the β-radiation has been deflected by the magnetic field (see also Figure 12.13b on page 134). Quantitative experiments with magnetic and electric fields confirm that β-radiation consists of fast-moving streams of negatively charged **electrons**. It is therefore more precisely designated as β^--radiation or β^--particles.

β^--particles, being charged, cause a fair amount of ionisation, but not anywhere near as much as α-particles as they are much less massive (an electron has about 1/8000th of the mass of a helium nucleus). As β^--particles cause far less ionisation than α-particles, they can penetrate matter much further before their energy is used up.

Gamma radiation

All attempts to deflect gamma radiation in the most powerful electric and magnetic fields were to no avail. Eventually Rutherford confirmed that γ-radiation was a high-energy form of electromagnetic radiation, like X-rays, ultraviolet, light, microwaves and radio waves.

Gamma radiation is composed of minute bundles of energy called **photons** (which we talked about in Unit 2). Only when one such photon happens to collide directly with an atom of matter does ionisation occur. This means that γ-radiation is only weakly ionising.

Tip

Remember that the essentials for experiments on radioactivity are:

- safety – always follow strict guidelines, e.g. handling sources with tongs
- detector – usually a G-M tube
- counter – either a counter and stopwatch or a rate-meter
- background count – take for a sufficiently long time to get an average.

Like all electromagnetic radiation, γ-rays obey an inverse-square law, meaning that their intensity falls off as the square of the distance in air. Gamma rays easily pass through aluminium and are not entirely absorbed (or 'attenuated') even by several centimetres of lead.

Summary of the properties of α, β and γ-radiation

Table 13.1 summarises the properties of the three types of radioactive emissions. We saw in Section 9.3 that it is common practice in atomic physics to quote masses in **unified mass units**, symbol **u**, where 1 u is defined as exactly 1/12th of the mass of the carbon-12 nucleus (= 1.66×10^{-27} kg).

Property	α-particle	β-particle	γ-radiation
mass	4 u (6.64×10^{-27} kg)	about $\frac{1}{2000}$ u (9.11×10^{-31} kg)	0
charge	$+2e$	$-e$	0
speed	up to about $\frac{1}{20}c$	up to about $0.99c$	c
typical energy	0.6 to 1.3 pJ (4 to 8 MeV)	0 up to 2.0 pJ (0 to 12 MeV)	0.01 to 1.0 pJ (0.06 to 6 MeV)
relative ionising power	10 000	100	1
penetration	few cm of air	few mm of aluminium	few cm of lead
deflection in electric and magnetic fields	small deflection	large deflection	no deflection
nature	helium nucleus	electron	high-frequency electromagnetic radiation

Table 13.1 ▲

The relative ionising power – and hence penetrating power – of α-particles and β-particles is illustrated in the cloud chamber photographs shown in Figure 13.12a, b.

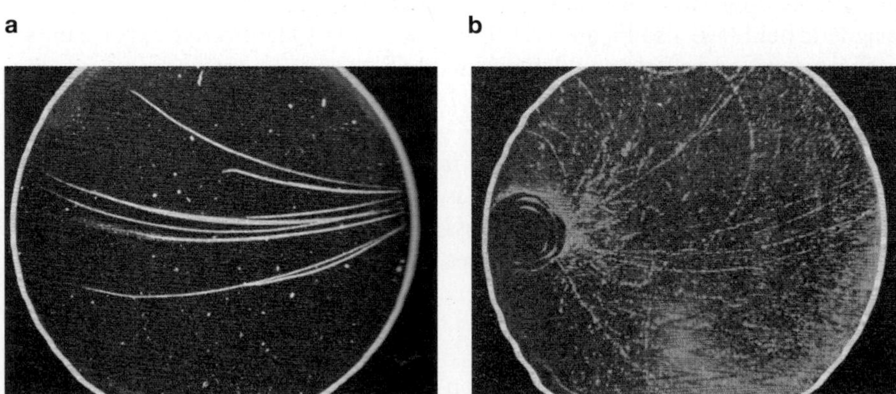

Figure 13.12 ▶
a) α-particle tracks and b) β-particle tracks

Figure 13.12a shows the strong, thick tracks produced by the heavily ionising α-particles, while Figure 13.12b shows the much thinner tracks of the less ionising β-particles. In each case, there is a magnetic field directed up out of the diagram (you should check this for yourself from the direction in which the tracks curve).

In the case of the α-particles, this field had to be very strong – over 4 teslas. Can you explain why the shorter α-particle tracks are more curved than the longer ones? The large-angle deflection near the end of one track is probably the result of a close encounter with a nucleus.

It can be deduced from the different curvatures of the β-particle tracks that the β-particles must have a range of different energies (or velocities). We will discuss the significance of this on pages 136–137.

We now know that these ionising radiations come from the unstable nuclei of radioactive elements, which, on emission of an alpha particle or a beta particle, decay into another, lower-energy nuclide.

Tip

The energy of ionising particles is usually expressed in electron-volts (eV), where 1 eV ≡ 1.60×10^{-19} J. Thus an energy of 1 pJ is equivalent to

$$\frac{1 \times 10^{-12}\,\text{J}}{1.60 \times 10^{-19}\,\text{J eV}^{-1}} = 6.25 \times 10^{6}\,\text{eV}$$

$$= 6.25\,\text{MeV}$$

As can be seen from Table 13.1, nuclear energies are in the order of MeV. Note that this means that nuclear energies are about a million times greater than energies encountered in chemistry, which are typically in the order of eV.

13.5 Disintegration processes

Let's start by revising some basic definitions.

From Unit 4 you should be familiar with the **proton number** (symbol Z) of an element being the number of protons in one atom. Remember that a neutral atom will also have the same number of electrons, for example an atom of oxygen will contain 8 protons in its nucleus, surrounded by 8 electrons. Note that the proton number is sometimes called the **atomic number** as this defines the element, e.g. $Z = 1$ is hydrogen, $Z = 2$ is helium, ... $Z = 8$ is oxygen, and so on.

The particles making up the nucleus of an atom, namely the protons and neutrons, are called nucleons. The **nucleon number** or **mass number** (symbol A) of an atom is the total number of protons and neutrons in the nucleus of that atom.

Isotopes are different forms of the *same element* (i.e. they have the same proton number, Z) that have a different number of *neutrons* in the nucleus (and therefore a different nucleon number, A). A **nuclide** is the nucleus of a particular isotope and is represented by the following symbol:

nucleon number → A
X ← symbol for element
proton number → Z

Worked example

Uranium occurs naturally in the form of two isotopes, $^{235}_{92}$U and $^{238}_{92}$U.

a) Explain what is meant by 'isotopes'.

b) Copy and complete the table to show the proton number, nucleon number and number of neutrons for each isotope.

Isotope	Proton number	Nucleon number	Number of neutrons
$^{235}_{92}$U	i)	ii)	iii)
$^{238}_{92}$U	iv)	v)	vi)

Table 13.2 ▲

Answer

a) Isotopes are different forms of the same element that have a different number of neutrons in the nucleus.

b)

Isotope	Proton number	Nucleon number	Number of neutrons
$^{235}_{92}$U	92	235	235 − 92 = 143
$^{238}_{92}$U	92	238	238 − 92 = 146

Table 13.3 ▲

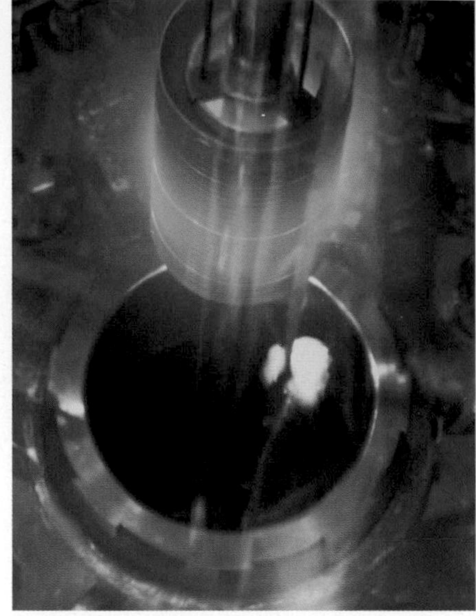

Figure 13.13 ▲
Removal of a fuel rod from a reactor

The proton number determines the number of electrons in each atom, which in turn gives rise to the chemical properties. Isotopes therefore have the *same chemical properties* because they have the same proton number. However, they will have different atomic masses and therefore *different densities*. The latter property of isotopes enables the separation of uranium-235 and uranium-238 by a diffusion process in the manufacture of 'enriched' uranium fuel rods for nuclear reactors.

We can now look at what happens in α-decay and β-decay.

α-emission

As an α-particle is a helium nucleus, it has the symbol ${}^{4}_{2}He$. A typical α-emitter is americium-241, for which the decay process is

$$ {}^{241}_{95}Am \rightarrow {}^{237}_{93}Np + {}^{4}_{2}He $$

americium-241 → neptunium-237 + α-particle

Neptunium-237 is formed because conservation laws decree that **proton numbers and nucleon numbers must total the same on each side of the equation.**

This process can be represented on a plot of nucleon number (mass number), A, against proton number (atomic number), Z. This is shown in Figure 13.14.

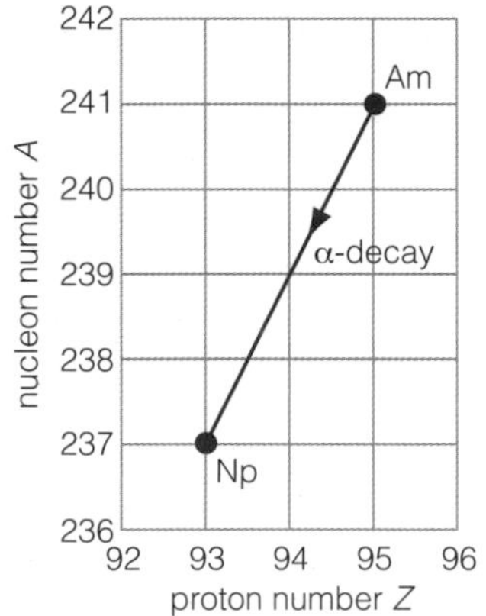

Figure 13.14 ▲
Graphical representation of α-decay

Worked example

Neptunium-237 decays further by α-emission to form an isotope of protactinium (Pa). Write the equation for this decay process.

Answer

Start by writing down what we know:

$$ {}^{237}_{93}Np \rightarrow Pa + {}^{4}_{2}He $$

Now balance the proton numbers, giving Z = (93 − 2) = 91 for protactinium, and then the nucleon numbers, giving A = (237 − 4) = 233 for this isotope of protactinium. The equation is thus:

$$ {}^{237}_{93}Np \rightarrow {}^{233}_{91}Pa + {}^{4}_{2}He $$

β^--emission

Since a β^--particle is an electron, we write it as ${}^{0}_{-1}e$. A typical β^--emitter is strontium-90. The decay process for this is:

$$ {}^{90}_{38}Sr \rightarrow {}^{90}_{39}Y + {}^{0}_{-1}e $$

strontium-90 → yttrium-90 + β^--particle

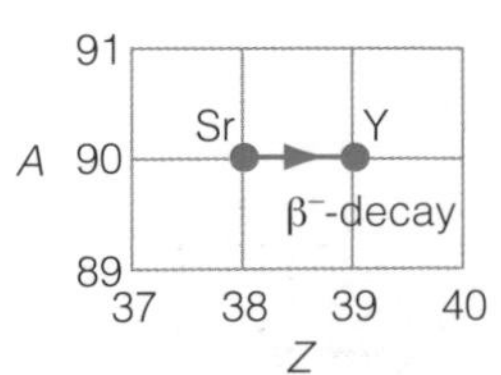

Figure 13.15 ▲
Graphical representation of β^--decay

Once again, the proton and nucleon numbers must balance on each side of the equation. As with α-decay, we can show this on an A–Z plot (Figure 13.15).

We saw in the cloud chamber photograph in Figure 13.12b on page 134 that β-particles are emitted with a range of energies. If a graph of the number of β-particles having a particular energy is plotted against the energy, a curve like that in Figure 13.16 is obtained.

This continuous distribution of energy could not be explained at first – a particular decay should provide the β-particle with a definite amount of energy. However, most had a great deal less than this, and where the rest of the energy went to was a mystery. In order to account for this 'missing' energy, Wolfgang Pauli, in 1930, put forward the idea of a new particle, having no

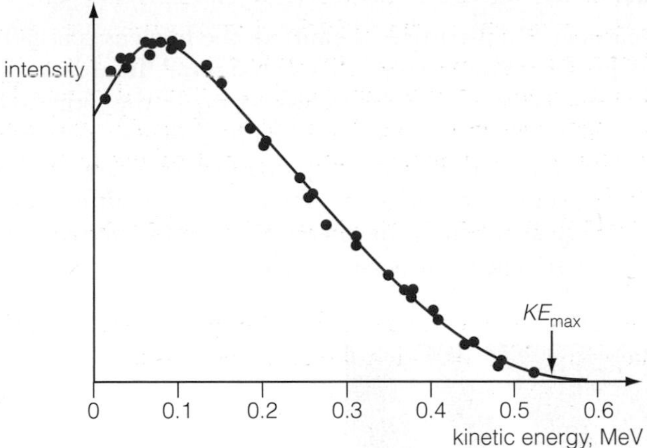

Figure 13.16 ◀
Energy spectrum of β-decay electrons from strontium-90

mass or charge but a range of possible energies. This imaginary particle, which was necessary for conservation of momentum, as well as energy, in β-decay, was given the name **neutrino** by Enrico Fermi. By 1934 Fermi had developed a theory of β-decay to include the neutrino, which some 20 years later led to the concept of the so-called 'weak interaction'. Neutrinos have now been detected and are considered to have no charge but a very small mass.

So, β-decay is accompanied by the emission of a neutrino. In the case of β^--decay, the neutrino is of a particular type called an 'electron antineutrino' (symbol $\bar{\nu}_e$). In the decay of strontium-90, the total energy released from the decay is 0.546 MeV. This is mostly shared by the emitted electron and the antineutrino, with a small amount being taken by the recoiling yttrium nucleus. It follows that the maximum kinetic energy KE_{max} in Figure 13.16 will be about 0.546 MeV.

Worked example

Yttrium-90 decays further by another β^--emission to form an isotope of zirconium (Zr). Write down the equation for this process.

Answer

As before, write down the known information:

$$^{90}_{39}\text{Y} \rightarrow \text{Zr} + {}^{0}_{-1}\text{e} + \bar{\nu}_e$$

Balancing the proton numbers gives Z = (39 − (−1)) = 40 for zirconium and balancing the nucleon numbers gives A = (90 − 0) = 90 for this particular isotope of zirconium. The equation is therefore:

$$^{90}_{39}\text{Y} \rightarrow {}^{90}_{40}\text{Zr} + {}^{0}_{-1}\text{e} + \bar{\nu}_e$$

β^+-emission

We discussed in Section 9.2 that the electron has an antiparticle, called the **positron**. Some of the lighter elements have isotopes, mostly artificially made, that decay by emitting a positron, or β^+-particle. A typical example is potassium-38, which decays to argon-38 by β^+-emission:

$$^{38}_{19}\text{K} \rightarrow {}^{38}_{18}\text{A} + {}^{0}_{1}\text{e} + \nu_e$$

potassium-38 → argon-38 + β^+-particle + electron neutrino

Note that in this decay process an 'electron neutrino' (ν_e) is also produced. This decay can be represented on an A–Z curve as shown in Figure 13.17.

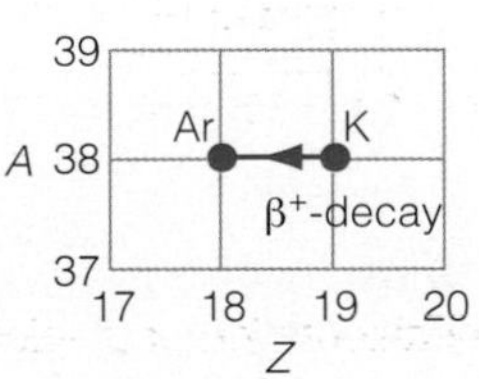

Figure 13.17 ▲
Graphical representation of β^+-decay

a

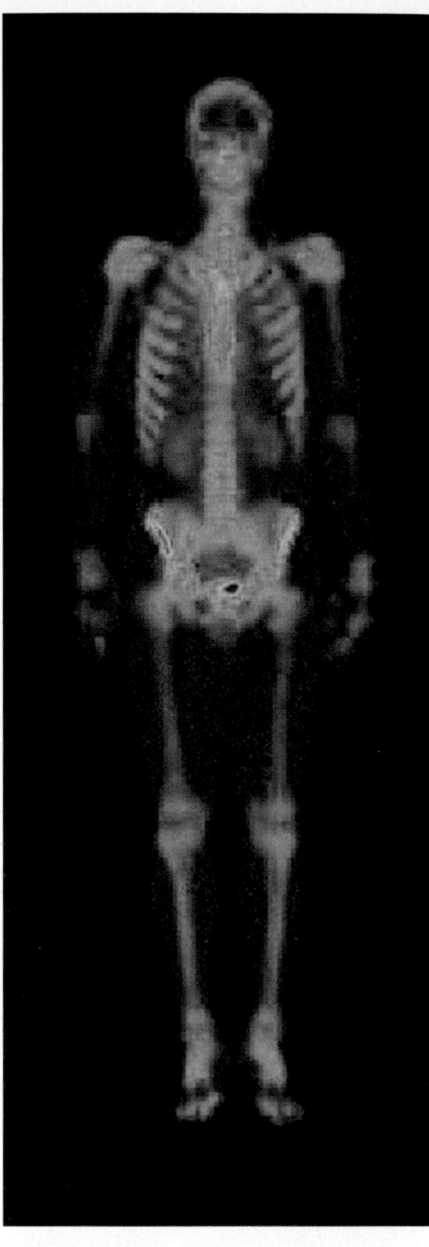

γ-emission

When an α-particle or a β-particle is emitted, the nucleus is usually left with a surplus of energy – it is said to be in an 'excited state'. In order to achieve a stable state the nucleus gives out the excess energy in the form of a quantum of γ-radiation. Thus α- and β-emission is nearly always accompanied by γ-radiation. Gamma rays have identical properties to X-rays; they differ from X-rays only in respect of their origin. Gamma rays are emitted from an excited *nucleus*, while X-rays come from the energy released by excited *electrons* returning to lower energy levels outside the nucleus.

b

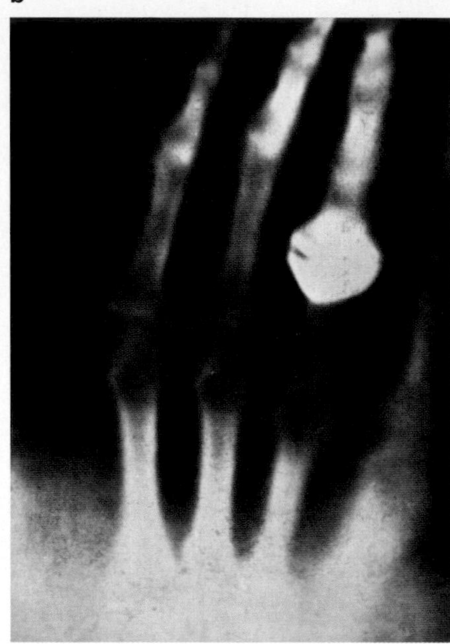

Figure 13.18 ▲
a) A modern coloured gamma scan of a human skeleton and b) the first X-ray photograph taken of human bones

Worked example

The mass of a radium-226 nucleus is 225.9771 u. It decays into radon-222, which has a nuclear mass of 221.9703 u, by emitting an α-particle of mass 4.0015 u and energy 4.8 MeV.

a) Show that the energy released in this decay is about 5 MeV.
b) Suggest what happens to the difference between this energy and the 4.8 MeV taken away by the α-particle.

Answer
In Chapter 9 (page 84) we used Einstein's equation $\Delta E = c^2 \Delta m$ in situations involving the creation and annihilation of matter and antimatter particles. We need to use the same principle here.

a) Adding up the mass of the radon-222 nucleus and the α-particle gives us

$$221.9703\,\text{u} + 4.0015\,\text{u} = 225.9718\,\text{u}$$

Taking this away from the mass of the radium-226 nucleus gives us

$$\Delta m = 225.9771\,\text{u} - 225.9718\,\text{u} = 0.0053\,\text{u}$$
$$\Delta m = 0.0053\,\text{u} = 0.0053 \times 1.66 \times 10^{-27}\,\text{kg} = 8.80 \times 10^{-30}\,\text{kg}$$

So

$$\Delta E = c^2\Delta m = (3.00 \times 10^8\,\mathrm{m\,s^{-1}})^2 \times 8.80 \times 10^{-30}\,\mathrm{kg} = 7.92 \times 10^{-13}\,\mathrm{J}$$

$$\Delta E = \frac{7.92 \times 10^{-13}\,\mathrm{J}}{1.6 \times 10^{-19}\,\mathrm{J\,eV^{-1}}} = 4.95 \times 10^6\,\mathrm{eV} \approx 5\,\mathrm{MeV}$$

b) The α-particle takes away only 4.8 MeV of this energy. Some of the remainder will be taken by the radon-222 nucleus, which recoils in order to conserve momentum. As the radon-222 nucleus is much more massive than the α-particle, its recoil velocity, and hence its kinetic energy, will be very small. The remaining energy is taken away by a quantum of γ-radiation, which is also emitted in the decay.

13.6 Spontaneous and random nature of radioactive decay

The emission of radiation is both **spontaneous** and **random**. By this we mean that the nuclei disintegrate independently – we cannot tell which nucleus will decay next, nor when it will decay. It is an attempt by an unstable nucleus to become more stable and is unaffected by physical conditions such as temperature and pressure.

As there is usually a *very* large number of nuclei involved, we can employ statistical methods to the process. We can therefore make the assumption that the rate of decay of a particular isotope at any instant will be proportional to the number of nuclei of that isotope present at that instant. In other words, the more radioactive nuclei we have, the more likely it is that one will decay. This is a fundamental concept of radioactive decay and can be expressed in simple mathematical terms as:

$$\text{rate of decay (or activity)} = -\frac{\mathrm{d}N}{\mathrm{d}t} = \lambda N$$

where N = number of nuclei
and λ = decay constant

The unit of **activity** is the **becqerel** (Bq), which is a rate of decay of one disintegration per second. Therefore in base SI units it is equivalent to $\mathrm{s^{-1}}$.

From our equation $-\mathrm{d}N/\mathrm{d}t = \lambda N$, we can see that, as $\mathrm{d}N/\mathrm{d}t$ has units of $\mathrm{s^{-1}}$ and N is just a number, the decay constant λ must also have units of $\mathrm{s^{-1}}$.

Tip

You need to know that the expression $\mathrm{d}N/\mathrm{d}t$ is a shorthand way of saying 'the rate of decay at a particular instant' and that the negative sign means that N *decreases* with time t. A mathematical treatment using calculus is *not* needed for examination purposes.

Worked example

The following is taken from an article about potassium:

> Potassium, symbol K, is an element with atomic number 19. Potassium-40 is a naturally occurring radioactive isotope of potassium. Two stable (non-radioactive) isotopes of potassium exist, potassium-39 and potassium-41. Potassium-39 comprises most (about 93%) of naturally occurring potassium, and potassium-41 accounts for essentially all the rest. Radioactive potassium-40 comprises a very small fraction (about 0.012%) of naturally occurring potassium. Potassium-40 decays to an isotope of calcium (Ca) by emitting a beta particle with no attendant gamma radiation (89% of the time) and to the gas argon by electron capture with emission of an energetic gamma ray (11% of the time). Potassium-40 is an important radionuclide in terms of the dose associated with naturally occurring radionuclides.

a) Explain, with reference to potassium, what is meant by *isotopes*.

b) Write down the nuclear equation for the decay of potassium-40 by emission of a beta particle.

c) You have about 130 g of potassium in your body.
   i) A mass of 39 g of naturally occurring potassium contains 6.0×10^{23} nuclei. Show that there would be about 2×10^{20} nuclei of potassium-40 in your body.
   ii) Given that the decay constant for potassium-40 is $1.7 \times 10^{-17}\,\text{s}^{-1}$, calculate the dose (activity) caused by this amount of potassium-40.
   iii) Comment on this amount of activity compared with the value for background radiation.

Answer

a) Isotopes are different forms of the same element and so have the same atomic number. For example, potassium has an atomic number of 19, meaning that it has 19 protons in its nucleus. Isotopes have differing numbers of neutrons in the nucleus and therefore a different nucleon number. Potassium-39 has 20 neutrons, potassium-40 has 21 neutrons and potassium-41 has 22 neutrons.

b) ${}^{40}_{19}\text{K} \rightarrow {}^{40}_{20}\text{Ca} + {}^{0}_{-1}\text{e}$

c) i) If 39 g of naturally occurring potassium contains 6.0×10^{23} nuclei, 130 g will contain

$$6.0 \times 10^{23} \times \frac{130}{39} \text{ nuclei} = 2.0 \times 10^{24} \text{ nuclei}$$

But potassium-40 is only 0.012% of naturally occurring potassium, so number of potassium-40 nuclei is

$$2.0 \times 10^{24} \times \frac{0.012}{100} = 2.4 \times 10^{20} \text{ nuclei} \approx 2 \times 10^{20} \text{ nuclei}$$

ii) $$\frac{dN}{dt} = \lambda N \Rightarrow \frac{dN}{dt} = 1.7 \times 10^{-17}\,\text{s}^{-1} \times 2.4 \times 10^{20} \text{ nuclei}$$

$$= 4.1\,\text{kBq}$$

iii) An activity of 4.1 kBq is equivalent to $4.1 \times 10^{3}\,\text{s}^{-1} \times 3600\,\text{s} \approx$ 15 000 000 counts in an hour. From the information given in Figure 13.3 on page 129, this is of the order of magnitude of background radiation from external sources, so we are being equally bombarded with radiation from without and within!

13.7 Half life

We can model radioactive decay with the following experiment using dice.

Experiment

Modelling radioactive decay

Take 24 dice (or small cubes of wood with one face marked with a cross). Shake the dice in a beaker and tip them onto the bench. Remove all the dice that have landed with the 'six' uppermost (or the cubes of wood with the 'cross' uppermost) – these are deemed to have 'decayed'. Count the number N of dice remaining. Repeat this process until all the dice have 'decayed'.

Tabulate your results as follows:

Number of throws, *t*	0	1	2	3	4	5	6	7	8	9
Dice remaining, *N*	24									

Table 13.3 ▲

Plot a graph of *N* against *t*. You should obtain a curve something like that in Figure 13.19.

You will observe that the curve is not smooth, because whether or not a 'six' is thrown is a **random** event, just like radioactivity. It is instructive to repeat the experiment and average your results, or else combine your results with those of one or more classmates. You will find that the more results you combine, the smoother your curve will become.

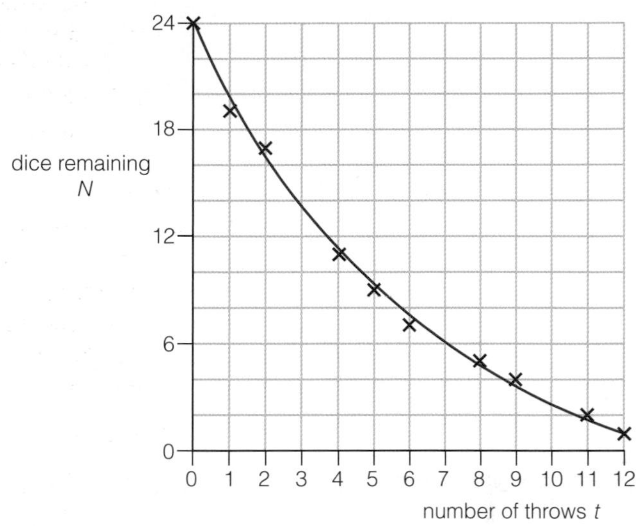

Figure 13.19 ▲

Radioactivity is a random process similar to throwing dice, although it differs from our model insomuch as it is a *continuous* process and the numbers involved are *much, much larger*.

Worked example

In a dice-throwing experiment the following data were obtained for 48 dice:

Number of throws, *t*	0	1	2	3	4	5	6	7	8
Dice remaining, *N*	48	41	32	27	23	18	15	14	10

Table 13.4 ▲

Plot a graph of *N* against *t* and use your graph to estimate how many throws it would take for *half* of the initial 48 dice to 'decay' (i.e. 24 left).

Answer

Having plotted your points, draw a *smooth* curve of best fit through them, as in Figure 13.19. You should find that 24 dice are left after about '3.7' throws, so we can say as an estimation that on average it would take 4 throws for half the dice to 'decay'.

As we said before, radioactivity differs from our dice-throwing model as it is *continuous*. The **half life** of a particular isotope is **the time for a given number of radioactive nuclei of that isotope to decay to half that number**. This is illustrated in Figure 13.20.

As radioactivity is an entirely *random* process, we can only ever determine an *average* value of the half life because it will be slightly different each time we measure it (just as you will get slightly different curves each time you do the 'dice' experiment). However, the very large number of nuclei involved in radioactive decay means that our defining equation, $-dN/dt = \lambda N$, is statistically valid.

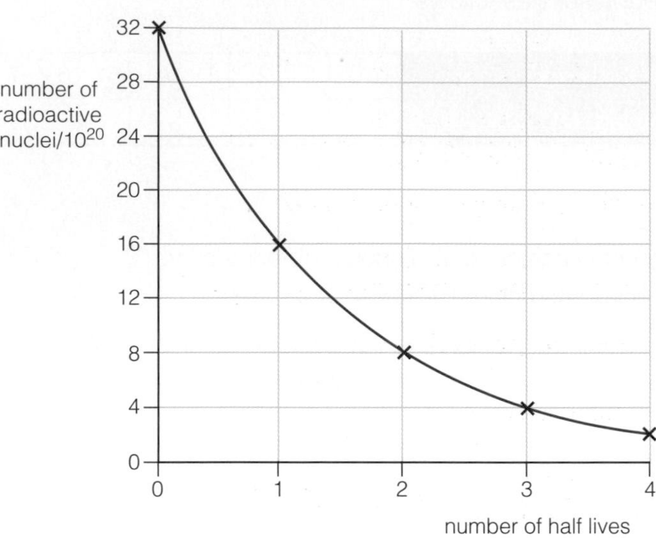

Figure 13.20 ▲
Radioactive decay curve

Tip

Remember that for a graph of any exponential decay the *ratio* of the *y*-values is constant for equal increments in the *x*-values. This is how you can check whether or not a curve is exponential.

An example of this was given on page 45 and there is another example on page 166.

The graph obtained for radioactive decay, such as that in Figure 13.20, is an **exponential** curve. We came across exponential curves in Chapter 6 when we observed the discharge of a capacitor through a resistor – do you remember $Q = Q_0 e^{-t/RC}$? The equivalent equation for radioactive decay is

$$N = N_0 e^{-\lambda t}$$

From this equation it can be deduced mathematically that the half life, $t_{\frac{1}{2}}$, is related to the decay constant, λ, by the expression

$$t_{\frac{1}{2}} = \frac{\ln 2}{\lambda}$$

If you are studying A-level mathematics, you might like to try to deduce this for yourself, but as far as the physics examination is concerned you only need to recognise and use the expression. You will, of course, need to know which button to press on your calculator to show that $\ln 2 = 0.693$!

Half lives can vary from the very long to the very short. For example, the half life of uranium-238 is 4.5×10^9 years, while that of polonium-214, one of its decay products, is a mere 1.6×10^{-4} seconds!

A long half life means that the radioactive substance has probably been around for a long time if it is a naturally occurring isotope such as uranium-238. More significantly, it will be around for a very long time to come! This is particularly important when considering nuclear reactors; some of the fission products in the nuclear fuel rods have half lives of millions of years, which means that the re-processing or safe disposal of these 'spent' fuel rods is a very dangerous and costly business. This major disadvantage of nuclear power stations has to be weighed up against the advantages of not using the world's precious supplies of coal and oil and not producing the atmospheric pollution associated with burning fossil fuels.

Worked example

1 In an earlier example we saw that the isotope potassium-40 has a decay constant of $1.7 \times 10^{-17}\,s^{-1}$. Use this to calculate the half life, in years, of potassium-40.

2 Calculate the decay constants for uranium-238 and polonium-214 from the values of their half lives given above.

Answer

1 $t_{\frac{1}{2}} = \frac{\ln 2}{\lambda} = \frac{0.693}{1.7 \times 10^{-17}\,\mathrm{s}^{-1}} = 4.08 \times 10^{16}\,\mathrm{s}$

$$= \frac{4.08 \times 10^{16}\,\mathrm{s}}{365 \times 24 \times 60 \times 60\,\mathrm{s\,y}^{-1}} = 1.3 \times 10^{9}\,\mathrm{y}$$

2 From $t_{\frac{1}{2}} = (\ln 2)/\lambda$ we get, for uranium-238:

$$\lambda = \frac{\ln 2}{t_{\frac{1}{2}}} = \frac{0.693}{4.5 \times 10^{9} \times 365 \times 24 \times 60 \times 60}\,\mathrm{s}^{-1}$$

$$= 4.9 \times 10^{-18}\,\mathrm{s}^{-1}$$

and for polonium-214:

$$\lambda = \frac{\ln 2}{t_{\frac{1}{2}}} = \frac{0.693}{1.6 \times 10^{-4}}\,\mathrm{s}^{-1} = 4.3 \times 10^{3}\,\mathrm{s}^{-1}$$

Worked example

The following is taken from an article about iodine:

> Iodine-131, which decays by beta and gamma emissions, is used in a number of medical procedures, including monitoring and tracing the flow of thyroxine from the thyroid. With its short half life of 8 days, it is essentially gone in three months.

a) Calculate the decay constant for iodine-131.
b) Justify the statement that 'it is essentially gone in three months' by showing that the fraction remaining after this time is only about 1 part in 4000.
c) Suggest reasons why iodine-131 is 'used in a number of medical procedures'.

Answer

a) From $t_{\frac{1}{2}} = \frac{\ln 2}{\lambda}$

$$\Rightarrow \lambda = \frac{\ln 2}{t_{\frac{1}{2}}} = \frac{0.693}{8.0 \times 24 \times 60 \times 60\,\mathrm{s}} = 1.0 \times 10^{-6}\,\mathrm{s}^{-1}$$

b) The half life is 8 days, so one month is, to a reasonable approximation, 4 half lives, and three months will be about 12 half lives. The fraction of iodine-131 remaining after three months is therefore $(\frac{1}{2})^{12}$, which is 1/4096, or about 1 part in 4000.

c) Iodine-131 is a beta and gamma emitter. As these radiations can penetrate the body, they can be detected outside the body, which enables the flow of thyroxine to be monitored easily. The half life of 8 days is long enough for the patient to be monitored for several days, but, as the calculation shows, short enough that very little remains in the body after two or three months.

13.8 Experimental determination of half life

As the activity of a radioactive isotope, dN/dt, is proportional to the number of nuclei N, we can express the half life as the time for a given *activity* to reduce to half that rate. If we use the symbol A for activity, our equation becomes

$$A = A_0 e^{-\lambda t}$$

Since we can easily measure activity, or count-rate, this gives us a convenient way of determining half lives.

Tip

Make sure you learn the experiment to *model* radioactive decay as described on pages 140–141.

You may get the opportunity to see half life measured in the laboratory but you will not be asked to describe any such experiment in an examination. What *is* important is that you know how the results of such an experiment would be analysed to determine the half life.

Exercise

A typical set of data from an experiment to determine the half life of protactinium-234, which is a β^--emitter, is shown in Table 13.5.

Background count: 135 counts in 5 minutes

Average: 27 min^{-1}

t/s	0	20	40	60	80	100	120	140	160	180
Count-rate/min^{-1}	527	437	367	304	262	216	183	157	136	122
Corrected count-rate/min^{-1}	500	410								

Table 13.5 ▲

Complete the data by calculating the corrected count-rates and plot a graph of this count-rate against time. You should get a characteristic exponential decay curve.

Find an average value for the half life by taking the average of the time to decay from, for example 500 to 250 counts min^{-1} and 250 to 125 counts min^{-1}. You should get a value of about 72 s.

Tip

It doesn't matter at which point you start when finding a half life. In the example given, you could find the time to decay from 400 to 200 min^{-1}, or from 300 to 150 min^{-1}. Check that these points also give 72 s for the half life.

Taking it further

If we take our equation for radioactive decay, $A = A_0 e^{-\lambda t}$, and take logarithms to base 'e' on both sides of the equation, we get:

$$\ln A = -\lambda t + \ln A_0$$

This is of the form:

$$y = mx + c$$

Therefore, if we plot a graph of $\ln A$ (on the y-axis) against t (on the x-axis) we will get a straight line of negative gradient $(-\lambda)$ and intercept $\ln A_0$.

Exercise

You need to be able to analyse an exponential function like this for your Practical Assessment, so you should practise plotting a 'log' graph, or more correctly a 'ln' graph, using the data in the Exercise above. Determine a value for the decay constant λ from the gradient of your graph and hence calculate the half life $t_{\frac{1}{2}}$ from the equation $t_{\frac{1}{2}} = (\ln 2)/\lambda$.

Tip

Remember, if you are plotting a logarithmic graph of an exponential, you must take logs to base 'e' (the 'ln' button on your calculator).

13.9 Radioactive dating

One practical application of radioactive decay is radioactive dating. This enables us to determine the age of rocks and other geological features, including the age of the Earth itself, and also of a wide range of natural materials and artefacts made from natural materials.

Carbon-14 dating

Through fixing atmospheric carbon dioxide in the process of photosynthesis, living plants have almost the same ratio of radioactive carbon-14 to stable carbon-12 as in the atmosphere. After plants die, the fraction of carbon-14 in their remains decreases exponentially due to the radioactive decay of the carbon-14. From the measurement of the ratio of carbon-14 to carbon-12 in an organic archaeological artefact – say a wooden ship – the age of the artefact can be estimated. This method is not entirely accurate for a number of reasons, such as the change in composition of the carbon dioxide in the atmosphere that has taken place over the last few thousand years. Also, as the half life of carbon-14 is 5730 years, the method is not suitable beyond about 60 000 years Nevertheless carbon-14 dating remains a useful tool for archaeologists.

Worked example

Figure 13.21 shows the Anglo-Saxon burial ship discovered at Sutton Hoo, in Suffolk, in 1939. The many priceless treasures found at the site are now in the British Museum. Carbon-14 dating has helped establish the age of the burial ground. The half life of carbon-14 is 5730 years.

a) Explain why carbon-14 dating is not suitable for dating things older than about 60 000 years.

b) Show that the decay constant for carbon-14 is about $1.2 \times 10^{-4}\,y^{-1}$.

c) The count-rate for 1 g of carbon in equilibrium with the atmosphere is $15.0\,min^{-1}$. If samples from the Sutton Hoo burial site were found to have a count-rate for 1 g of $12.7\,min^{-1}$ when measured in 1998, estimate the date of the burial site.

Answer

a) 60 000 years is about 10 half lives. This means that after 60 000 years a fraction of only about $(\frac{1}{2})^{10} = 1/1024$, or 1 part in 1000, of the carbon-14 will remain. This would be too small to measure with any degree of accuracy.

b) From $t_{\frac{1}{2}} = (\ln 2)/\lambda$

$$\Rightarrow \lambda = \frac{\ln 2}{t_{\frac{1}{2}}} = \frac{0.693}{5730\,y} = 1.21 \times 10^{-4}\,y^{-1} \approx 1.2 \times 10^{-4}\,y^{-1}$$

c) From $N = N_0 e^{-\lambda t}$

$$\Rightarrow \frac{N}{N_0} = e^{-\lambda t} \Rightarrow \frac{12.7}{15.0} = e^{-\lambda t} \Rightarrow 0.847 = e^{-\lambda t}$$

Taking natural logs on both sides of the equation gives

$$\ln(0.847) = -0.166 = -\lambda t$$

$$\Rightarrow -0.166 = -1.21 \times 10^{-4}\,y^{-1} \times t \Rightarrow t = \frac{-0.166}{-1.21 \times 10^{-4}\,y^{-1}} = 1376\,y$$

The date of the burial site is therefore about 1998 − 1376 = 622. The best we can really say is that it is likely to be early 7th century.

Figure 13.21 ▲
The Sutton Hoo burial ship

Age of the Earth

Long-lived radio-isotopes in minerals, such as potassium-40 which decays to the stable argon-40 with a half life of 1.3×10^9 years, provide the means for determining long time scales in geological processes. Various methods are used, which provide data for modelling the formation of the Earth and the solar system, and for determining the age of meteorites and of rocks from the Moon. The techniques used are complex and beyond the scope of

A-level. The average of a large number of independent methods gives the age of the Earth as 4.54 billion years (4.54×10^9 years), with an uncertainty of less than 1%. This is consistent with the age of the universe determined from measurements of the Hubble constant (see page 207), which give a value for the age of the universe of between 13 and 14 billion years.

13.10 Nuclear medicine

No discussion of nuclear decay would be complete without looking more closely at some of its applications in medicine. Ionising radiation is used both for diagnosis (finding out what is wrong) and therapy (trying to put it right). For diagnostic imaging, the radioactive isotope is concentrated in the organ under investigation and the emissions are detected outside the body. For therapy, the emissions are usually designed to destroy cancerous tissue. In both cases, it is important that minimal damage is done to healthy cells.

For diagnosis, the isotope must be absorbed by the organ without affecting the organ's function in any way. The isotope must have a half life that is long enough to complete the diagnostic examination, but short enough so that the patient is subjected to the minimum dose of radiation.

For therapy, external sources of high-energy gamma rays, for example from cobalt-60, are now being replaced by high-energy (MeV) X-rays, because X-rays are more easily switched off and their energy can be altered as required. Alternatively, high doses of a radio-isotope accumulated in an organ can be used to kill cells from within the body, for example doses of the order of 400 MBq are used to treat overactive thyroids, while doses of several GBq are used for cancer therapy. Since cancerous cells are rapidly dividing, they are more susceptible to radiation damage.

Table 13.6 summarises some of the isotopes used in nuclear medicine.

Isotope	Emission	Energy	Half life	Use
Technetium-99m (^{99m}Tc)	γ	140 keV	6.0 h	*Diagnosis:* Localisation of tumours Monitoring blood flow in heart and lungs Kidney investigations
Iodine-123 (^{123}I)	γ	160 keV	13 h	*Diagnosis:* Localisation of tumours Assessing thyroid function
Iodine-131 (^{131}I)	β^-, γ	360 keV	8 days	*Therapy:* Thyroid function and tumours

Table 13.6 ▲

Meta-stable technetium-99m is the most versatile and commonly used radio-isotope and is worth special consideration. It is a decay product of molybdenum-99, which is produced by neutron bombardment in a nuclear reactor. The relevant nuclear equations are:

$$^{99}_{42}\text{Mo} \rightarrow {}^{99m}_{43}\text{Tc} + \beta^- \text{ (half life 67 hours)}$$

$$^{99m}_{43}\text{Tc} \rightarrow {}^{99}_{43}\text{Tc} + \gamma \text{ (half life 6.0 hours)}$$

The technetium-99m is chemically separated from the molybdenum by dissolving it in a saline solution and flushing it out. The principle of this process, which is called 'elution', is outlined in Figure 13.22.

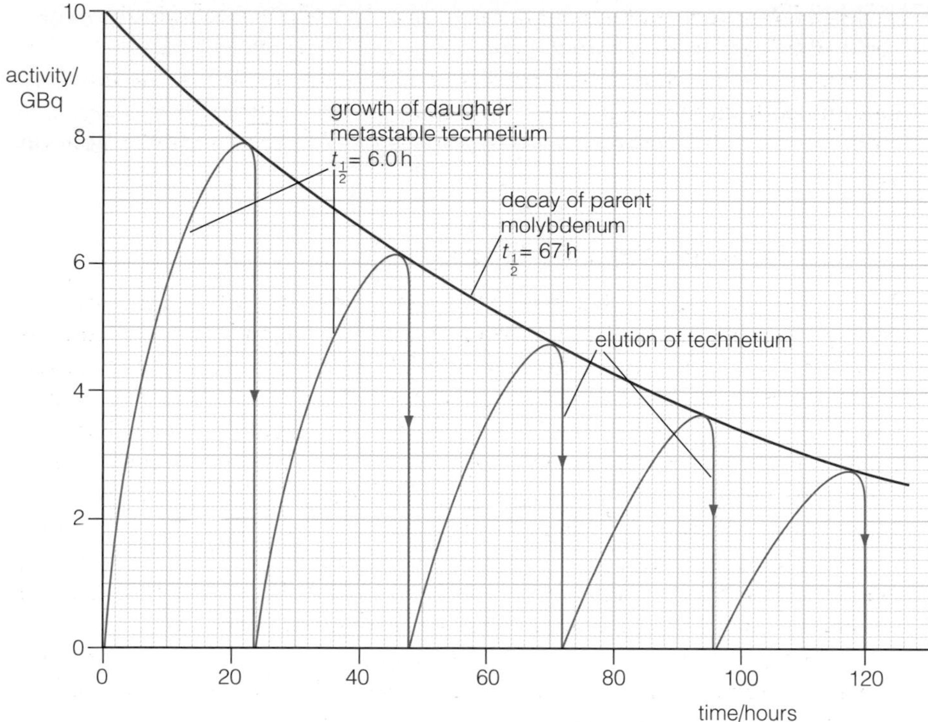

Figure 13.22 ▲

The half life of the molybdenum is short enough to allow fresh supplies of technetium to be removed daily, and long enough so that the elution cell will last for over a week (2 to 3 half lives) before it needs to be recharged. The 6-hour half life of the technetium-99m is ideal for most medical examinations.

Tip

To calculate $e^{-1.73}$, insert 1.73 into your calculator, press +/− to get −1.73, and then press the e^x key. You should get 0.177.

Worked example

a) The half life of molybdenum-99 is 67 hours. What percentage of a sample will remain after one week?

b) Technetium-99 actually decays by β^--emission, but the half life is very long (2.2×10^5 years) and its activity is too small to have much effect on the body cells. Explain why its activity is very small.

Answer

a) Using $N = N_0 e^{-\lambda t}$ where $\lambda = \dfrac{0.693}{t_{\frac{1}{2}}} = \dfrac{0.693}{67\,\text{h}} = 0.0103\,\text{h}^{-1}$

$\Rightarrow N/N_0 = e^{-0.0103 \times 24 \times 7} = e^{-1.73} = 0.177$

This means that 18% (to the nearest percent) of the sample remains after one week.

b) The half life is 2.2×10^5 years, so the decay constant λ will be

$$\lambda = \frac{0.693}{2.2 \times 10^5 \times 365 \times 24 \times 60 \times 60\,\text{s}} = 1.0 \times 10^{-13}\,\text{s}^{-1}$$

As activity $= \lambda N$ and λ is so small, the activity will be very small.

Tip

In calculations involving half lives, like that in part a), it is a good idea to do a rough check.

A week is $(24 \times 7) \div 67 = 2.5$ half lives.

After 1 half life 50% remains.

After 2 half lives 25% remains.

After 3 half lives $12\frac{1}{2}$% remains.

So after a week, or 2.5 half lives, 18% would seem reasonable as it is somewhere between $12\frac{1}{2}$% and 25%.

REVIEW QUESTIONS

1 Which of the following types of radiation does **not** emanate from the nucleus?

A alpha B beta C gamma D X-rays

2 Which of the following types of radiation would cause the most ionisation when passing through air?

A alpha B beta C gamma D X-rays

3 The range of α-particles in air is typically:

A 0.5 mm B 5 mm C 50 mm D 500 mm

4 When an isotope decays by α-emission, which of the following takes place?

A Its mass number and its atomic number both increase.

B Its mass number stays the same and its atomic number increases.

C Its mass number stays the same and its atomic number decreases.

D Its mass number and its atomic number both decrease.

5 When an isotope decays by β^--emission, which of the following takes place?

A Its mass number and its atomic number both increase.

B Its mass number stays the same and its atomic number increases.

C Its mass number stays the same and its atomic number decreases.

D Its mass number and its atomic number both decrease.

6 Bromine-83 has a half life of 2.4 hours. The fraction of a sample of bromine-83 remaining after one day has elapsed would be about:

A 0.1 B 0.01 C 0.001 D 0.0001

7 a) What is meant by *background radiation*? State two sources of background radiation.

b) If you were given a glass beaker of soil which was thought to be radioactively contaminated, how would you determine whether α, β or γ-radiation was present in the soil sample? You may assume that you have normal laboratory apparatus available.

8 A particular type of smoke detector uses an ionising chamber containing a radioactive isotope of americium-241 ($^{241}_{95}$Am), which is an alpha-emitter and has a half life of 460 years. A schematic diagram of the arrangement is shown in Figure 13.23.

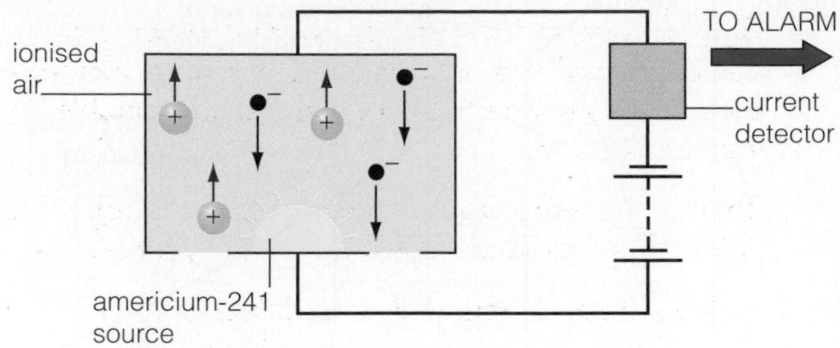

Figure 13.23 ▲

The alpha particles ionise the air, thus enabling it to conduct a small current. If smoke particles enter the chamber they reduce the amount of ionisation and the drop in current triggers an alarm.

a) Explain what is meant by *ionisation*.

b) What is an *alpha particle*?

c) An alpha particle has a range of a few centimetres in air.

i) What happens to its energy as it slows down?

ii) What eventually happens to the alpha particle?

d) i) What is meant by an *isotope*?

ii) How many protons and how many neutrons are there in a nucleus of the isotope americium-241?

iii) The americium-241 decays to an isotope of neptunium (Np). Write down a nuclear equation for this process.

e) i) What is meant by the term *half life*?

ii) Show that the decay constant for americium-241 is approximately $4.8 \times 10^{-11}\,s^{-1}$.

iii) The mass of americium-241 in a typical source is 1.6×10^{-8} g. Given that 241 g of americium-241 contains 6.0×10^{23} nuclei, calculate the number of radioactive nuclei in the source.

iv) Hence calculate the activity of the source.

f) Suggest three reasons why americium-241 is a suitable source for such a smoke detector.

9 Iodine-131 ($^{131}_{53}$I) is a β^--emitter and has a half life of 8.0 days. A sample has an activity of 32 kBq.

a) Explain why the activity will fall to about 2 kBq after one month.

b) Show that its decay constant is approximately $1 \times 10^{-6}\,s^{-1}$.

c) How many nuclei would be needed to give an activity of 32 kBq?

d) Given that 131 g of iodine-131 contains 6.0×10^{23} nuclei, what would be the mass of iodine-131 in a source having this activity?

e) This isotope, in the form of a solution of sodium iodide, is used as a radioactive 'tracer' in medicine. Suggest two reasons why it is particularly suitable for this purpose.

f) Iodine-131 decays to an isotope of xenon (Xe). Write down a nuclear equation for this process.

g) Iodine-131 also emits some γ-radiation.

i) In what ways does γ-radiation differ from α and β-radiation?

ii) Why does the iodine-131 nucleus emit γ-radiation as well as β^--radiation?

10 Complete these nuclear equations:

a) Plutonium-239 decaying by α-emission to an isotope of uranium (U).

$_{94}\text{Pu} \rightarrow$

b) Cobalt-60 decaying by β^--emission to an isotope of nickel (Ni).

$_{27}\text{Co} \rightarrow$

c) Magnesium-23 decaying by β^+-emission to an isotope of sodium (Na).

$_{12}\text{Mg} \rightarrow$

d) Show each of these decays on a plot of nucleon number against proton number.

11 The 'Big Bang' was thought to have created mainly hydrogen and helium. In the hearts of stars the formation of carbon is possible through the so-called *triple alpha* reaction, in which three helium nuclei (alpha particles) fuse to make a nucleus of carbon-12.

Rather than recreate the scorching conditions inside stars, physicists at CERN watched the reaction in reverse. To do this they created nitrogen-12, which is transformed into carbon-12 by beta-plus decay; the carbon-12 then breaks into three alpha particles.

Write down the nuclear equations for the reactions described in each of the paragraphs above.

12 The diagram shows a protactinium generator in which a layer of β^--emitting protactinium-234 has been created.

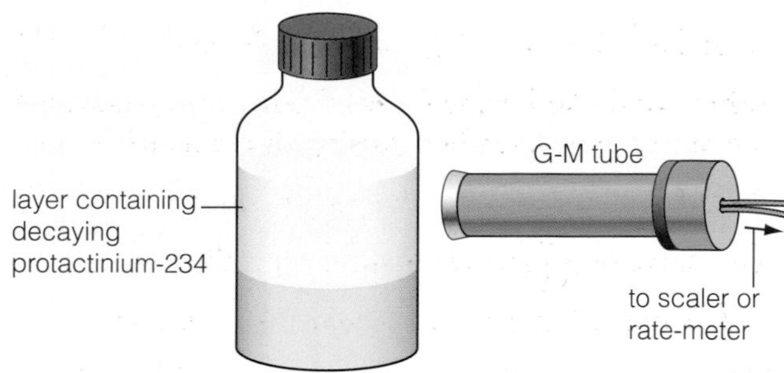

Figure 13.24 ▲

a) Define the term *half life*.

b) Explain how you could use the arrangement shown above to find the half life of protactinium-234, which has a value of about one minute. Your answer should include:

- a description of the readings you would take;
- a sketch of the graph you would plot;
- an explanation of how you would find the half life from your graph.

c) A bottle of milk is contaminated with a small amount of strontium-90, which is also a β^--emitter and has a half life of 28 years. Give two reasons why the method you have described for the protactinium-234 would not be suitable for determining the half life of strontium-90.

14 Oscillations

As you learnt in Unit 2 of your AS course, waves occur in a variety of forms – from shock waves generated by earthquakes to γ-radiation emanating from the nuclei of atoms. The study of wave motion is therefore very important, as it has so many applications in everyday life. It would be a good idea for you to spend a few minutes revising your AS work on waves, because we will be building on some of the ideas in this chapter. In particular, we will be looking in detail at a special form of oscillatory motion, called simple harmonic motion. Many of the waves you looked at (or heard!) in Unit 2 were caused by oscillations approximating very closely to simple harmonic motion. Its study is of great importance in a wide range of areas of physics and engineering – from the design of bridges to the development of MRI scanners.

14.1 Simple harmonic motion

We saw the importance of mechanical oscillations when we studied waves in Unit 2. Oscillating springs, vibrating rules, pendulums, sound waves, water waves and shock waves from earthquakes are all examples of mechanical oscillations caused by some vibrating source.

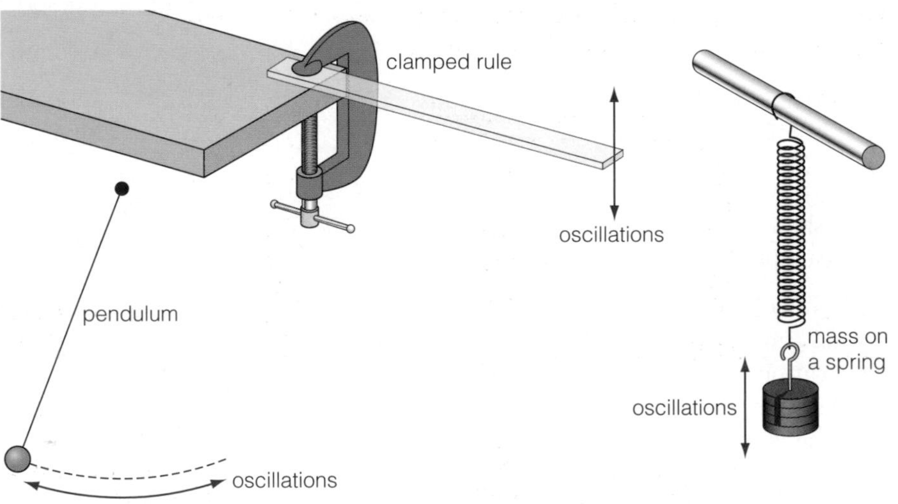

Figure 14.1 ▲
Oscillating systems

Definitions

Amplitude, ***A***, is the maximum displacement from the mean (equilibrium) position (in metres, m).

Period, ***T***, is the time taken to complete one oscillation (in seconds, s).

Frequency, ***f***, is the number of complete oscillations per second (in hertz, Hz).

In oscillations, we observe that the motion is repetitive about a fixed point, with the object at rest at the extremes of the motion and moving with maximum speed in either direction at the midpoint. Three properties can be used to describe an oscillation. They are: the amplitude, A, the period, T, and the frequency, f. Remember that period and frequency are related by $f = 1/T$ or, conversely, $T = 1/f$, and that we measure frequency in Hz ($\equiv s^{-1}$).

Worked example

1 What is the period of the mains alternating current supply, which has a frequency of 50 Hz? Give your answer in milliseconds.

2 A spring is found to complete 20 oscillations in 13.3 s. Calculate
  a) the period, and
  b) the frequency of the oscillations.

Answer

1 $T = \frac{1}{f} = \frac{1}{50\,\text{s}^{-1}} = 0.02\,\text{s} = 20\,\text{ms}$

2 **a)** $T = \frac{13.3\,\text{s}}{20} = 0.665\,\text{s}$

b) $f = \frac{1}{T} = \frac{1}{0.665\,\text{s}} = 1.50\,\text{Hz}$

To a good approximation, all the oscillations described above show two common characteristics:

- the force acting on the oscillating body, and therefore its acceleration, is *proportional to the displacement* of the body (i.e. its distance from the mean, or equilibrium, position); and
- the force, and therefore the acceleration, always acts in a direction *towards the equilibrium position.*

An oscillating body that satisfies these conditions is said to have **simple harmonic motion**, or s.h.m. for short. We can combine these conditions into a simple equation:

$$F = -kx$$

where k is a constant, and the minus sign means that the force, F, is always opposite in direction to the displacement, x, that is, *towards* the equilibrium position (see Figure 14.2). Remember, force, acceleration and displacement are *vectors*, so their direction is important.

Note

F and x are vectors. As can be seen in Figure 14.2, F (upwards) is in the opposite direction to x (downwards).

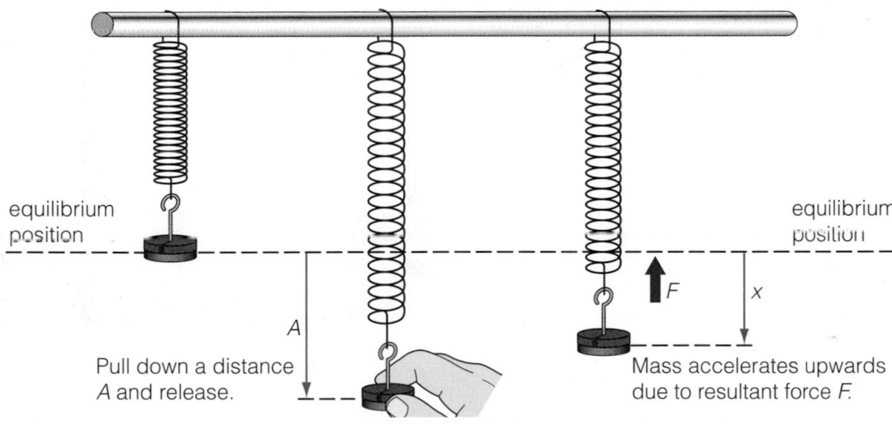

Figure 14.2 ▲

Since force is proportional to acceleration, we can express the equation in terms of acceleration. By convention we call the constant ω^2 so the simple harmonic equation becomes:

$$a = -\omega^2 x$$

Now let's look at some examples of simple harmonic motion.

Note

Later on we will see that $\omega = 2\pi f$.

14.2 Spring

Note

The introduction of the minus sign is to show the difference in direction of F and x.

In Unit 1 we saw that the equation for a spring that obeys Hooke's law is $F = kx$, where k is the spring constant, or 'stiffness', of the spring, and x is the extension. So

$$\text{from } F = ma \quad \Rightarrow \quad a = F/m, \text{ where } F = -kx$$

$$\Rightarrow a = -\frac{kx}{m} = -\frac{k}{m}x$$

This equation is of the form

$$a = -\omega^2 x$$

where $\omega = \sqrt{\frac{k}{m}}$.

As shown later (page 160), $\omega = 2\pi f$, and so $T = 2\pi/\omega$. Therefore a mass m oscillating on a light vertical spring that obeys Hooke's law will execute s.h.m. with an oscillation period given by the equation

$$T = 2\pi\sqrt{\frac{m}{k}}$$

Experiment

Finding the spring constant of a spring from Hooke's law

Suspend a 100 g mass hanger from a vertical spring as shown in Figure 14.3.

Use a set-square against the vertical rule to determine the initial position, h_0, of the bottom of the spring. Now carefully add extra masses, Δm, in 50 g increments up to about 300 g. After each mass has been added, record the position h of the bottom of the spring. The extension of the spring for each mass is then $\Delta x = h - h_0$. Tabulate your results as in Table 14.1 below.

Now plot a graph of F (in newtons) against Δx (in metres). You should get a straight line through the origin if the spring obeys Hooke's law. The graph may curve a bit at the bottom end, as some springs need a small force to separate the coils before the spring starts to stretch. Take the gradient of the straight part of your graph to find the spring constant, k.

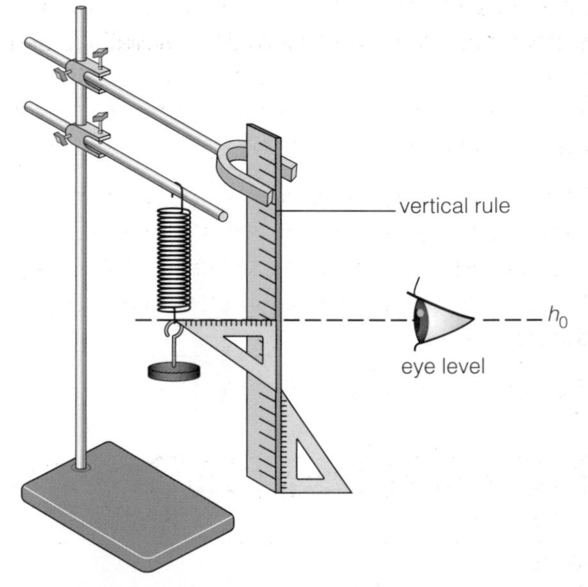

Figure 14.3 ▲

Δm/kg	0.000	0.050	0.100	0.150	0.200	0.250	0.300
F/N (= Δmg)	0.00	0.49	0.98	1.47	1.96	2.45	2.94
h/mm							
Δx/mm (= $h - h_0$)							

Table 14.1 ▲

Exercise

The following data were obtained in an experiment to stretch a spring.

Δm/kg	0.000	0.050	0.100	0.150	0.200	0.250	0.300
F/N ($= \Delta mg$)	0.00	0.49	0.98	1.47	1.96	2.45	2.94
h/mm	412	430	449	468	486	504	523
Δx/mm ($= h - h_0$)	0	18	37				

Table 14.2 ▲

Copy and complete Table 14.2 by adding the rest of the Δx values ($h_0 = 412$ mm) and then plot a graph of F (in newtons) against Δx (in metres).

Draw a large triangle to determine the gradient. This should give a value for the spring constant, k, of about 27 N m^{-1}.

Experiment

Finding the spring constant of a spring from simple harmonic motion

Use the same arrangement and the same spring as in the previous experiment. Time 20 small vertical oscillations of masses in the range 100 g to 350 g, in 50 g increments. Note that in this experiment the mass, m, is the *total* mass, including that of the mass hanger. Repeat each timing.

To help you judge the start and stop positions, put a marker at the *centre* of the oscillations, for example a pin secured to the vertical rule with Blu-Tack as shown in Figure 14.4. This is called a 'fiducial' mark – a mark by which you can 'judge' the position of the mass.

Tabulate your results, as in Table 14.3.

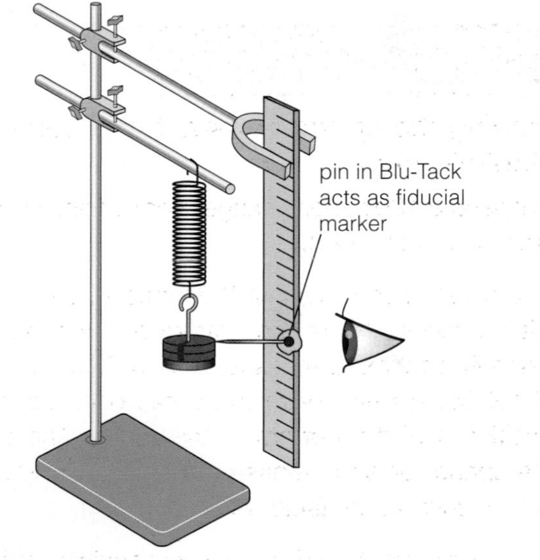

Figure 14.4 ▲

m/kg	$20T$/s		Mean T/s	T^2/s^2
0.100				
0.150				

Table 14.3 ▲

Why do we need values of T^2? Our equation is $T = 2\pi\sqrt{\frac{m}{k}}$, so squaring both sides of the equation gives:

$$T^2 = \frac{4\pi^2 m}{k} \quad \Rightarrow \quad T^2 = \frac{4\pi^2}{k}m$$

This means that if you plot a graph of T^2 against m you should get a straight line through the origin of gradient $4\pi^2/k$.

Plot your graph, determine the gradient and hence find the value of k. How does this compare with your previous value for k?

Exercise

The data in Table 14.4 were obtained by timing small vertical oscillations of masses on a spring. Copy and complete the table by adding the rest of the values for T and T^2.

Plot a graph of T^2 against m, which should be a straight line through the origin.

Draw a large triangle to determine the gradient, which you should find to be about 1.48 ($s^2\,kg^{-1}$).

The gradient $= \frac{4\pi^2}{k}$ so $k = \frac{4\pi^2}{\text{gradient}} \approx 27\,N\,m^{-1}$

m/kg	20*T*/s		Average *T*/s	T^2/s^2
0.100	7.79	7.83	0.390	0.152
0.150	9.41	9.35	0.469	0.220
0.200	11.01	11.07		
0.250	12.41	12.33		
0.300	13.38	13.42		
0.350	14.42	14.36		

Table 14.4 ▲

Tip

Always remember to divide by the number of oscillations when finding the period.

The quickest way here to find the average value for T is to add the two values for $20T$ and then divide by 40.

14.3 Simple pendulum

In this context, 'simple' means

- a small, dense, pendulum bob, e.g. lead or brass sphere, and
- a light, inextensible string.

In the case of the simple pendulum, the force causing oscillation is provided by a component of the weight of the pendulum bob as shown in Figure 14.5.

The required force is the component $mg\sin\theta$. If θ is small, and in radians, then $\sin\theta \approx \theta$ and so the force is proportional to the displacement. It can then be shown, for oscillations of small amplitude (less than about 10°), that the period, to a good approximation, is given by:

$$T = 2\pi\sqrt{\frac{l}{g}}$$

where l is the distance from the point of suspension to the centre of gravity of the bob.

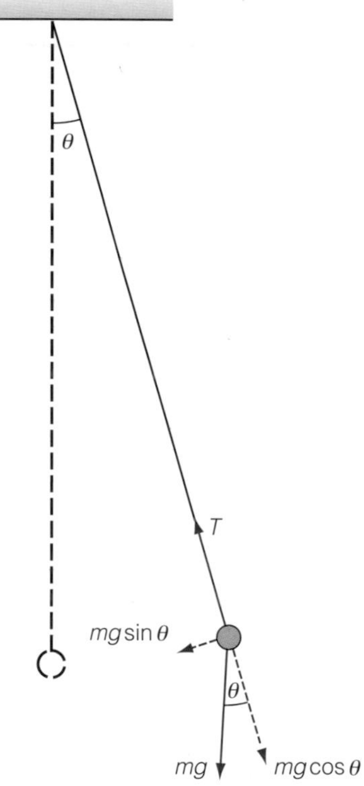

Figure 14.5 ▲

Exercise

We can use measurements of the period of a simple pendulum to practise plotting a logarithmic graph to test a proposed relationship. We can see from simple observation that the period depends on the length of the pendulum, so let's assume that they are related by an equation of the form

$T = al^b$

where a is a constant and b is a power.

If we take logarithms (to base 'e') on both sides of the equation we get

$\ln T = b \ln l + \ln a$

Tip

For investigating powers, you could use ln *or* log. For exponential equations you *must* use ln. If you use ln for *both*, you can't go wrong!

The data in Table 14.5 were obtained for the period T of a simple pendulum. Note that if you were doing this yourself, which you might like to do, you should find T from the average of two lots of 20 oscillations, as in the spring experiment (page 153).

Copy and complete Table 14.5 by adding the rest of the values of ln (l/m) and ln (T/s), and then plot a graph of ln (T/s) against ln (l/m). Think carefully when choosing your axes – the ln (l/m) values are negative!

You should get a graph like Figure 14.6.

l/m	T/s	ln (l/m)	ln (T/s)
0.500	1.42	−0.693	0.351
0.600	1.55		
0.700	1.68		
0.800	1.80		
0.900	1.90		
1.000	2.01		

Table 14.5 ▲

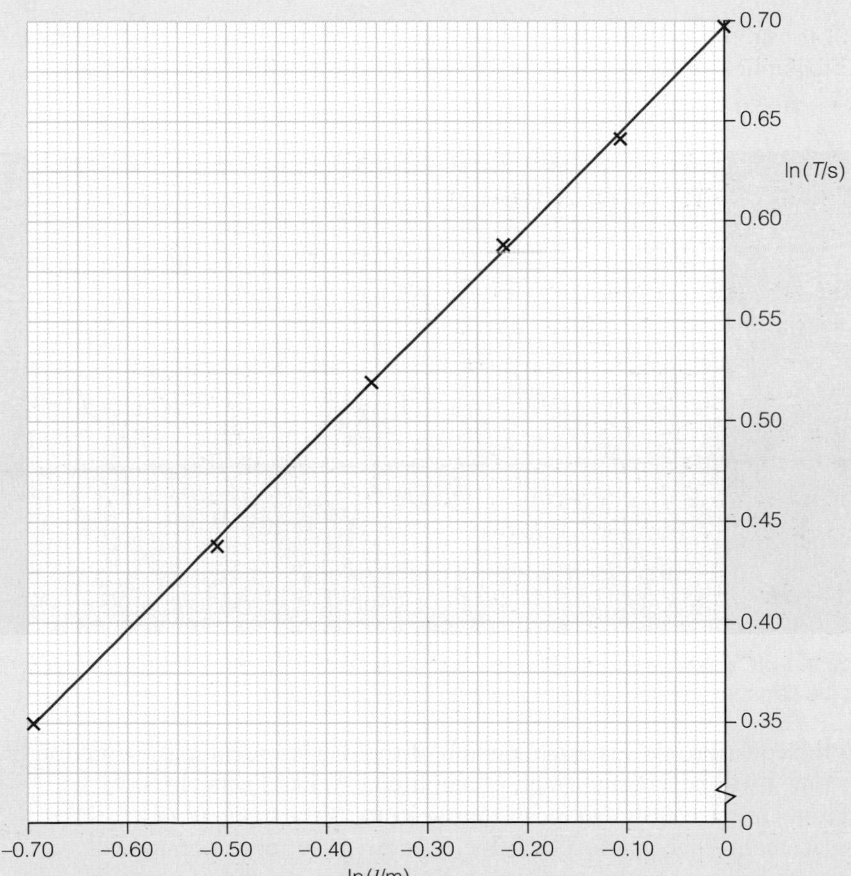

Figure 14.6 ▲

Draw a large triangle on your graph to determine the gradient. This should be close to 0.50, or $\frac{1}{2}$.

This suggests that $T = al^{1/2}$ or $T = a\sqrt{l}$ for a simple pendulum.

The intercept on the ln T axis is 0.700. So ln a = 0.700, giving a = 2.01. From the equation for a pendulum this should be equal to $\frac{2\pi}{\sqrt{g}}$. Is it?

Tip

To get a large scale, it may be necessary to choose a graph scale that does not start at the origin. This is quite often the case when plotting a 'log' graph.

Note

This is typical of the sort of analysis you will need to be familiar with for the Practical Assessment.

14.4 Equations of simple harmonic motion

Before we go any further, we should mention that in perfect simple harmonic motion the frequency of the oscillations *does not depend on the amplitude of oscillation*. We say the motion is **isochronous** – from the Greek 'iso' (the same) 'chronos' (time). In real situations, the motion only approximates to simple harmonic motion, but the following equations will still be more or less valid. The basic defining equation for s.h.m. is

$$a = -\omega^2 x$$

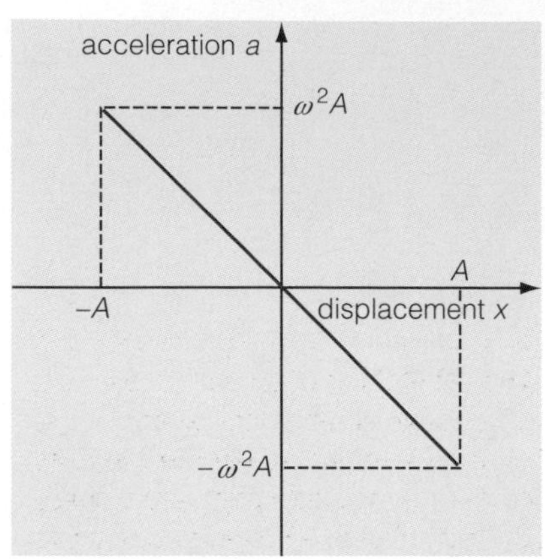

Figure 14.7 ▲

The graph of this equation is shown in Figure 14.7. Note that the graph line is of finite length, determined by the maximum displacement, that is the amplitude $\pm A$. The corresponding maximum acceleration is $\pm\omega^2 A$.

This equation has an infinite number of solutions for x, depending on the exact point in the motion at which we decide to start timing. An obvious place to start is at one extreme of the motion – after all, this is how we start a pendulum or spring oscillating, by displacing it and then letting go.

If $t = 0$ at one end of the motion, x will at that time be equal to the amplitude, A. The solution of the equation is then

$$x = A\cos\omega t$$

This is beautifully illustrated in Figure 14.8. A vertically oscillating spring is photographed stroboscopically, with the camera moving horizontally at a steady speed.

Figure 14.8 ▲
Simple harmonic motion illustrated with an oscillating spring

For the equation $x = A\cos\omega t$, a graph of the displacement x as a function of the time t will be a cosine graph of amplitude A, like Figure 14.9a.

From Unit 1 you should recall that velocity is the gradient of a displacement–time graph. The velocity at any point of the s.h.m. will therefore be equal to the gradient of the cosine curve at that point. If you are doing A-level mathematics, you will know that this can be found by differentiating the above equation, which gives

$$v = -\omega A\sin\omega t$$

For A-level physics you just need to be able to recognise and use this expression. The velocity–time graph therefore looks like Figure 14.9b, which is a negative sine wave.

Why is there a negative sign in the above equation? Well, think of pulling a pendulum to your right and letting go at $t = 0$, as in Figure 14.10. During the first half-cycle of its motion, the velocity of the pendulum will be back

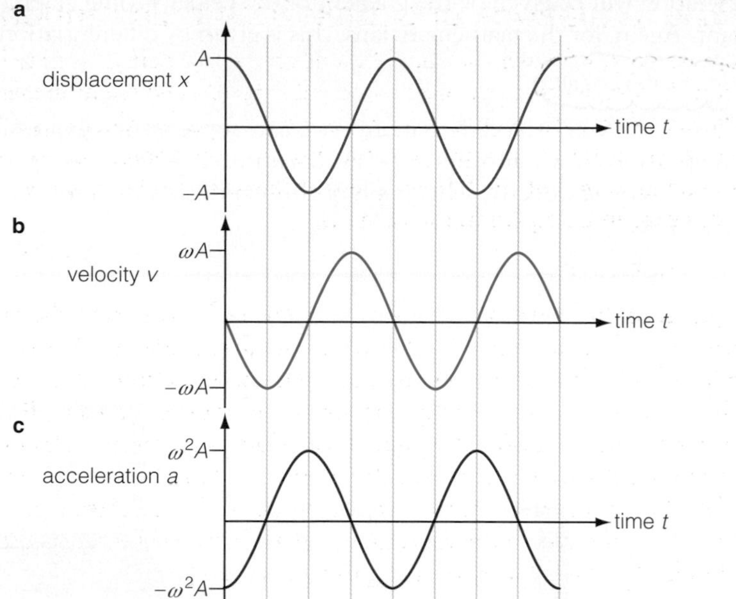

Figure 14.9 ▲
Graphical representation of s.h.m

Tip

When sketching graphs such as these, draw vertical lines every ¼ cycle. Mark where each line has its maximum, minimum and zero values, and then sketch the sine or cos wave between these points.

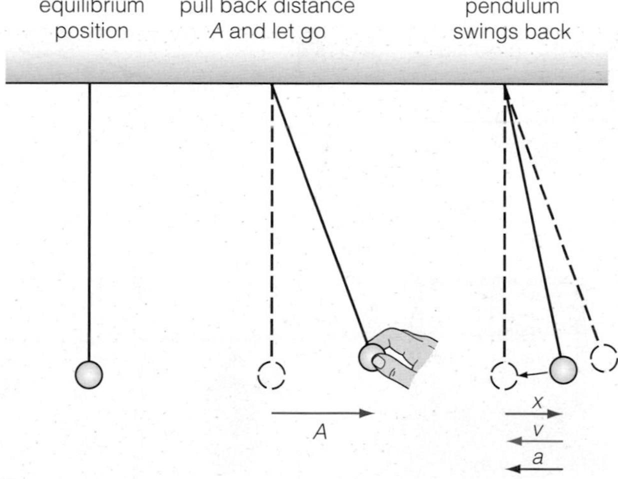

Figure 14.10 ▲

towards your left – in other words in the opposite sense to the displacement you have given it. Hence the negative sign. You can also see from the graph for x that the gradient during the first half-cycle is negative.

There are two key points in the motion:

- at each end of the oscillation, when the velocity is momentarily zero, the displacement curve is a maximum, positive or negative (A or $-A$), and its gradient $= 0$;
- at the centre of the motion, corresponding to $x = 0$, the velocity has its maximum value – you can see from the graph for x that the gradient, and hence the velocity, is a maximum at each point where the curve crosses the t-axis; furthermore, the gradient alternates between being positive and negative as the body moves first one way, and then back again, through the midpoint of its oscillations.

We have seen how the displacement and velocity vary with time. What about the *acceleration*? As acceleration is defined as the rate of change of velocity, the acceleration will be given by the gradient of the velocity–time graph at any instant. Again, for the mathematicians, this is given by differentiation and yields

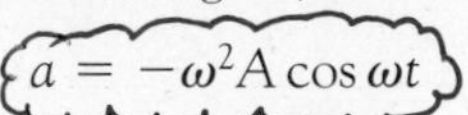

$$a = -\omega^2 A \cos \omega t$$

Don't worry if you can't differentiate – we can arrive at this expression in another way. If we substitute $x = A\cos\omega t$ into $a = -\omega^2 x$, we get $a = -\omega^2 A\cos\omega t$ directly! The graph of this expression is shown in Figure 14.9c.

Experiment

Obtaining the graphs for simple harmonic motion using a motion sensor

Figure 14.11a shows the arrangement for a spring. A card is attached to the masses to give a good reflective surface, but may not be needed if the base of the masses is large enough. Typically the data logger might be set to record for 10 seconds at a sampling rate of 100 per second. The computer can be programmed to give displays of displacement, velocity and acceleration against time, as shown in Figure 14.12 opposite.

In Figure 14.11b, a pendulum is shown connected to a rotary sensor. Such a sensor is useful for investigations of rotational motion – in this case it will measure the *angular* displacement of the pendulum.

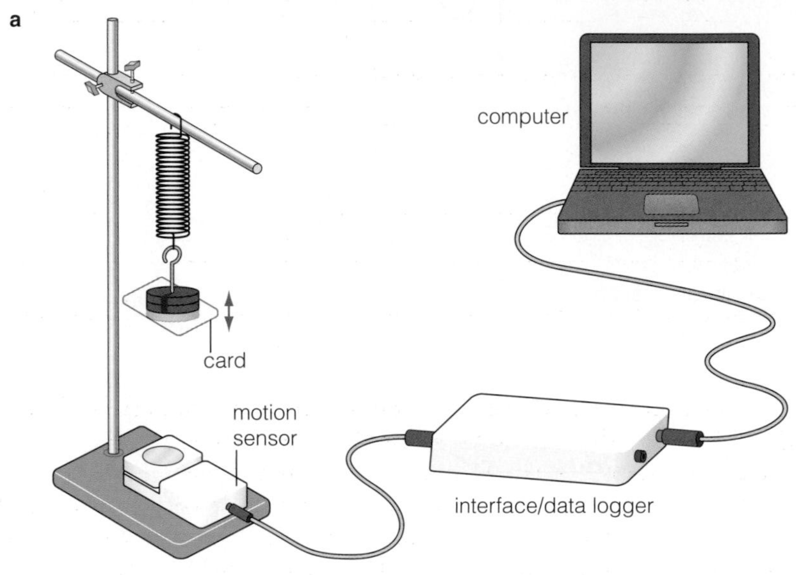

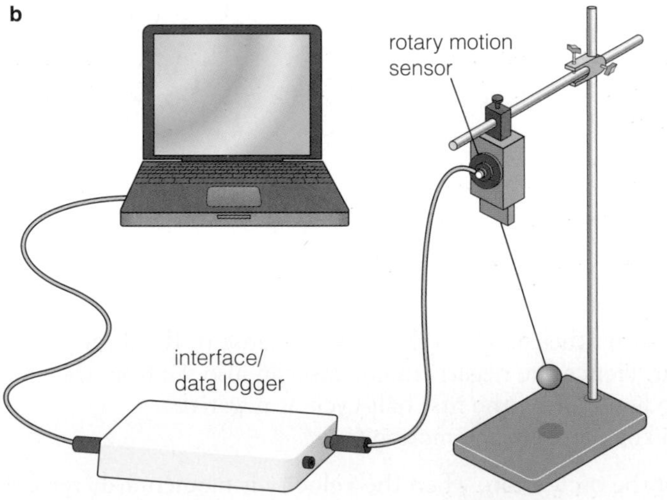

Figure 14.11 ▶

Worked example

A printout from an experiment to investigate the motion of a mass oscillating vertically on a spring is shown in Figure 14.12.

a) Use the plot of displacement against time to determine the amplitude A, the period T and the frequency f of the motion.

b) Calculate the spring constant (stiffness) k, given that $T = 2\pi\sqrt{\frac{m}{k}}$ and $m = 200\,\text{g}$.

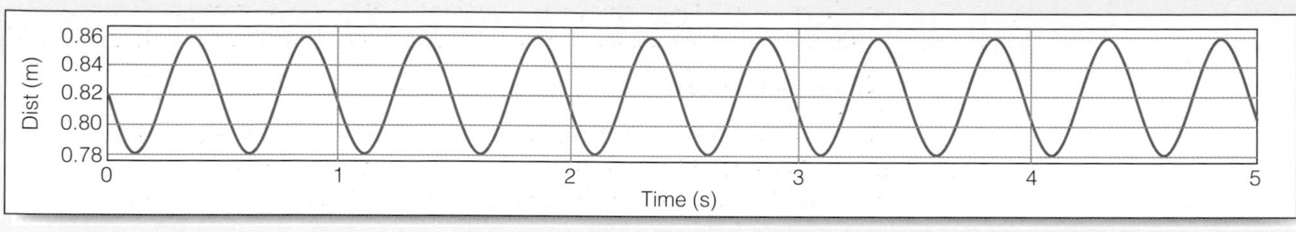

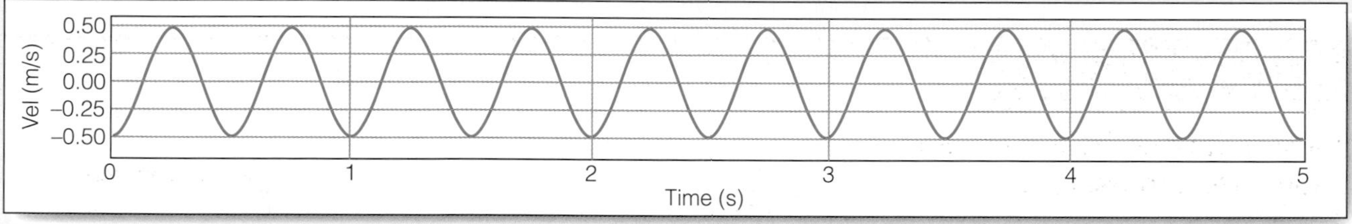

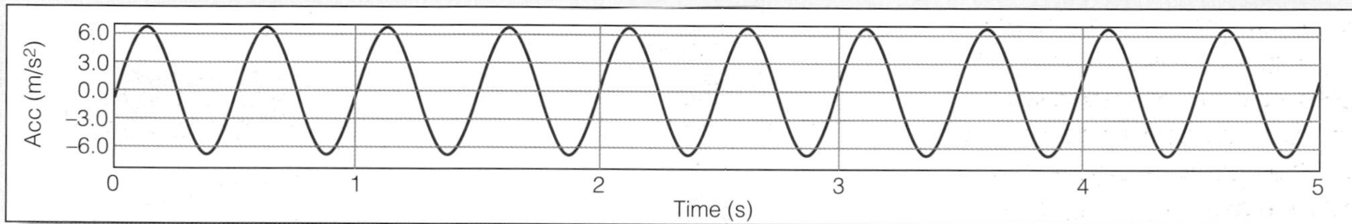

Figure 14.12 ▲

c) The equation for velocity is $v = -\omega A \sin \omega t$. Use your answers from part a) to show that the constant ω is $4\pi\,\mathrm{s}^{-1}$, given that $\omega = 2\pi f$.
d) Use the equation $v = -\omega A \sin \omega t$ to calculate the maximum value of the velocity and compare your value with that from the velocity–time plot.
e) Use the equation $a = -\omega^2 x$ to calculate the maximum acceleration and compare your value with that from the acceleration–time plot.
f) Sketch a graph of acceleration against displacement for the motion.

Answer

a) The amplitude is the maximum displacement in either direction from the centre of the oscillations, $(0.86 - 0.82)\,\mathrm{m}$ and $(0.78 - 0.82)\,\mathrm{m}$, so $A = \pm 0.040\,\mathrm{m}$.

The period is the time for one complete oscillation. There are two complete oscillations in 1 s, so $T = 0.50\,\mathrm{s}$.

From $f = \frac{1}{T} \Rightarrow f = \frac{1}{0.50\,\mathrm{s}} = 2.0\,\mathrm{Hz}$

b) From $T = 2\pi\sqrt{\frac{m}{k}} \Rightarrow T^2 = 4\pi^2\frac{m}{k}$

$$\Rightarrow k = \frac{4\pi^2 m}{T^2} = \frac{4\pi^2 \times 0.200\,\mathrm{kg}}{(0.50\,\mathrm{s})^2} = 32\,\mathrm{N\,m^{-1}}$$

c) $\omega = 2\pi f = 2\pi \times 2.0\,\mathrm{s}^{-1} = 4\pi\,\mathrm{s}^{-1}$

d) From the equation $v = -\omega A \sin \omega t$, the maximum velocity will be when $\sin \omega t$ has its maximum value, as ω and A are constants. The maximum value that a sine can have is 1, giving:

$$v_{max} = \pm\omega A = 4\pi\,\mathrm{s}^{-1} \times 0.040\,\mathrm{m} = 0.50\,\mathrm{m\,s^{-1}}$$

From the velocity–time plot we can read off that the maximum velocity is, indeed, $0.50\,\mathrm{m\,s^{-1}}$.

e) From the equation $a = -\omega^2 x$, the maximum acceleration will be when the displacement x is a maximum, that is when $x = A$.

$$\Rightarrow a_{max} = \omega^2 A = (4\pi\,s^{-1})^2 \times 0.040\,m = 6.3\,m\,s^{-2}$$

From the acceleration–time plot we can see that the maximum acceleration is just over $6.0\,m\,s^{-2}$, which agrees with the calculated value.

f) The graph of acceleration against displacement is the graph of $a = -\omega^2 x$. It is therefore a straight line of negative slope passing through the origin, as shown in Figure 14.13.

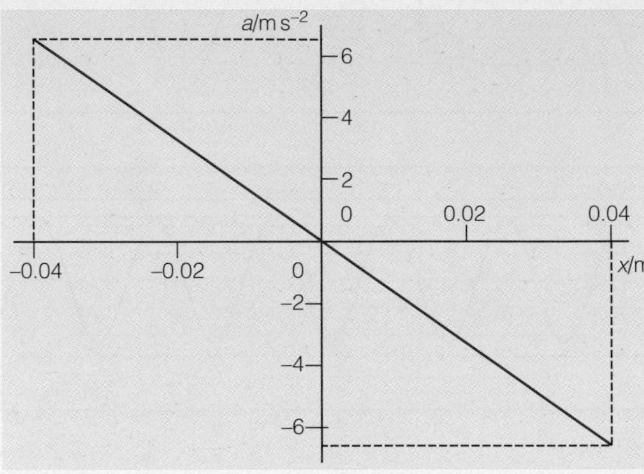

Figure 14.13 ▲

Tip

If you are asked to *sketch* a graph, the axes must always be labelled, with units, and also given a scale if you know any numerical values. Your line should then be related to the scale. You do *not* however need to draw the graph accurately on graph paper.

In this example, your scales should be labelled to show the limits of the line, in this case $a = \pm 6.3\,m\,s^{-2}$ when $x = \pm 0.040\,m$.

The meaning of ω

We have used ω^2 as the constant in the equation that defines s.h.m, and we said that we would explain later what ω was. So, let's start by plotting a graph of $\cos\omega t$ against ωt: Figure 14.14.

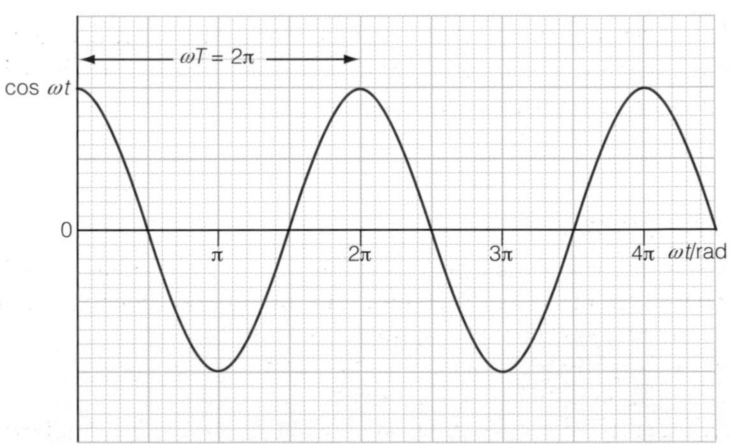

Figure 14.14 ▲

Note that ωt is in *radians*. For one complete cosine wave, $\omega t = 2\pi$ radians. One complete cosine wave represents one oscillation, which takes a time T, the period. This means that $t = T$. We therefore have that $\omega T = 2\pi$, or

$$\omega = \frac{2\pi}{T}$$

This means that ω has units of rad s^{-1}, which are the units for angular velocity. As the frequency $f = 1/T$ we also have

$$\omega = 2\pi f \qquad \text{and} \qquad f = \frac{\omega}{2\pi}$$

Worked example

Figure 14.15 shows one of the cylinders of a car engine. To a good approximation, the piston oscillates up and down with simple harmonic motion. In this engine, the length of the stroke is 80 mm and the piston has a mass of 600 g.

a) What is the amplitude of the piston's motion?
b) What is the frequency of the motion when the engine is running at 3000 revolutions per minute?
c) Show that the maximum acceleration experienced by the piston is about $4000\,\text{m}\,\text{s}^{-2}$. At which point in its motion does this occur?
d) Calculate the maximum force exerted on the piston.
e) Aluminium alloys, which are both light and strong, are being developed for manufacturing pistons. Suggest why.

Answer

a) The stroke is the distance from the top to the bottom of the piston's motion. The amplitude is the maximum displacement, in either direction, from the midpoint of the motion, so the amplitude is *half* the stroke. The amplitude is therefore 40 mm.
b) The frequency will be $3000/60\,\text{s} = 50\,\text{s}^{-1}$ (or Hz).
c) From the s.h.m. defining equation $a = -\omega^2 x \Rightarrow a = -(2\pi f)^2 x$.

   The piston will have maximum acceleration when x has its maximum value, that is when x is equal to the amplitude, A. Its magnitude will be

   $$a_{max} = (2\pi f)^2 A = (2\pi \times 50\,\text{s}^{-1})^2 \times 0.040\,\text{m} = 3948\,\text{m}\,\text{s}^{-2} \approx 4000\,\text{m}\,\text{s}^{-2}$$

   This will occur at the top and bottom of the piston's motion.

d) The maximum force acting on the piston will be given by $F = ma$.

   $$F_{max} = 0.600\,\text{kg} \times 4000\,\text{m}\,\text{s}^{-2} = 2400\,\text{N}$$

e) If the piston is light, it will require much less force to accelerate it, giving greater engine efficiency. It needs to be strong to withstand the very large forces acting on it without distorting.

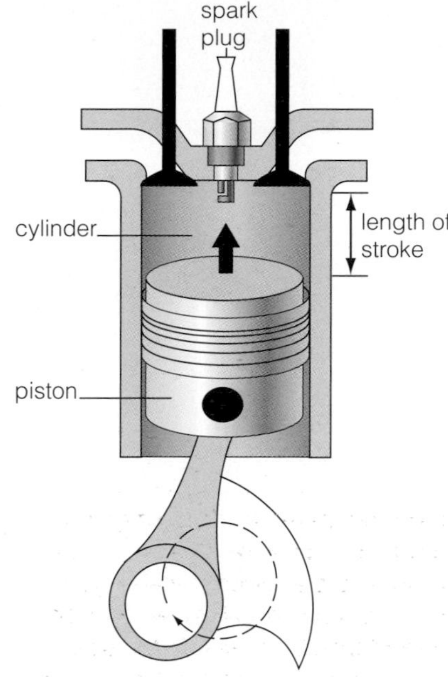

Figure 14.15 ▲

Tip

There are two points to remember in a calculation like this:

- the amplitude is *half* the distance between the two extremes of the motion, and
- care must be taken to put brackets round $2\pi f$ when squaring it, $(2\pi f)^2$, to ensure that each term is squared.

Worked example

In a harbour, the water is 4.0 m deep at low tide and 10.0 m deep at high tide. The variation in water level with time is, to a good approximation, simple harmonic motion, with two high tides per day.

a) What is the period, in hours, of this motion?
b) What is the amplitude of the motion?
c) Sketch a graph of the depth of water against time for one day, beginning at high tide.
d) Calculate the depth of water i) 2.0 hours after high tide, and ii) 1.5 hours after low tide. Show these points on your graph.

Answer

a) If there are two tides per day (24 hours), the period will be 12 hours.
b) The difference in depth between high tide (10.0 m) and low tide (4.0 m) is 6.0 m. The amplitude is the maximum displacement from the midpoint, in either direction, that is ±3.0 m from the mid-depth of 7.0 m.

c)

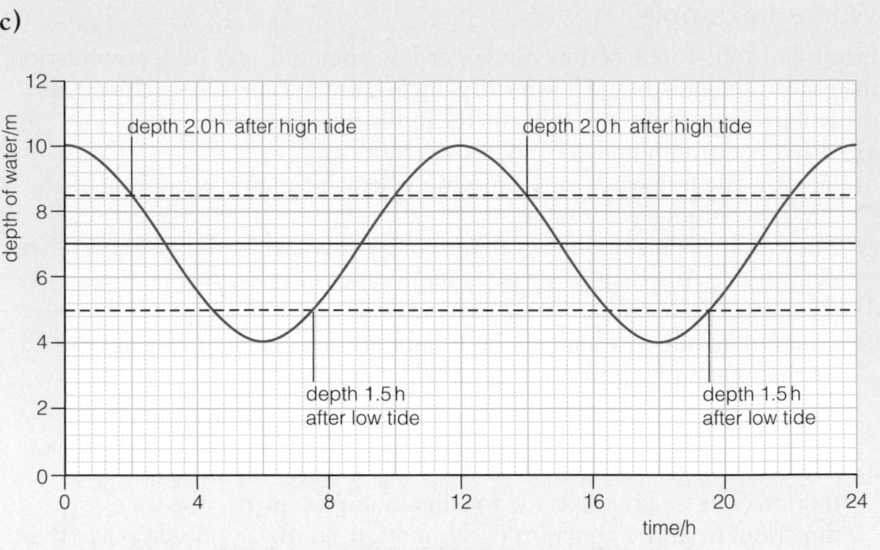

Figure 14.16 ▲

d) It might be helpful to put the two points on the graph first, as shown, so that you can check that you get sensible answers for the depths of the water.

i) We know that $x = A\cos\omega t$, where $A = 3.0\,\text{m}$ and $t = 2.0\,\text{h}$. Therefore to find x, we must first of all find ω.

$$\omega = \frac{2\pi}{T} = \frac{2\pi\,\text{rad}}{12.0\,\text{h}} = \frac{\pi}{6}\,\text{rad}\,\text{h}^{-1}$$

$$x = A\cos\omega t \quad\Rightarrow\quad x = 3.0\,\text{m} \times \cos\left(\tfrac{\pi}{6}\,\text{rad}\,\text{h}^{-1} \times 2.0\,\text{h}\right)$$

$$= 3.0\,\text{m} \times \cos\left(\tfrac{\pi}{3}\,\text{rad}\right) = 3.0\,\text{m} \times 0.50 = 1.5\,\text{m}$$

This means that the water level is 1.5 m *above* its midpoint, so the water is 7.0 m + 1.5 m = 8.5 m deep.

ii) Low tide will occur 6.0 hours after high tide (taken as $t = 0$), so 1.5 hours after low tide means that $t = 7.5\,\text{h}$.

$$x = A\cos\omega t \quad\Rightarrow\quad x = 3.0\,\text{m} \times \cos\left(\tfrac{\pi}{6}\,\text{rad}\,\text{h}^{-1} \times 7.5\,\text{h}\right)$$

$$= 3.0\,\text{m} \times \cos\left(\tfrac{5\pi}{4}\,\text{rad}\right) = 3.0\,\text{m} \times -0.707$$
$$= -2.1\,\text{m}$$

This means that the water level is 2.1 m *below* the midpoint, so the water is 7.0 m − 2.1 m = 4.9 m deep.

Tip

Three things to remember:

- x is the *displacement*, which means it is a *vector* – so its sign represents its *direction* relative to the equilibrium position;
- ωt must be in *radians* – you must therefore make sure you know how to put your calculator into radian mode and remember to do so; and
- ω and t must have *consistent* units – in this example there was no need to convert ω from rad h^{-1} into rad s^{-1} because t was in hours.

14.5 Energy in simple harmonic motion

As a pendulum swings to and fro there is a continuous interchange of kinetic and (gravitational) potential energy. At one end of its swing, the pendulum momentarily comes to rest and so its kinetic energy is zero. At this point it will have maximum potential energy because the bob is at its highest point. As the bob swings down, it loses gravitational potential energy and gains kinetic energy. At the bottom of the swing, the midpoint of the motion, the bob will have maximum velocity and thus maximum kinetic energy. As this is the lowest point of its motion, its gravitational potential energy will have its minimum value. See Figure 14.17. This cyclic interchange of energy is repeated twice every oscillation.

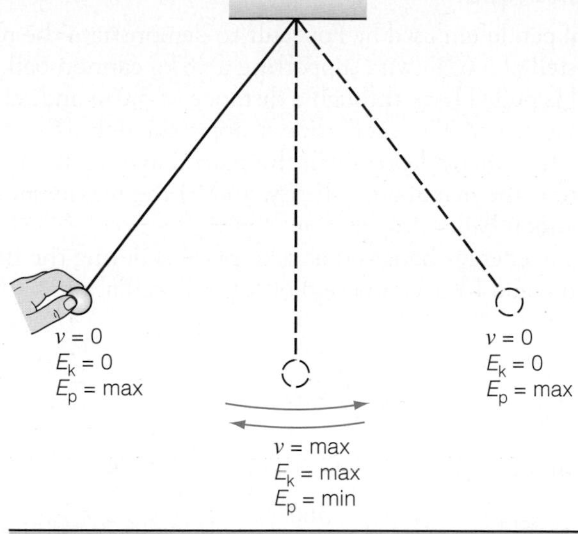

Figure 14.17 ▲

We saw in the Worked example on page 159 that, from the equation $v = -\omega A \sin \omega t$, the maximum velocity will be when $\sin \omega t = \pm 1$, giving $v_{max} = \pm \omega A$. As kinetic energy $E_k = \frac{1}{2}mv^2$,

$$\Rightarrow (E_k)_{max} = \frac{1}{2}m(v_{max})^2 = \frac{1}{2}m\omega^2 A^2$$

This occurs at the equilibrium position, where the potential energy has its minimum value. If we define that the potential energy is zero at the equilibrium position, then the energy of the system is $\frac{1}{2}m\omega^2 A^2$ at the equilibrium position. If there is no damping (that is, no energy is transferred to the surroundings), by the law of conservation of energy the *total* energy $(E_k + E_p)$ of the system must be constant and equal to $\frac{1}{2}m\omega^2 A^2$. This is shown in Figure 14.18.

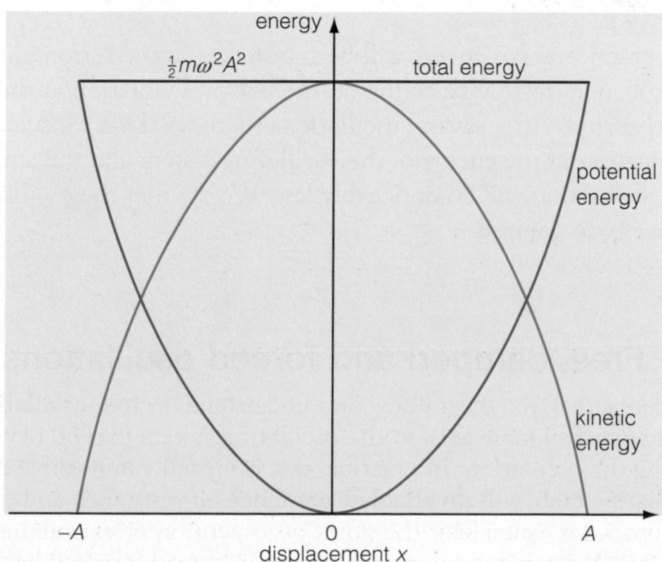

Figure 14.18 ▲
Energy interchange in s.h.m.

Worked example

The original pendulum used by Foucault to demonstrate the rotation of the Earth consisted of a 67 m wire supporting a 28 kg cannon ball. Imagine this cannon ball is pulled back through a distance of 3.0 m and released.

a) Show that its period of oscillation is approximately 16 s.
b) Through how many degrees will the Earth have rotated in this time?
c) Calculate i) the maximum velocity, and ii) the maximum kinetic energy of the cannon ball.
d) Discuss the energy changes that take place i) during the first half oscillation, and ii) over a period of several oscillations.

Answer

a) $T = 2\pi\sqrt{\left(\frac{l}{g}\right)} = 2\pi\sqrt{\left(\frac{67\,\text{m}}{9.8\,\text{m s}^{-2}}\right)} = 16.4\,\text{s} \approx 16\,\text{s}$

b) The Earth rotates through 360° in one day ($= 24 \times 60 \times 60 = 864\,000\,\text{s}$).

In 16 s it will have rotated $\frac{360°}{864\,000\,\text{s}} \times 16\,\text{s} = 6.7 \times 10^{-3}$ degree

c) i) From $v = -\omega A \sin \omega t \Rightarrow v_{\text{max}} = \pm\omega A$

$$\omega = 2\pi f = \frac{2\pi}{T} = \frac{2\pi}{16.4\,\text{s}} = 0.383\,\text{s}^{-1}$$

$$v_{\text{max}} = \pm\omega A = \pm 0.383\,\text{s}^{-1} \times 3.0\,\text{m} = \pm 1.15\,\text{m s}^{-1}$$

ii) $(E_k)_{\text{max}} = \frac{1}{2}m(v_{\text{max}})^2 = \frac{1}{2} \times 28\,\text{kg} \times (1.15\,\text{m s}^{-1})^2 = 19\,\text{J}$

d) i) At the point of release, the cannon ball will have 19 J of gravitational potential energy. As it starts swinging, this potential energy will be gradually changed into kinetic energy as the ball falls and gains velocity. At the bottom of the swing, the ball will have maximum velocity and all the potential energy will have been converted into 19 J of kinetic energy. As the ball starts to rise again, its kinetic energy will gradually be converted back into potential energy. When it comes to rest at the end of its swing, its kinetic energy will be zero and the potential energy will once again be 19 J.

ii) In practice, some energy will be transferred to the surroundings through air resistance acting on the bob and vibration at the point of suspension. After several oscillations there will be a considerable reduction in the energy of the oscillating system and the amplitude of the motion will be noticeably less. We say that the oscillations have been **damped**.

14.6 Free, damped and forced oscillations

These are terms that you must know and understand. A **free oscillation** is one in which no external force acts on the oscillating system except, of course, the force causing the oscillation. In practice, this is virtually impossible to achieve as the oscillating body will invariably experience air resistance and other frictional forces, for example at the point of suspension of a pendulum or spring. If a small, dense pendulum bob on a thin thread is pulled back and released, the ensuing oscillations will approximate to free oscillations.

An oscillating system does work against the external forces acting on it, such as air resistance, and so uses up some of its energy. This transfer of energy from the oscillating system to internal energy of the surrounding air causes the oscillations to slow down and eventually die away – the oscillations are **damped**, as in the previous Worked example on Foucault's pendulum. In a pendulum clock, this energy is gradually restored to the oscillating pendulum by means of a coiled spring, which has to be re-wound from time to time.

When you set your mobile phone on 'vibration mode' and it rings, the part of your body in contact with the phone 'feels' the vibrations. This is because your body has been made to vibrate by the phone. You are experiencing **forced oscillations**, which have the *same frequency* as the vibrating source – in this case your mobile phone. Similarly, you hear sound because the mechanical oscillations of the air particles forming the sound wave force your eardrum to vibrate at the frequency of the sound wave. Your ear converts these mechanical vibrations into an electrical signal, which your brain interprets as a sound.

Experiment

Investigating damped oscillations

We can investigate damped oscillations by means of a long pendulum with a paper cone attached to the bob to increase air resistance.

This experiment is best carried out in pairs, with one person observing the amplitude at one end of the oscillation and the second person at the other end, as shown in Figure 14.19. Alternatively use a motion sensor to record the displacement of the oscillations. It works best if the pendulum can be suspended from the ceiling to give a longer period, although reaching the ceiling can be hazardous!

Figure 14.19 ▲

Initially, find the period T of the pendulum *without* the cone by timing 10 oscillations twice and finding the average. Then attach the cone.

Now pull the pendulum back so that the cone is level with the end of one of the metre rules – that is an initial amplitude of 1.00 m. Let go, and record the amplitude at the end of each swing until it is too small to measure. You may have to experiment with the size of the cone so that you can get about four or five complete oscillations.

Tabulate your results as in Table 14.6.

$10T$/s	$10T$/s	Average T/s

No. of swings	0	1	2	3	4	5	6	7	8
Amplitude/m	1.00								

Table 14.6 ▲

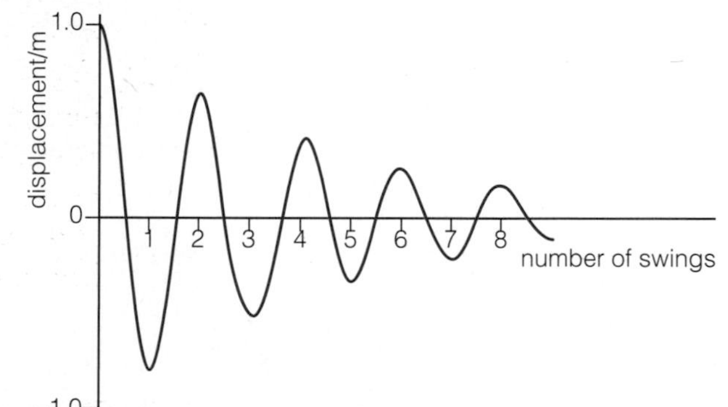

Figure 14.20 ▲

Plot a sketch graph of displacement against time. You can plot the points corresponding to the amplitude at the end of each swing, remembering that each swing is half a period and that the displacement will be alternately positive and negative. You now have to use some skill in joining the points with a cosine curve. Your graph should look like Figure 14.20.

Note

The 'number of swings' means from one extreme to the other and so is the number of *half* periods.

The damping looks as though it could be exponential. Can you remember how to test this? If not, you will have to look at the next Worked example.

Worked example

The following data were obtained for a damped pendulum.

$10T$/s	$10T$/s	Average T/s
29.81	30.17	3.00

No. of swings	0	1	2	3	4	5	6	7	8
Amplitude/m	1.00	−0.80	0.64	−0.51	0.41	−0.33	0.26	−0.21	0.17

Table 14.7 ▲

To check whether the damping is exponential, we have to look at successive *ratios* of the amplitude. We can do this from the data – see Table 14.8.

Swings	0 → 1	1 → 2	2 → 3	3 → 4
Ratio of amplitudes	1.00/0.80 = 1.25	0.80/0.64 = 1.25	0.64/0.51 = 1.25	0.51/0.41 = 1.24
Swings	4 → 5	5 → 6	6 → 7	7 → 8
Ratio of amplitudes	0.41/0.33 = 1.24	0.33/0.26 = 1.27	0.26/0.21 = 1.24	0.21/0.17 = 1.24

Table 14.8 ▲

We can see that the ratio is very nearly constant, which shows that the damping is exponential.

Exercise

Copy and complete the data in Table 14.9 below to show the ratios of successive positive amplitudes, using data from the Worked example above.

Swings	0 → 2	2 → 4	4 → 6	6 → 8
Ratio of positive amplitudes	1.00/0.64 = 1.56			

Table 14.9 ▲

Now test your own results from the experiment.

14.7 Resonance

A simple way of demonstrating, and understanding, what is meant by 'resonance' is to use the arrangement shown in Figure 14.21, which is known as Barton's pendulums.

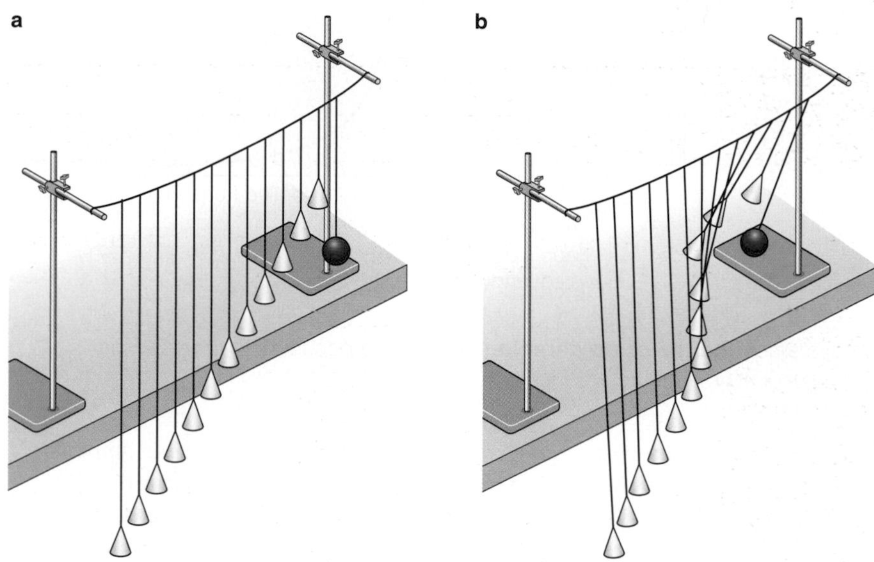

Figure 14.21 ◀ Barton's pendulums

Figure 14.21a shows the initial set-up. The heavy, dense pendulum bob, at the right-hand end, is pulled back a few centimetres and released. This pendulum will oscillate at its **natural frequency** f_0, determined by its length and given by the equation $T = 2\pi\sqrt{(l/g)}$. All the other, light pendulums are coupled to this driver pendulum by the string to which they are tied. They experience **forced oscillations**, equal in frequency to that of the driving pendulum, as some of the energy of the driving pendulum will be transferred to each of the other pendulums. Although they will all be set in motion, the pendulums of length nearest to the length of the driving pendulum will absorb more energy because their natural frequency of vibration is close to that of the driving pendulum. These pendulums oscillate with large amplitude as shown in Figure 14.21b. A pendulum having the *same* natural frequency (same length) as the driving pendulum will absorb by far the most energy and will be forced to oscillate with very large amplitude. This is called **resonance**.

Experiment

Investigating resonance

You can investigate resonance using the arrangement shown in Figure 14.22.

Give the mass a small vertical displacement and determine the period of oscillations, T_0, by timing 20 oscillations and repeating. Hence calculate the natural frequency of oscillation, f_0.

Set the frequency, f, of the signal generator to about $\frac{1}{2}f_0$ and switch on. Gradually increase the frequency until the mass vibrates with maximum amplitude – you should find this happens when $f \approx f_0$. Repeat the experiment, this time starting with $f \approx 2f_0$ and reducing the frequency until resonance occurs. You can further repeat the experiment using different masses.

A graph of the amplitude of oscillation of a particular mass as a function of the driving frequency would look like Figure 14.23 on the next page.

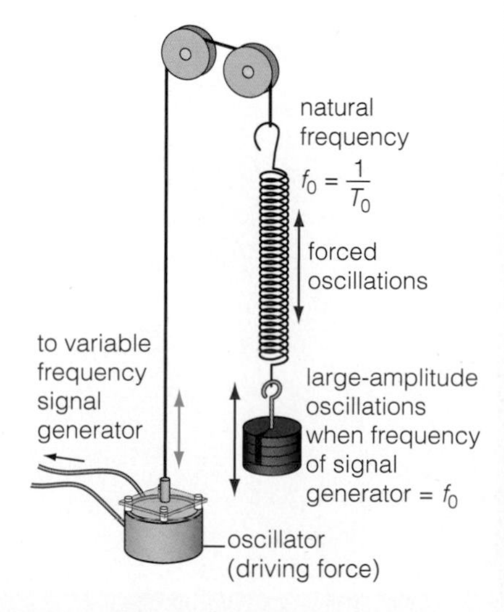

Figure 14.22 ▲

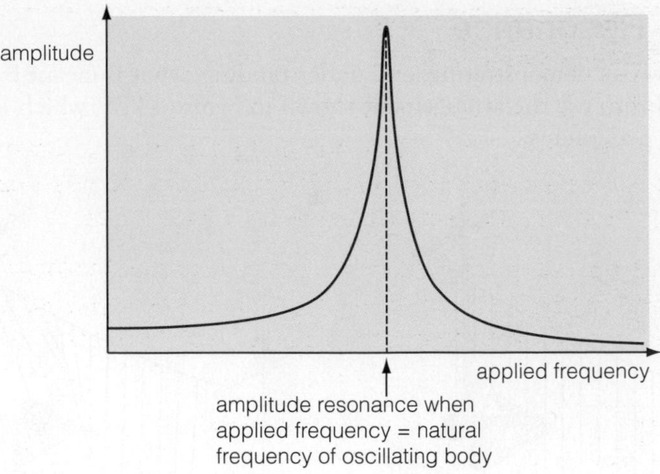

Figure 14.23 ▲
For a system **with little or no damping**, resonance occurs when the applied frequency equals the natural frequency

Tip

When you explain what happens in resonance always consider the *energy transfer* that takes place.

Tip

Be careful to distinguish between *amplitude* resonance and *energy* resonance.

In Figure 14.23 we can easily see that the maximum amplitude, i.e. *amplitude resonance*, occurs when the driving frequency f is equal to the natural frequency of vibration of the mass, f_0. This is only the case if there is very little damping. As the mass oscillates up and down with large amplitude, there is clearly *energy transfer* from the signal generator to the mass.

For a damped system, the situation is somewhat complex. All you need to know is that, if there is damping, then the resonant frequency at which the *amplitude* is a maximum is lower than the natural frequency, and that this difference increases as the degree of damping increases. This is shown in Figure 14.24. However, the maximum energy transfer, or *energy resonance*, always occurs at the natural frequency.

Figure 14.24 also shows two other features of damped resonance – as the amount of damping increases, the resonant peak is much lower, and the resonance curve broadens out.

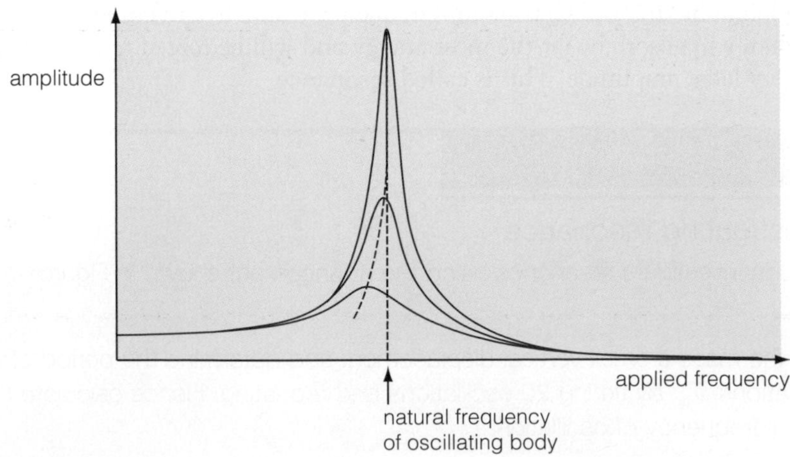

Figure 14.24 ▲
For a **damped system**, amplitude resonance occurs at a frequency that is lower than the natural frequency

Experiment

Investigating a damped system

Damped oscillations can be investigated using a mass on a spring as described in the previous experiment. Damping is provided by a cardboard disc held in between the masses, as shown in Figure 14.25.

Printouts for a small disc (light damping) and for a large disc (heavier damping as more air resistance) are shown in Figure 14.26. You might like to investigate how the damping depends on the area of the disc.

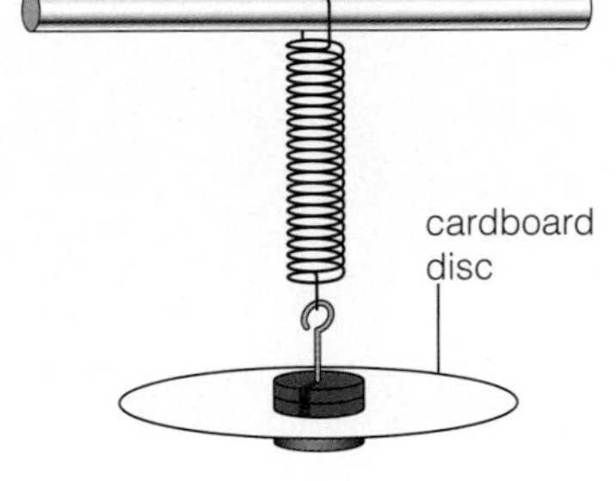

Figure 14.25 ▲

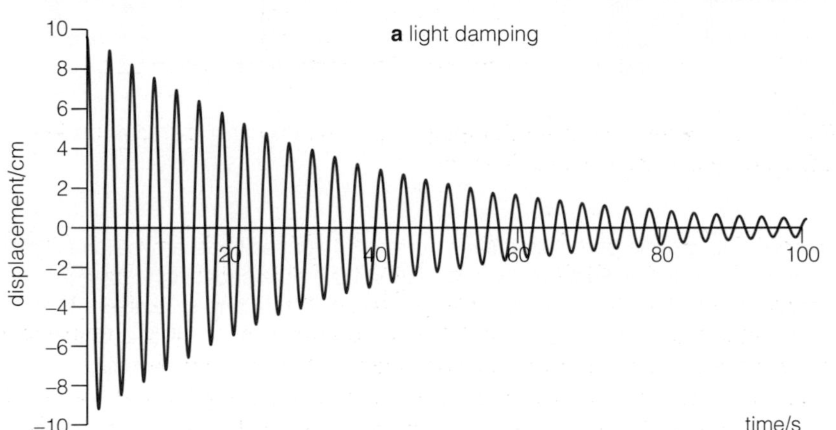

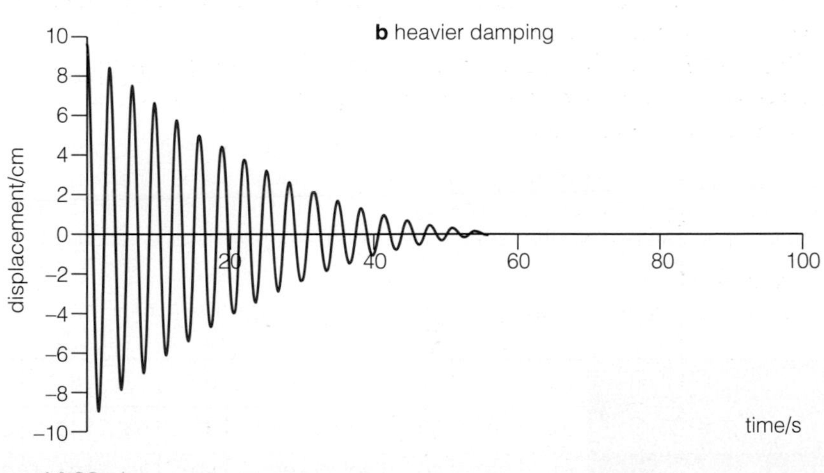

Figure 14.26 ▲

Exercise

Resonance can be demonstrated very elegantly with an electrical circuit containing a capacitor, coil and resistor connected in series with a signal generator giving an alternating current. This is shown in Figure 14.27.

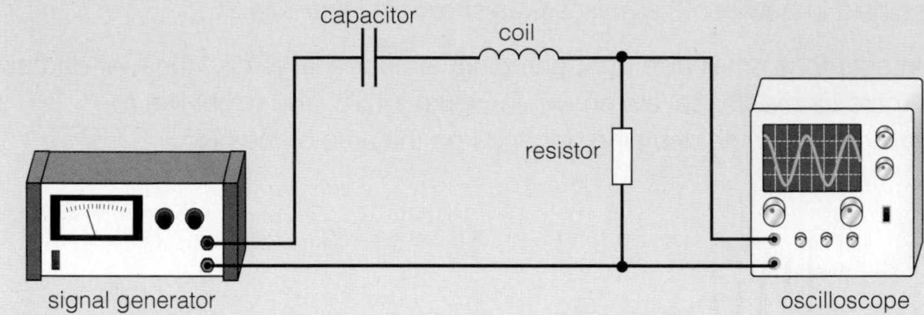

Figure 14.27 ▲

In one quarter-cycle of the alternating current, the capacitor charges up and stores energy in the electric field between its plates. When the capacitor is fully charged, the current in the circuit (and therefore in the coil) is zero and so there is no magnetic field in the coil. In the next quarter-cycle, the capacitor discharges, charge flows in the circuit and the current in the coil causes energy to be stored in the magnetic field it creates.

This process is repeated during the next half-cycle, but in the opposite sense. There is thus a continuous interchange of energy stored in the electric field between the plates of the capacitor and energy stored in the magnetic field of the coil.

The amount of energy stored in each component depends on the frequency of the alternating current. At a certain frequency, the energy stored in the capacitor is exactly equal to that stored in the coil, and the energy drawn from the power supply (i.e. the current) is a maximum – in a word, resonance.

In Figure 14.27, the oscilloscope measures the potential difference across a resistor of known value, from which the circuit current can be calculated. The oscilloscope can also be used to check the frequency calibration of the signal generator. The resonance is 'damped' by the resistor. If the value of the resistor is increased, the damping is greater.

Data from such an experiment are shown in Table 14.10.

f/kHz	3.0	4.0	5.0	5.5	6.0	6.5	7.0	8.0	9.0
I/mA (R = 10 Ω)	20	33	83	125	143	91	63	40	30
I/mA (R = 22 Ω)	19	28	60	80	80	56	45	31	24

Table 14.10 ▲

Plot a graph of I against f for each resistance value and comment on the shape of the curves that you obtain. What is the value of the resonant frequency?

You should obtain graphs like those shown in Figure 14.28 opposite.

Note that a little bit of imaginative, careful drawing needs to be done to get the best curves – the peak of the 10 Ω curve is somewhere between 5.5 kHz and 6.0 kHz and *not* at 6.0 kHz, although the maximum *recorded* current is at 6.0 kHz. The resonant frequency is about 5.8 kHz.

The graphs are characteristic resonance curves, with the flatter 22 Ω curve showing a greater degree of damping and a slightly lower resonant frequency.

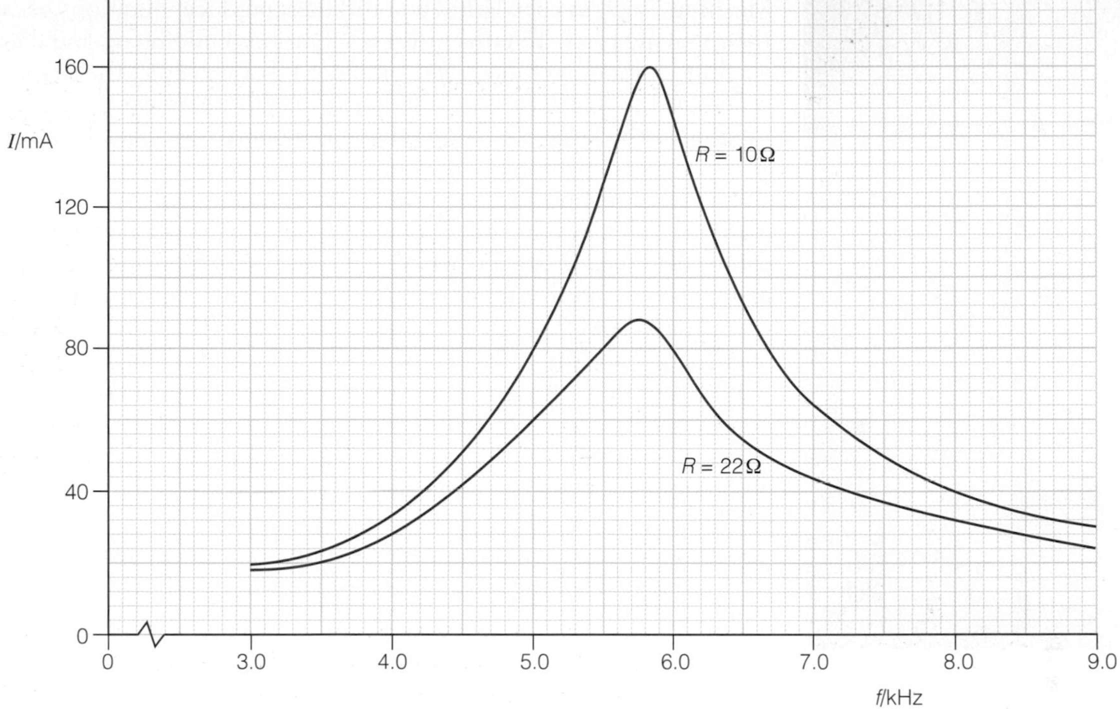

Figure 14.28 ▲

Damping is important in the design of machines and buildings to prevent unwanted vibrations, which if they built up to large amplitude through resonance could cause severe damage, witness the famous Tacoma Narrows Bridge collapse in the USA in 1940. Video clips of this spectacular event can be found on numerous sites by putting 'Tacoma Bridge collapse' into a search engine – well worth the effort!

Sixty years later, on 10 June 2000, the Millennium Bridge crossing the River Thames in London was opened. It was nicknamed the Wobbly Bridge as pedestrians felt an unexpected swaying motion on the first two days after the bridge opened. The natural sway motion of people walking caused small sideways oscillations in the bridge, which in turn caused people on the bridge to sway in step, increasing the amplitude of the bridge and producing resonance. The bridge had to be closed and the problem was tackled by retrofitting dampers to dissipate the energy. The bridge re-opened on 22 February 2002 and has not been subject to any significant vibration since.

Figure 14.29 ▲
The collapse of Tacoma Narrows Bridge

Figure 14.30 ▲
a) The Millenium Bridge, London, and b) one of the dampers that had to be fitted

Figure 14.31 ▲
Burj Dubai skyscraper

Machines, such as cars, lathes and turbines, produce vibrations because of their moving parts. To prevent these vibrations building up and damaging the machine, various techniques are employed. For example, the shape of a lathe is designed so that its resonant frequency is nowhere near the frequency of rotation of the lathe, and a car engine is mounted on special dampers to absorb the energy of the vibrations.

The latest technology is to coat turbine blades with a special ductile material. You may remember from Unit 1 that a ductile material is one that can be deformed plastically without fracture, which means that ductile materials can absorb a lot of energy. If vibrations occur in a ductile material, the material goes through a hysteresis loop each vibration. This absorbs the energy and prevents vibrations of large amplitude building up.

Buildings in earthquake zones are now being designed using ductile construction materials as well as different types of damping mechanisms. Recent research shows that ductility may be a more important factor in the absorption of the energy from earthquake shock waves than damping.

The Burj Dubai, which is the world's tallest building, has a structural system designed and engineered for seismic conditions and extensive wind-tunnel testing has enabled the tower to resist high wind loads while minimising vibration.

Making use of resonance

It's not all bad news. We do actually make use of resonance – every day if you have a radio, television or microwave oven! When you 'tune in' a radio (or a TV, but you don't have to do this very often once the channels are set), you alter the capacitance of a variable capacitor in the tuning circuit (see Figure 6.11 on page 44). When the natural frequency of oscillations in the tuning circuit matches the frequency of the incoming radio signal, resonance occurs and the tuning circuit absorbs energy strongly from the radio waves.

We saw in Unit 4 (page 34) that the oscillating electric field in a microwave oven excites the water molecules in the food, and that energy transferred from the electric field is dissipated as internal energy in the food so that it heats up. The frequency of the microwaves is critical if the energy is to transfer effectively – the microwave frequency must be close to the natural frequency of vibration of the water molecules, and then resonance will take place and the water molecules will strongly absorb energy from the electric field. A frequency of 2.45 GHz is used, because this means the time it takes for the electromagnetic wave to change the electric field from positive to negative is just the right amount of time for the water molecules to rotate. Hence the water molecules can rotate at the fastest possible rate. In addition, this frequency is not used for communications, so microwave ovens won't interfere with mobile phones, televisions, and so forth.

So why do rotating water molecules heat food? The answer has to do with the nature of internal energy and temperature. As we saw in Chapter 11, internal energy is the random kinetic energy of the individual atoms and molecules. As the water molecules rotate, they bump other molecules causing them to begin moving randomly. The process is like frictional heating (see page 107). Microwave energy converts to internal energy by causing the molecules in food to increase the average speed of their random motions – your meal gets nice and hot!

MRI (magnetic resonance imaging) is a medical diagnostic technique that has been used since the beginning of the 1980s. One advantage of an MRI scan is that it uses magnetic and radio waves, so that there is no exposure to X-rays or any other damaging forms of radiation.

How does an MRI scanner work? The patient lies inside a large, cylinder-shaped magnet (see Figure 7.15 on page 63). The strong magnetic field, 10 000 to 30 000 times stronger than the magnetic field of the Earth, exerts a force on the protons within the hydrogen atoms of the patient's body. All the protons, which

Figure 14.32 ▲
MRI scan of the brain and spinal cord

normally lie in random directions, line up parallel to the magnetic field. Then, short bursts of radio waves, of frequency between 1 MHz and 100 MHz, are sent from the scanner into the patient's body. The protons absorb energy from the radio waves (this is the 'resonance' bit of MRI) and are knocked out of alignment. When the burst of radio waves stops, the protons re-align parallel to the magnetic field. As the protons re-align, they emit tiny radio signals. These are detected by a receiving device in the scanner, which transmits the signals to a computer. Most of the hydrogen atoms in our bodies are in the form of water molecules. As each type of tissue has a different water content, the strength of the signal emitted from different body tissues varies. The computer creates a picture based on the strength and location of the radio signals emitted from the body, with a different colour or shade corresponding to the different strength of signal.

MRI magnetic fields are incredibly strong – typically 0.5 to 2.0 T. A watch flying off an arm and into a MRI machine is entirely possible and it has been known for a vacuum cleaner to be sucked into a scanner – it needed a winch to pull it out! How do we get such a strong magnetic field? In a word – superconductivity – superconducting electromagnets, cooled to 4 K by liquid helium (see page 110). What would we do without physicists?

REVIEW QUESTIONS

1 In simple harmonic motion, which of the following does **not** depend on the amplitude of oscillation?

A acceleration B energy C frequency D velocity

2 When a pendulum passes through its midpoint it has maximum:

A acceleration B energy C frequency D velocity

3 At the end of a swing a pendulum has zero:

A acceleration B energy C frequency D velocity

4 A pendulum makes 20 oscillations in 16.0 s. Its equation of motion is of the form $x = A\cos\omega t$, where ω is equal to:

A 0.80π B 1.60π C 1.25π D 2.50π

5 A well designed suspension system in a car can help prevent unwanted resonance in various parts of the car such as the bodywork and the exhaust. Each part has its own particular frequency of vibration.

a) What name is given to this frequency?

b) Explain what is meant by resonance.

c) Sketch a graph to show how the amplitude varies when different frequencies of vibration are applied to a system. Mark the resonant frequency on your graph.

d) A car has shock absorbers to dampen the vibrations. Add a second curve to your graph to show the effect of damping on the system.

e) With reference to your two curves, explain how 'a well designed suspension system in a car can help prevent unwanted resonance'.

6

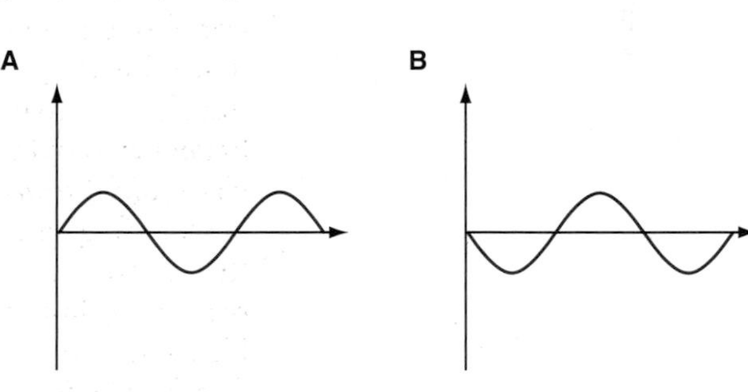

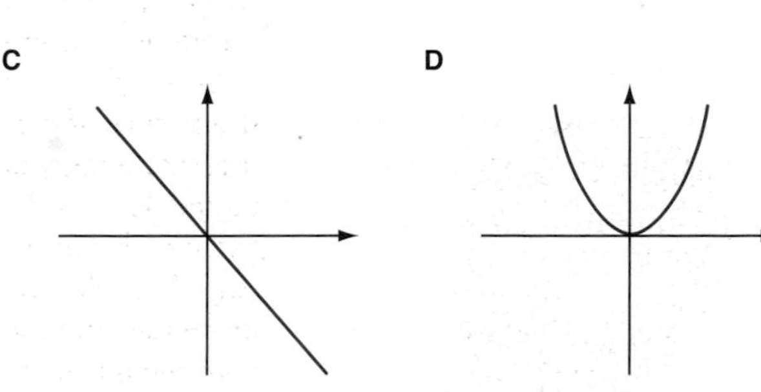

Figure 14.33 ▲

Which of the above graphs associated with simple harmonic motion could be a plot of

a) acceleration against displacement,

b) potential energy against displacement?

7 A child has a bouncy ball attached to a length of rubber cord.

a) Sketch velocity–time graphs for the ball when it

i) makes vertical oscillations when suspended from the rubber cord, and

ii) bounces up and down on a hard surface.

b) Explain, with reference to your graphs, why the bouncing ball does not have simple harmonic motion.

8 The piston in a motor cycle engine moves up and down with simple harmonic motion. The distance from the bottom of the piston's motion to the top of its motion, called the stroke, is 80 mm. The engine is running at 6000 rpm.

a) Explain why the motion of the piston can be represented by the equation

$x = 4.0 \times 10^{-2} \cos 628t$

b) Calculate

i) the maximum acceleration, and

ii) the maximum speed of the piston.

c) Draw sketch graphs of

i) the displacement against time, and

ii) the velocity against time for two cycles of the motion.

Draw your graphs under each other with the same time scale and add suitable numerical values to your scales.

9 A spring is suspended vertically with a mass of 400 g attached to its lower end. The mass is pulled down a distance of 60 mm and released. It is then found to make 20 oscillations in 11.4 s. The displacement x of the spring varies with the time t according to the equation $x = A \cos \omega t$.

a) What is the value of A in this equation?

b) Show that ω is approximately equal to $11\,\text{s}^{-1}$.

c) What is i) the maximum acceleration, and ii) the maximum speed of the mass?

d) Show that the maximum kinetic energy of the mass is about 0.09 J.

e) Show that the maximum force acting on the mass is about 3 N.

f) Hence calculate the spring constant (stiffness) k for the spring.

g) Calculate the maximum potential energy stored in the spring from the relationship $E_p = \frac{1}{2}kx^2$. Comment on your answer.

h) Sketch a graph showing the kinetic, potential and total energy of the mass as a function of time for a half-cycle of the motion.

10 In an earthquake, waves radiate outwards from the epicentre through the Earth, causing the particles of the Earth to vibrate with simple harmonic motion and energy to be transmitted.

a) State what conditions must occur for the motion of the particles to be simple harmonic.

b) Figure 14.34 shows the variation of potential energy E_p with displacement x of a particle at a distance of 100 km from the epicentre.

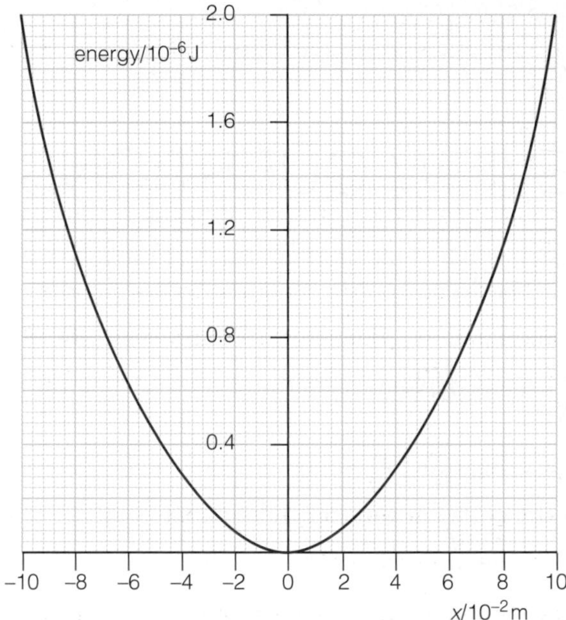

Figure 14.34

Sketch a copy of the graph and add labelled lines to show the variation with displacement of

i) the kinetic energy E_k, and

ii) the total energy E_T of the particle.

c) Calculate the stiffness k of the 'bonds' between particles vibrating within the Earth.

d) Explain two ways in which buildings in earthquake zones can be designed to be safer in the event of an earthquake.

15 Universal gravitation

We all experience a gravitational attraction to the Earth: our weight. Gravitational forces act over very large distances – the Sun keeps the planets in orbit by its gravitational pull. We considered electric forces and fields in Chapter 5; gravitational forces and fields are in many ways analogous, and the way physicists describe them are very similar.

In this chapter, after revising some AS work on gravitational potential energy, you will meet Isaac Newton's formula for the interaction between two masses and learn how to predict the motion of satellites.

15.1 Uniform gravitational fields

Near the Earth's surface, the gravitational field (or *g*-field) is a uniform field. The lines of force are parallel, equally spaced and the arrows are downwards – Figure 15.1. The size of the gravitational force on an object in this field is the same at every place.

The strength of the gravitational field *g* is defined by the equation

$$g = \frac{F_g}{m}$$

where F_g is the size of the gravitational force experienced by a body of mass *m*. You will be familiar with this relationship in the form $W = mg$. The unit of *g* is $\mathrm{N\,kg^{-1}}$ or, as a newton $\mathrm{N \equiv kg\,m\,s^{-2}}$, the unit of *g* is also $\mathrm{m\,s^{-2}}$.

All objects fall with the same **acceleration** *g* in the same **gravitational field** *g*.

Gravitational field strength *g* is a vector quantity. The direction of *g* is the same as that of the gravitational force F_g. At the Earth's surface the size of *g* is about $9.8\,\mathrm{N\,kg^{-1}}$ or $9.8\,\mathrm{m\,s^{-2}}$.

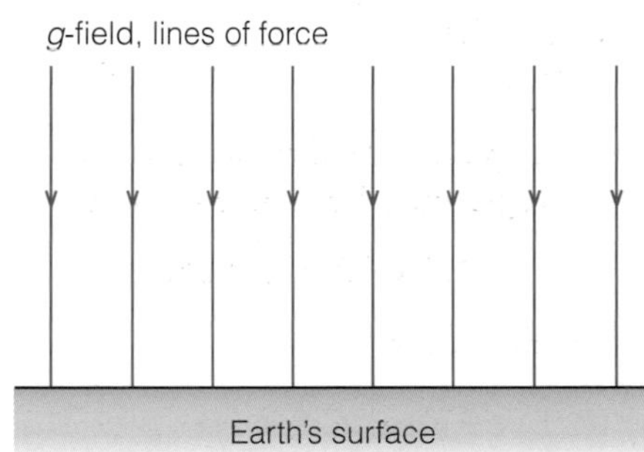

Figure 15.1 ▲
A uniform gravitational field

Worked example

The values of *g* on the surface of the planets Mars and Jupiter are $3.7\,\mathrm{N\,kg^{-1}}$ and $23\,\mathrm{N\,kg^{-1}}$ respectively.

Figure 15.2 ▲
Mars is much smaller than Jupiter, therefore its gravitational field strength is much smaller

a) Calculate the force needed to support a rock of 16 kg on each planet.

b) How much gravitational potential energy (*GPE*) does the rock gain when raised 2.0 m on each planet?

Answer

a) As $g = F_g/m$ then the gravitational force on the rock is $F_g = mg$. The force F needed to support the rock is equal to F_g (Newton's first law).

$\therefore \quad F = (16\,\text{kg})(3.7\,\text{N kg}^{-1}) = 59\,\text{N}$ on Mars
and $F = (16\,\text{kg})(23\,\text{N kg}^{-1}) = 370\,\text{N}$ on Jupiter

b) The gain in *GPE* of the rock is equal to the work done $F_{av}\,\Delta x$, that is $mg\,\Delta x$, in lifting it.

On Mars the gain in $GPE = (59\,\text{N})(2.0\,\text{m}) = 120\,\text{J}$
On Jupiter the gain in $GPE = (370\,\text{N})(2.0\,\text{m}) = 740\,\text{J}$

Variation of *g*

We have said that the Earth's field is uniform, with parallel field lines, near the surface, but this is an approximation. The field is actually a radial field (see Figure 15.9) so the strength decreases with height. The value of g at the top of the Earth's highest mountain is about 0.3% less than its value at sea level.

There is also some variation around the globe. The measured value of the free fall acceleration varies from $9.83\,\text{m s}^{-2}$ at the poles to $9.78\,\text{m s}^{-2}$ at the equator. Some of this variation is the result of the fact that the Earth is not an exact sphere, and some is the result of the Earth's rotation.

Worked example

Figure 15.3 shows a girl standing on a set of bathroom scales. She is at the Earth's equator.

a) What is her centripetal acceleration? Take the Earth's radius to be 6400 km.

b) Show that the g-field needed to produce this acceleration is less than $0.05\,\text{N kg}^{-1}$ and comment on the result.

Tip

Look back to Chapter 4 if you need to revise centripetal acceleration.

Answer

a) The girl's speed as the Earth rotates is

$$v = 2\pi \times \frac{6.4 \times 10^6\,\text{m}}{24 \times 3600\,\text{s}} = 465\,\text{m s}^{-1}$$

Her centripetal acceleration is

$$\frac{v^2}{r} = \frac{(465\,\text{m s}^{-1})^2}{6.4 \times 10^6\,\text{m}} = 0.034\,\text{m s}^{-2}$$

Figure 15.3 ▲

b) To produce this acceleration needs a gravitational field of $0.034\,\text{N kg}^{-1}$. This g-field is only a small fraction (less than 0.5%) of the Earth's g-field at the equator.

Tip

When asked to comment, an answer that includes some quantitative statement (numbers) is often better than offering only words.

On a local scale, tiny variations of g of the order of 1 part in 10^8 can be detected by geologists. Such variations help them to predict what lies below the Earth's surface at that point and possibly, for example, to locate oil deposits.

Equipotential surfaces

The surface of the Earth, the ground, is often taken as a place where objects have zero gravitational potential energy: it is an **equipotential surface**. (See also page 33 in Chapter 5.) There is no change in the *GPE* of an object when it is moved from one place to another on this surface. Above the ground we can imagine a whole series of flat 'contour' surfaces – represented by the dashed lines in Figure 15.4. Each is an equipotential surface. No work is needed to move an object across such a surface, but to move it upwards between any two adjacent surfaces involves a change in gravitational potential. In the case of Figure 15.4, with 50 m intervals between adjacent surfaces, the change in potential is

$$g\,\Delta h = 9.8\,\text{N kg}^{-1} \times 50\,\text{m} = 490\,\text{J kg}^{-1}$$

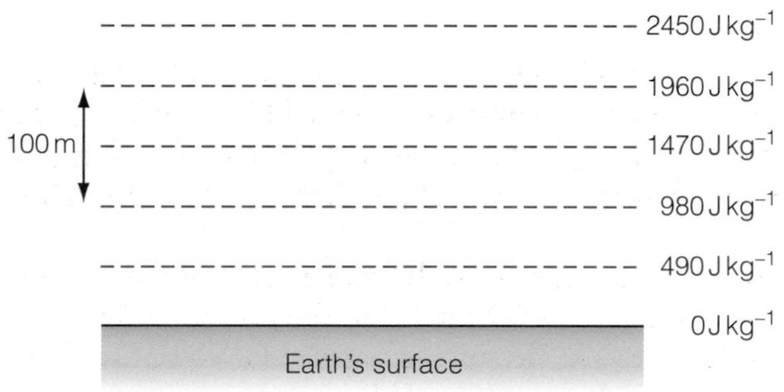

2450 J kg⁻¹ → $2450\,\text{J kg}^{-1}$

$1960\,\text{J kg}^{-1}$

100 m — $1470\,\text{J kg}^{-1}$

$980\,\text{J kg}^{-1}$

$490\,\text{J kg}^{-1}$

$0\,\text{J kg}^{-1}$

Earth's surface

Figure 15.4 ▲
Gravitational equipotential surfaces

Worked example

A multi-storey car park has six levels, G, 1, …, 5, each 3.0 m above the other. A car of mass 1600 kg is parked on level 2.

a) Draw a simple labelled sketch of the car park. Label the levels on your sketch with values for gravitational potential. Take $g = 10\,\text{N kg}^{-1}$ and give level G a potential of $0\,\text{J kg}^{-1}$.

b) What is the change in *GPE* of the car if it moves i) down to level 1, ii) up to level 5?

Answer

a) Gravitational potential difference between levels $= 10\,\text{N kg}^{-1} \times 3.0\,\text{m}$ $= 30\,\text{J kg}^{-1}$ (see Figure 15.5).

b) i) Change in *GPE* $= (1600\,\text{kg})(-30\,\text{J kg}^{-1}) = -48\,000\,\text{J}$

ii) Change in *GPE* $= (1600\,\text{kg})(+90\,\text{J kg}^{-1}) = +144\,000\,\text{J}$

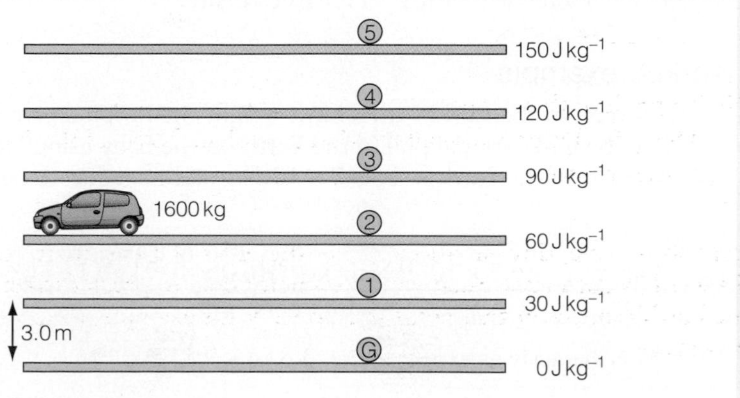

Figure 15.5 ▲

Figure 15.6 ▲

15.2 Newton's law of gravitation

Isaac Newton knew that bodies are accelerating when they are moving in a circle at a constant speed, and hence realised that the Moon is *continuously accelerating* (falling) towards the Earth. He is said to have conceived his law of gravitation by linking this observation with the fact that a falling apple also accelerates towards the Earth.

This led Newton to propose that:

Every particle in the universe attracts every other particle with a force F, given by

$$F = \frac{Gm_1m_2}{r^2}$$

where m_1m_2 is the product of the masses of the two particles, r is their separation and G ('big gee') is a constant called the **gravitational constant**.

Gravitational forces are very small unless one of the 'particles' is a planet or star, and G is *very* difficult to measure in the laboratory. It has a value, to three significant figures, of $6.67 \times 10^{-11}\,\text{N m}^2\,\text{kg}^{-2}$. (See question 6 on page 184 for a way of estimating a value for G.) This is an inverse-square law, very similar to Coulomb's law for electric charges (see page 36). The difference is that, so far as we know, all gravitational forces are attractions.

Worked example

The Moon has a mass of 7×10^{22} kg and is 4×10^8 m away from us – each to only one significant figure. Estimate the gravitational pull of the Moon on you and comment on your answer.

Answer

For a mass of 70 kg, the force $F = Gm_1m_2/r^2$ is

$$F = \frac{(7 \times 10^{-11}\,\text{N m}^2\,\text{kg}^{-2})(7 \times 10^{22}\,\text{kg})(70\,\text{kg})}{(4 \times 10^8\,\text{m})^2}$$

$$= 0.002\,\text{N}$$

This is a very, very small force compared to your body weight on Earth, which is approximately 700 N.

Tip

When putting awkward numbers into calculators to get values like *F* here, it is often helpful to start with the number on the bottom and square it, then use the inverse key before multiplying by the numbers on top to get the answer.

Newton's law, as expressed by the equation above, applies to *particles* and to *spherical objects* such as the Earth. If we use it to find the gravitational force between irregular objects, for example two elephants a certain distance apart, the calculated force will only be a rough estimate.

Worked example

A baby arrives in the world with a birth weight the hospital gives as '3.8 kg'. Calculate the gravitational pull of the Earth on the baby using Newton's law. Take the mass of the Earth to be 6.0×10^{24} kg and its radius to be 6.4×10^6 m.

Answer

$$F = \frac{Gm_1m_2}{r^2} = \frac{(6.7 \times 10^{-11}\,\text{N m}^2\,\text{kg}^{-2})(6.0 \times 10^{24}\,\text{kg})(3.8\,\text{kg})}{(6.4 \times 10^6\,\text{m})^2}$$

$= 37\,\text{N}$ (exactly as expected, i.e. $3.8\,\text{kg} \times 9.8\,\text{N kg}^{-1}$)

Satellites

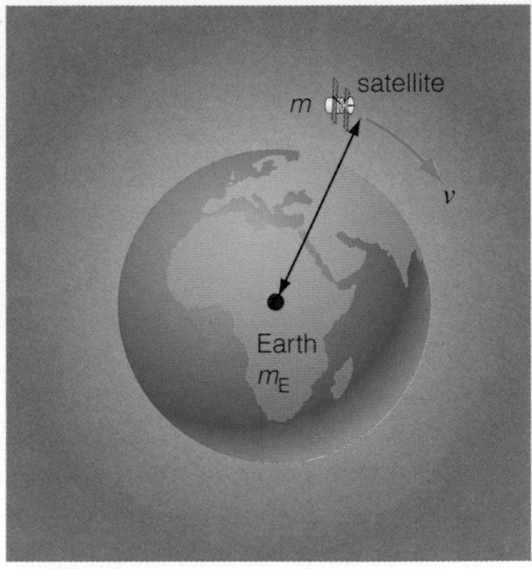

Figure 15.7 ▲

A body on which the only force acting is the pull of the Earth, its weight, is said to be in a state of **free fall** (see Section 4.3). Satellites remain in free fall for a long time as they circle the Earth. (A free-fall parachutist free-falls for only a few seconds before drag forces affect the motion. During most of the fall the parachutist is moving at a steady speed of about 125 m.p.h.)

The Earth has only one natural satellite – the Moon – but there are thousands of 'artificial' satellites placed in orbit above the Earth: communications satellites, weather satellites, military satellites, and so on. Figure 15.7 represents an artificial satellite orbiting the Earth. A free-body force diagram of the satellite would show there to be only one force acting on it, the gravitational attraction of the Earth towards its centre.

Applying Newton's second law and his law of gravitation to the satellite (of mass m moving with speed v) gives

$$\frac{mv^2}{r} = \frac{Gmm_E}{r^2}$$

and, as $v = \dfrac{2\pi r}{T}$

this leads to $r^3 = \dfrac{Gm_E}{4\pi^2} \times T^2$

As $Gm_E/4\pi^2$ is a constant, the result is that r^3 is proportional to T^2 (this is sometimes known as Kepler's third law), and so the further out the satellite, the longer it takes to orbit the Earth. For moons orbiting other planets in orbits of known radius, observations of their period of revolution enable us to calculate the mass of the planet as $m_P = 4\pi^2 r^3/GT^2$.

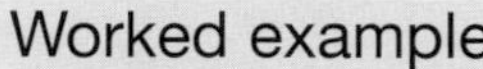

Worked example

Communications satellites are placed in geosynchronous orbits above the Earth's equator.

a) What is the angular velocity of a geosynchronous satellite?

b) Calculate the radius of the satellite's orbit.
Take $Gm_E = 4.0 \times 10^{14}\,\mathrm{N\,m^3\,kg^{-2}}$.

Answer

a) A geosynchronous satellite orbits the Earth once every 24 hours. Its angular velocity is $\omega = 2\pi\,\mathrm{rad}/(24 \times 3600\,\mathrm{s}) = 7.3 \times 10^{-5}\,\mathrm{rad\,s^{-1}}$

b) Using Newton's laws,

$$\frac{mv^2}{r} = mr\omega^2 = \frac{Gmm_E}{r^2}$$

$$\Rightarrow r^3\omega^2 = Gm_E = 4.0 \times 10^{14}\,\mathrm{N\,m^3\,kg^{-2}}$$

$$\therefore r^3 = \frac{(4.0 \times 10^{14}\,\mathrm{N\,m^3\,kg^{-2}})}{(7.3 \times 10^{-5}\,\mathrm{s^{-1}})^2}$$

and $r = 4.2 \times 10^7\,\mathrm{m}$ or 42 000 km (36 000 km above the Earth's surface)

Apparent weightlessness

Figure 15.8 ▲

Figure 15.8 shows an astronaut orbiting the Earth. He is in a continuous state of free fall, i.e. he is falling continuously towards the Earth with an acceleration g ($\mathrm{m\,s^{-2}}$) equal to the local gravitational field strength g ($\mathrm{N\,kg^{-1}}$).

From the photograph it is obvious that there is no 'supporting' force acting on the astronaut. When he returns to the space station the only force acting on him will continue to be his weight, and so he can 'float' around, as can other objects in the cabin that are not attached to the walls. The astronaut describes his condition as being one of **weightlessness**, but this is only apparently the case. His gravitational attraction to the Earth – his weight mg – is still acting on him all the time.

15.3 Radial gravitational fields

Figure 15.9 shows the lines of force (the blue lines), and some equipotential surfaces (the dashed blue lines), in the gravitational field of the Earth to a distance of about 20×10^3 km from the Earth's surface. (The radius of the Earth is 6.4×10^3 km.) On this scale the lines of force become wider apart as they spread out. Since the gravitational potential increases by equal amounts between the dashed lines, they too become further apart the greater the distance from the Earth. A similar diagram could be drawn for any isolated planet or star.

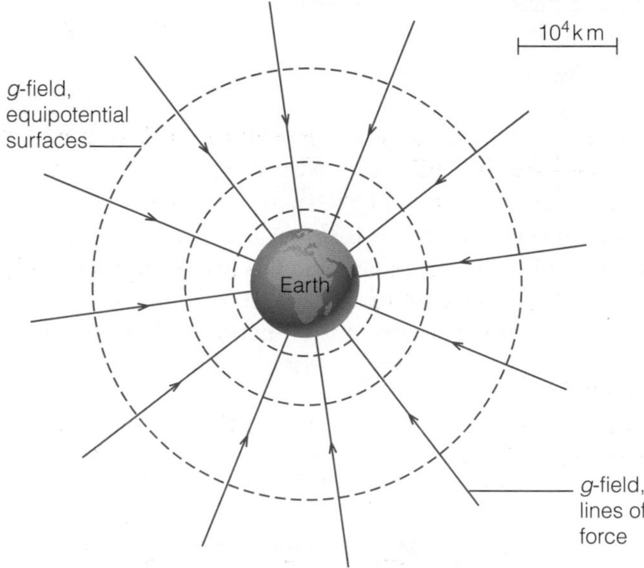

Figure 15.9 ▲
Field lines and equipotentials in a radial field

The gravitational force per unit mass, or the gravitational field strength, around the Earth is described by the equation

$$g = \frac{F_g}{m} = \frac{Gmm_E}{r^2} \div m$$

$$g = \frac{Gm_E}{r^2}$$

where m_E is the mass of the Earth, r is the distance from the centre of the Earth and G is the gravitational constant. In general $g = Gm/r^2$ where m is the mass of the moon, planet, star or other gravitationally attracting body.

Worked example

The mass of the Earth is 5.98×10^{24} kg and its radius is 6.37×10^3 km. Calculate the value of g at the Earth's surface.

Answer

$$g = \frac{Gm_E}{r^2} = \frac{(6.67 \times 10^{-11}\,\text{N m}^2\,\text{kg}^{-2})(5.98 \times 10^{24}\,\text{kg})}{(6.37 \times 10^6\,\text{m})^2}$$

$= 9.81\,\text{N kg}^{-1}$ (not surprisingly!)

Worked example

The mass of the Earth is 600×10^{22} kg and the mass of the Moon is 7×10^{22} kg, each to one significant figure. Explain why there is a point P somewhere between the Earth and the Moon where the gravitational field is zero. Give a rough estimate as to where P lies on the line between the two bodies.

Answer

Between Earth and Moon the Earth's *g*-field is towards E, $\mathbf{g}_E$, and the Moon's *g*-field is towards M, $\mathbf{g}_M$. At some point P, the vectors $\mathbf{g}_E + \mathbf{g}_M = 0$.

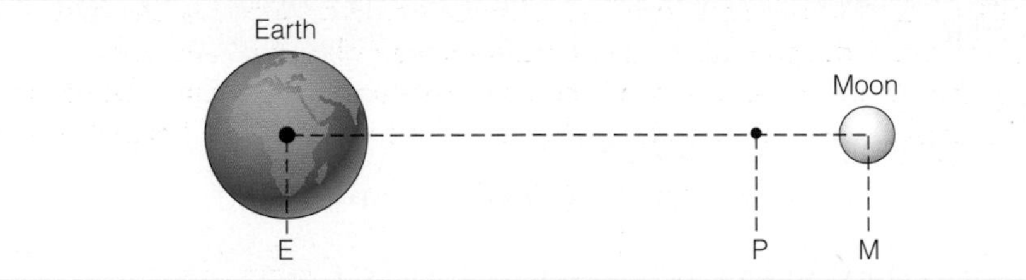

Figure 15.10 ▲

The point P where g_E is the same size as g_M will be much nearer to the Moon than to the Earth, i.e. EP > PM. This is because the Earth's mass 600×10^{22} kg is almost 100 times bigger than the Moon's mass 7×10^{22} kg and $g \propto m$.

As $g \propto 1/r^2$, EP will be about 10 times PM, i.e. P is approximately 90% of the way from the Earth to the Moon.

The Earth's *g*-field

Figure 15.11 shows the variation of *g* above the Earth's surface with distance *r* from the centre of the Earth.

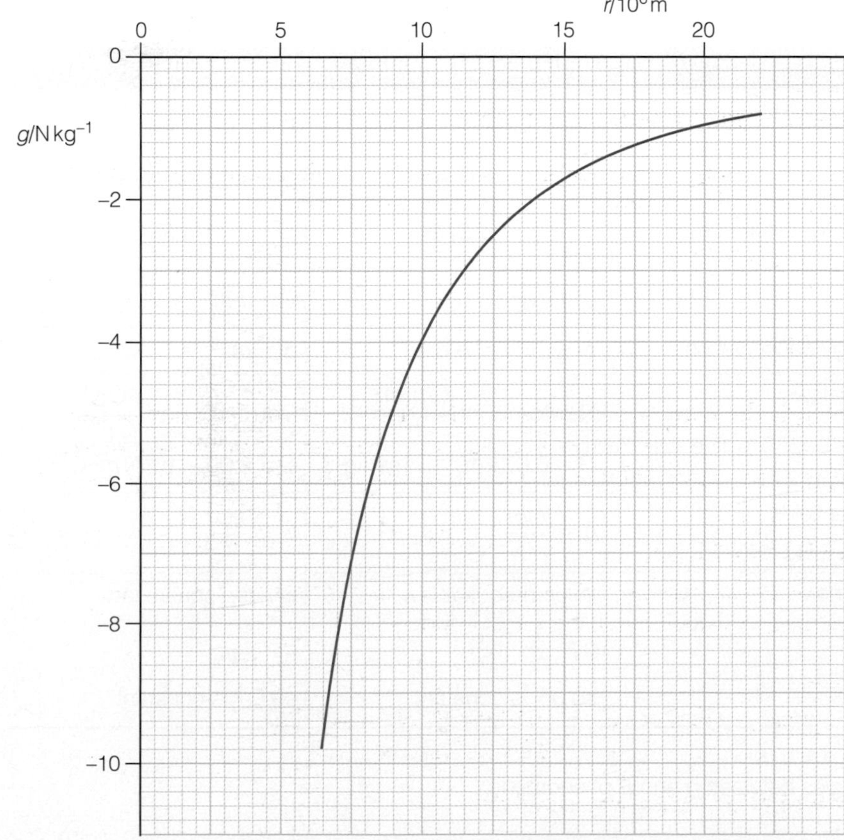

Figure 15.11 ◀
Variation of *g* above the Earth's surface

The graph line in Figure 15.11 shows the inverse-square relationship between g and r for values of r greater than 6.4×10^3 km (the radius of the Earth). The values of g are negative because r is measured away from the Earth but the g-field is towards the Earth. We do not know in detail how the Earth's gravitational field varies below the Earth's surface, but we can be sure it will fall to zero at the centre.

Worked example

A satellite is in a circular orbit of radius 7.5×10^3 km around the Earth. Use the data in the graph of Figure 15.11 to determine the speed of the satellite in its orbit.

Answer

At 7.5×10^6 m the size of g is $7.0\,\text{N kg}^{-1}$.
So the centripetal acceleration of the satellite v^2/r is $7.0\,\text{m s}^{-2}$.

$$\Rightarrow v^2 = (7.0\,\text{m s}^{-2})(7.5 \times 10^6\,\text{m}) = 5.3 \times 10^7\,\text{m}^2\,\text{s}^{-2}$$

So the speed of the satellite $v = 7200\,\text{m s}^{-1}$

Exercise

Test whether the graph in Figure 15.11 shows an inverse-square law relationship between *g* and *r*.

15.4 Comparing gravitational and electric fields

The formula, diagrams and graphs of this chapter show remarkable similarities with those of Chapter 5 – Electric fields. Table 15.1 summarises the similarities.

	Gravitational effects	Electrostatic effects
Field strength	$g = \frac{F_g}{m}$ unit N kg^{-1} or m s^{-2}	$E = \frac{F_e}{Q}$ unit N C^{-1} or V m^{-1}
Potential difference	$g\,\Delta h$, i.e. $\frac{\Delta(GPE)}{m}$ unit J kg^{-1}	$E\,\Delta x$, i.e. $\frac{\Delta(EPE)}{Q}$ unit J C^{-1} or V
Energy conservation	$\Delta(\frac{1}{2}mv^2) = mg\,\Delta h$	$\Delta(\frac{1}{2}mv^2) = Q\,\Delta V$
Force laws	Newton: $F = \frac{Gm_1m_2}{r^2}$ $G = 6.67 \times 10^{-11}\,\text{N m}^2\,\text{kg}^{-2}$	Coulomb: $F = \frac{kQ_1Q_2}{r^2}$ $k = 8.99 \times 10^9\,\text{N m}^2\,\text{C}^{-2}$
Radial fields	$g = \frac{Gm}{r^2}$	$E = \frac{kQ}{r^2}$
Graphs	Inverse-square law, $g \propto \frac{1}{r^2}$ so gr^2 is constant	Inverse-square law, $E \propto \frac{1}{r^2}$ so Er^2 is constant

Table 15.1 ▲
Analogies between *g* and *E* phenomena

There are, of course, differences between the two phenomena. One obvious one is that gravity is about masses while electricity is about charges. Other differences include:

- Gravitational forces affect all particles with mass, but electrostatic forces affect only particles that carry charge.
- Gravitational forces are always attractive, but electrostatic forces can be either attractive or repulsive.
- It is not possible to shield a mass from a gravitational field, but it is possible to shield a charge from an electrostatic field.

Worked example

Two protons of mass m_p in a helium nucleus each carry a charge e. Their centres are a distance d apart. The electrical force F_e between them is pushing them apart; the gravitational force F_g between them is pulling them together.

a) Show that the ratio F_e/F_g does not depend on the distance d.

b) Look up values of the relevant physical quantities (see the Data sheet, page 225) and show that the ratio F_e/F_g is about 10^{36}.

c) What units does this ratio have?

Answer

a) $F_e = \dfrac{kee}{d^2}$ and $F_g = \dfrac{Gm_pm_p}{d^2}$

$$\therefore \frac{F_e}{F_g} = \frac{kee}{Gm_pm_p} = \frac{ke^2}{Gm_p^{\,2}}$$

which does not depend on d.

b) $$\frac{F_e}{F_g} = \frac{(8.99 \times 10^9\,\text{N m}^2\,\text{C}^{-2})(1.60 \times 10^{-19}\,\text{C})^2}{(6.67 \times 10^{-11}\,\text{N m}^2\,\text{kg}^{-2})(1.67 \times 10^{-27}\,\text{kg})^2}$$

$$= 1.24 \times 10^{36}$$

c) There are no units. Both the C and the kg cancel, leaving $\text{N m}^2 \div \text{N m}^2$.

This huge number for the ratio F_e/F_g shows that it is not gravity that holds the nucleus together against electrical repulsion. There must be another force between nucleons.

REVIEW QUESTIONS

1 A high jumper of mass 83 kg raises his centre of gravity 1.3 m in crossing a high bar. The increase in GPE needed to achieve this:

A is less than 1000 J

B is about 100 J

C depends on how fast his approach run is

D is independent of his run-up speed.

2 A crane lifts a mass of 450 kg from ground level to the top of a building 30 m high.

a) What is its change of GPE?

b) What is the change in gravitational potential?

3 A stone of mass 3.0 kg is projected from ground level at a speed of 50 m s^{-1}.

a) What will be its height above the ground when it has a speed of 30 m s^{-1}?

b) Does your answer depend upon (i) the angle at which it is projected or (ii) its mass?

4 The planet Saturn has a radius of 60 × 10^{6} m and spins on its axis once every 3.7 × 10^{4} s (about 10 hours).

a) Calculate the centripetal acceleration of an object at rest on Saturn's equator.

b) The gravitational field strength on Saturn's equatorial surface is 10.6 N kg^{-1}. What would an object of mass 100 kg register on scales at Saturn's equator?

5 The planet Venus has a diameter of 12 × 10^{3} km and an average density of 5200 kg m^{-3}. Calculate the gravitational field strength at its surface.

6 Isaac Newton 'guessed' that the average density of the Earth was about 5000 kg m^{-3}. He used this guess to find a value for G, the gravitational constant. What value did he calculate?

7 A rifle bullet is fired horizontally around the Moon's equator at a speed v. It makes a full circle of the Moon in a time T. Calculate values for v and T given the following data:

mass of Moon = 7.3 × 10^{22} kg
radius of Moon = 1.7 × 10^{6} m

8 Table 15.2 gives the orbital period T of the four largest moons of the planet Jupiter. Their mean distance from the centre of Jupiter r is also given.

	Io	Europa	Ganymede	Callisto
T/days	1.8	3.6	7.3	17.0
r/10^{3} km	422	671	1070	1880

Table 15.2 ▲

a) Show that these data agree with the relationship $r^3 \propto T^2$.

b) Calculate the mass of the planet Jupiter.

9 The value of the gravitational field g at the surface of a planet of radius r and uniform density ρ is given by the relation $g = \frac{4}{3}\pi G\rho r$. Show that the right-hand side of this expression has the unit for g.

10 A satellite of mass m is moving at a speed v at a constant height h above the Earth's surface. The mass and radius of the Earth are m_E and r_E, and G is the gravitational constant.

a) Apply Newton's second law to the motion of this satellite.

b) Hence show that v decreases as h increases.

c) How does the mass of the satellite affect the speed v at a height h?

11 Figure 15.12 shows a body of mass m situated a distance R from the centre of the Earth and r from the centre of the Moon ($R + r$ = 3.8 × 10^{8} m).
Mass of Earth m_E = 598 × 10^{22} kg
Mass of Moon m_M = 7.3 × 10^{22} kg

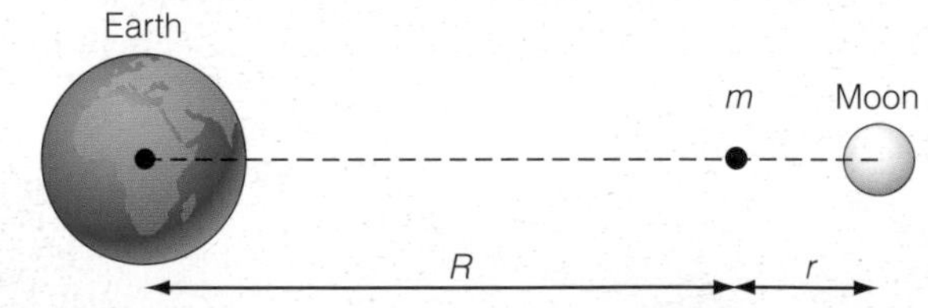

Figure 15.12 ▲

If at this point the body experiences zero net gravitational pull towards either body, show that

a) $R = \sqrt{\left(\frac{m_E}{m_M}\right)} \times r$

b) R = 3.4 × 10^{8} m

12 Show that the value of g hardly changes in the first 100 km above the Earth's surface, by calculating g_r and $g_{r\,+\,100\,\text{km}}$. Take r_E = 6.37 × 10^{6} m.

16 Astrophysics

'Astronomy' has been practised since very ancient times. Observations of the night sky were detailed and, since the early 17th century, increasingly powerful telescopes have been used. Observations and measurements were analysed and a mathematical picture of the 'Heavens' was developed. Physics has added a new dimension to this knowledge, and in this chapter you will learn about the life and death of stars and of our appreciation of the size of our universe.

'Astrophysics' is now a part of most university physics courses and a major field of research but, since it contributes little to helping us understand such earthly matters as global warming, we ask ourselves whether the money for astrophysics is well spent?

16.1 How far to the stars?

Many of us who live in towns or cities rarely see the night sky as our ancestors did. Look at page 95, the opening page of Unit 5. The photograph shows the Milky Way – the billions of stars in the middle of our own galaxy. Have *you* ever seen the Milky Way like this? Our own star – the Sun – is near the edge of the galaxy, so when we are away from city lights on a clear night and look towards the centre of the galaxy we have that amazing view. The whole galaxy is a swirling disc of several hundred billion ($>10^{11}$) stars. A similar spiral galaxy, called M51 or the 'Whirlpool', is shown in Figure 16.1. M51 is roughly 80 000 light years across and has a total mass about 150 billion times larger than the mass of the Sun!

Figure 16.1 ▲
A galaxy similar to our own, millions of light years away from us

Definition

A **light year** is the distance travelled by light in one year:
1 light year = 9.5×10^{15} m
or about 10^{16} m

The light year is a very, very large distance. Professional astronomers use other units such as 'parsecs' or 'astronomical units' for distance, but here we will only use metres and light years.

$$1 \text{ light year} = (3.00 \times 10^{8}\,\text{m s}^{-1})(365 \times 24 \times 3600\,\text{s}) = 9.5 \times 10^{15}\,\text{m}$$

Trigonometric parallax

'Nearby' stars – by near, we mean up to about 300 light years away – can be seen from ground-based observations to exhibit an annual 'wobble' relative to the 'fixed' distant stars in the background. From satellite observations, for example from the Hubble space telescope, the wobble of more distant stars can be detected. This wobble follows an annual pattern, and is the result of the Earth's movement round the Sun. Figure 16.2 illustrates the phenomenon that is known as **stellar** or **trigonometric parallax**.

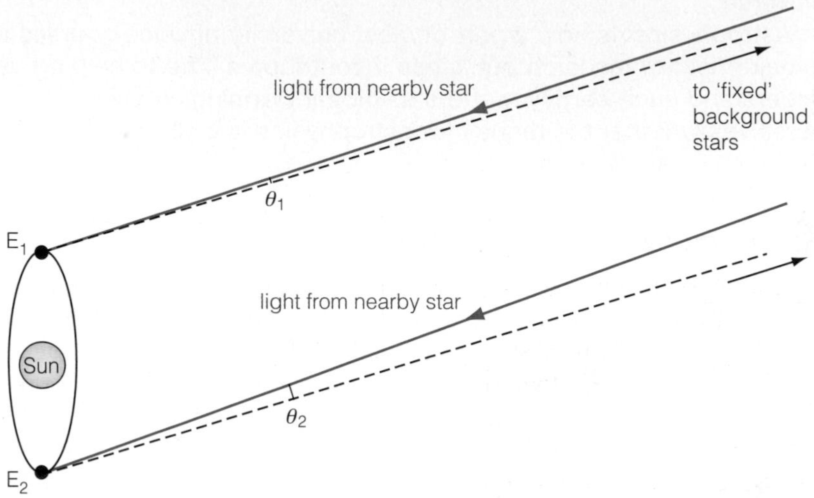

Figure 16.2 ▲
Trigonometric parallax

E_1 and E_2 are the positions of the Earth six months apart. The angles θ_1 and θ_2 to the nearby star measured from E_1 and E_2 are different, and the size of $\theta_2 - \theta_1 = \Delta\theta$ measures the size of its 'wobble'. A knowledge of the diameter of the Earth's orbit round the Sun plus some trigonometry enables us to calculate the distance from the Sun to the nearby star. Using different positions on the Earth's orbit allows the parallax angles to all the nearby stars to be measured. However, the distance to the *nearest* star is over 100 000 times the diameter of the Earth's orbit, so $\Delta\theta$ is tiny. It can be measured to a few millionths of a degree: 0.01 second of arc from the ground. There are only 50 stars for which $\Delta\theta > 0.25$ seconds of arc. Ground-based measurements of parallax only work for a few thousand stars, but measurements from satellites such as Hubble and Hipparcos above the atmosphere have greatly increased this number. (See also question 7 on page 196.)

Tip

There are 360° in a circle.

1 minute of arc $= \frac{1}{60}^{\circ}$

1 second of arc $= \frac{1}{3600}^{\circ}$

A second of arc is also called an **arcsec**. One arcsec is a very small angle – about the angle between two car headlights seen (if you could) from a distance of 200 km!

Worked example

Look out of the window and identify something like a chimney or a lamp post that is at least 50 m away. Extend one arm and hold your thumb so that your thumb is between your eyes and the chimney. Now close first one eye and then the other.

a) Describe what you see.

b) Sketch a diagram to explain what you see, giving lengths and angles where possible.

Answer

a) As one eye is closed and then the other one is closed, the chimney swaps over from lying on the left of the thumb when seen by the right eye, to lying on the right of the thumb when seen by the left eye.

b) For example:

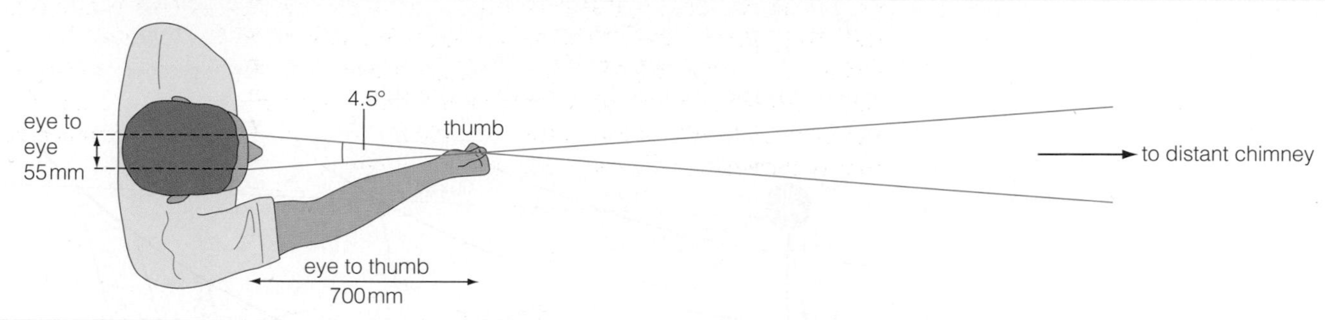

Figure 16.3 ▲

Using the lengths in the diagram, the parallax angle can be calculated as

$$\Delta\theta \approx \frac{55\,\text{mm}}{700\,\text{mm}} = 0.079\,\text{rad} = 4.5°$$

$$\text{or as } \tan\tfrac{1}{2}\Delta\theta = \frac{27.5\,\text{mm}}{700\,\text{mm}} \Rightarrow \Delta\theta = 4.5°$$

16.2 Luminosity and flux

Luminosity, symbol L, is the word astrophysicists use to describe the **total output power** of a star, unit W.

For example, the luminosity of the Sun is $L_{\odot} = 3.90 \times 10^{26}$ W – an incredible power output of 3.90×10^{26} joules per second. All we know about stars and our universe involves huge numbers.

Note

The suffix $_{\odot}$ is attached to values that refer to our Sun.

Worked example

The Sun's output power of $3.90 \times 10^{26}\,\text{J s}^{-1}$ is the result of mass–energy transfer within the Sun. Calculate the rate at which the Sun is transforming matter into electromagnetic wave energy.

Answer

Using $\Delta E = c^2\Delta m$ tells us that the mass loss needed to produce 3.90×10^{26} J of energy is

$$\Delta m = \frac{\Delta E}{c^2} = \frac{3.90 \times 10^{26}\,\text{J}}{(3.00 \times 10^{8}\,\text{m s}^{-1})^2} = 4.33 \times 10^{9}\,\text{kg}$$

which is over a million tonnes of matter – and this happens *each second*. The mechanism for this power production is described on page 198.

The electromagnetic wave energy per second per unit area from a star reaching us on Earth is called the **radiation flux** from the star, symbol F, unit W m^{-2}.

In non-astronomical situations radiation flux is usually referred to as the intensity of the light. You have probably met light intensity, perhaps in considering a surface illuminated by a 100 W light bulb, in your AS course.

The radiation flux received from the Sun at the Earth's surface is about $1000\,\text{W m}^{-2}$, depending on the state of the atmosphere and cloud cover. Above the atmosphere the radiation flux from the Sun, $F_{\odot}$, is $1350\,\text{W m}^{-2}$.

For more distant stars the flux is, of course, very much smaller. Most importantly, F and L are linked by the **inverse-square law**:

$$F = \frac{L}{4\pi d^2}$$

where d is the distance from Earth to the star.

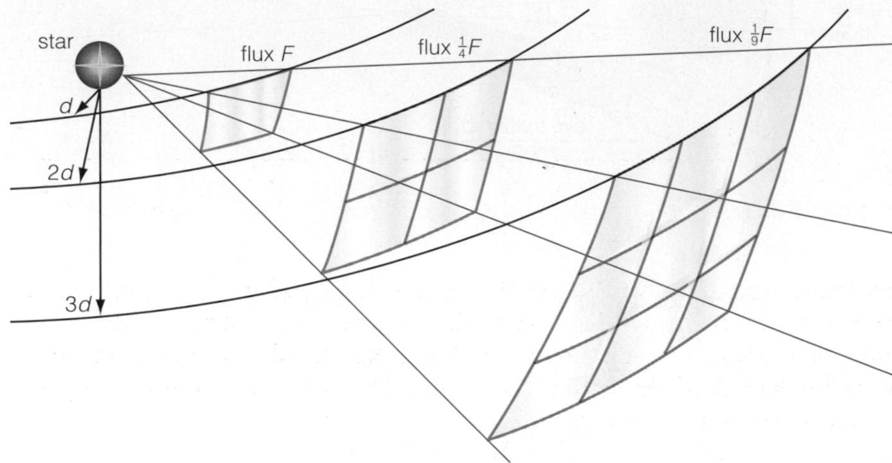

Figure 16.4 ▶ Radiation flux and distance: the inverse-square law

Referring to Figure 16.4:

- the radiation flux at $2d$ is one-quarter $(\frac{1}{2})^2$ that at d
- the radiation flux at $3d$ is one-ninth $(\frac{1}{3})^2$ that at d, etc.

When L of a given star is known and F is measured on Earth, the distance from Earth to the star can be determined, as

$$4\pi d^2 = \frac{L}{F} = \frac{\text{radiation flux}}{\text{luminosity}}$$

Worked example

Use the values of $L_\odot$ and $F_\odot$ given above to confirm that the Sun is 8.3 light minutes from the Earth.

Answer

$$4\pi d^2 = \frac{L_\odot}{F_\odot} = \frac{3.90 \times 10^{26}\,\text{W}}{1350\,\text{W}\,\text{m}^{-2}}$$

$$\Rightarrow d = 1.52 \times 10^{11}\,\text{m}$$

$$8.3 \text{ light minutes} = (3.0 \times 10^8\,\text{m}\,\text{s}^{-1})(8.3 \times 60\,\text{s}) = 1.49 \times 10^{11}\,\text{m}$$

Hence, to 2 sig. fig., the values are consistent.

16.3 Standard candles

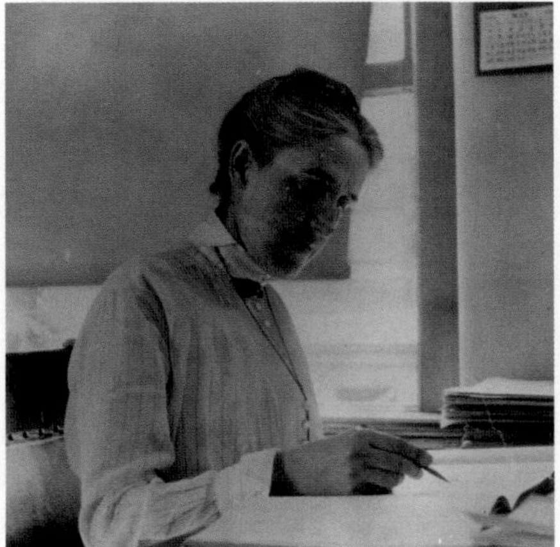

Figure 16.5 ▲ Henrietta Leavitt

The problem in using $4\pi d^2 = L/F$ to measure how far it is to a star (that is too far away to exhibit parallax) is how to determine the star's full power output – its luminosity L.

In the early 20th century an American astrophysicist, Henrietta Leavitt, working at Harvard, discovered that a type of star now called a **Cepheid** has a luminosity that varies with time. Such stars appear more bright and less bright with periods of the order of days. Further, she was able to establish that the maximum luminosity L of a Cepheid star was related to the period T of its

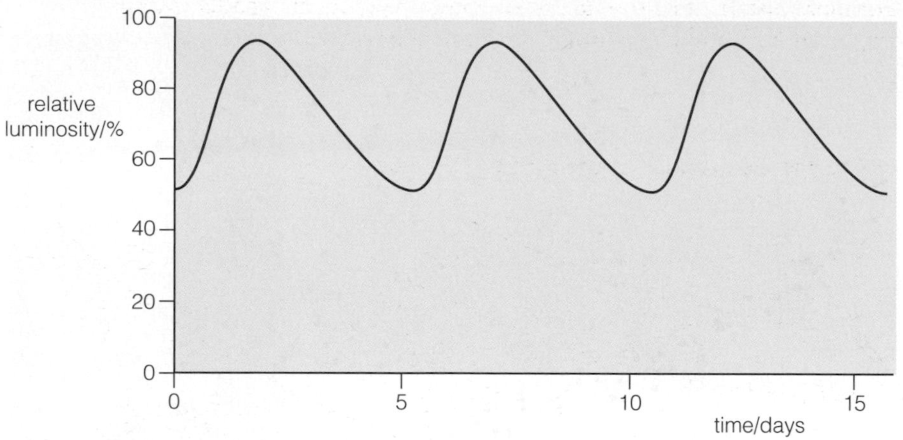

Figure 16.6 ◀ Typical luminosity variation for a Cepheid star

luminosity variation. For the first time it was thus possible, by measuring T for a Cepheid star that is too far away to show any parallax wobble, to know the star's luminosity. Such stars are valuable **standard candles**, which means that we can determine their absolute luminosity. The pole star Polaris is a Cepheid with a period of about 4 days.

The process is therefore:

- locate a Cepheid variable star
- measure its period (may be up to tens of days)
- find the star's luminosity using Leavitt's $T-L$ data
- measure the radiation flux F from the star at Earth.

This last measurement is not easy, but is what astronomers have spent hundreds of years perfecting using telescopes that track a chosen star at night as the Earth rotates.

Using this method the true scale of our galaxy became known: the Milky Way is about 150 thousand light years across. Furthermore, as Cepheid stars can be observed in galaxies beyond ours, this led to the earliest indication that our universe was huge. The distance to Andromeda, our nearest galaxy, is about 2.5 million light years. And there are millions of galaxies much, much further away. **Supernova** explosions can be used as standard candles to find the distance to even more distant galaxies, as we believe that the maximum luminosity of these exploding stars is the same all over the universe.

Figure 16.7 summarises the methods of measuring distance in the universe. (See page 206 for the red shift method.) For all this to work reliably it is important that there are stars that overlap the methods, for example a Cepheid that shows a parallax wobble, and a supernova close to a Cepheid.

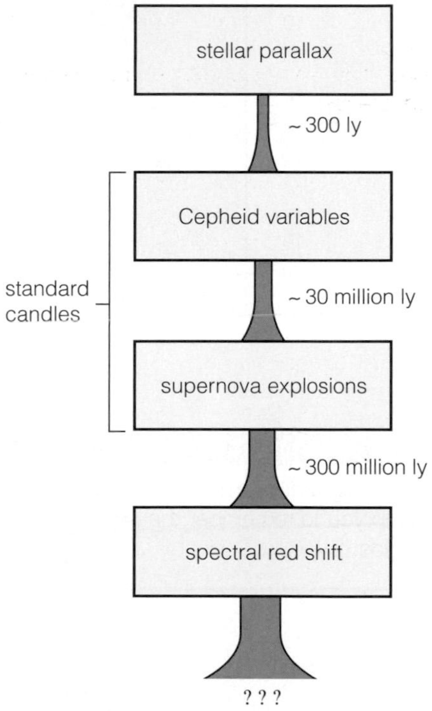

stellar parallax
~ 300 ly
Cepheid variables
standard candles
~ 30 million ly
supernova explosions
~ 300 million ly
spectral red shift
? ? ?

Figure 16.7 ▲ Measuring our universe

16.4 The Hertzsprung–Russell diagram

The **Hertzsprung–Russell** or **H–R diagram** is a plot of stellar luminosity and surface temperature. The one in Figure 16.8 is a simplified version which indicates clearly the regions in which 'main sequence' stars, white dwarf stars and red giant stars are located. Note it is a diagram not a graph – each dot represents a single star. To understand the diagram you need to remember what the two axes are telling you.

The vertical axis is luminosity, scaled in 'Sun-powers', i.e. multiples of $L_{\odot}$ (no units). The positions of the stars go up and down from 1 (where the Sun sits) in powers of 10, i.e. the scale is logarithmic. The Sun's luminosity is 3.90×10^{26} W.

The horizontal axis is the surface temperature T of the star in kelvin. This scale is also logarithmic and (an unfortunate historical blip) goes from high

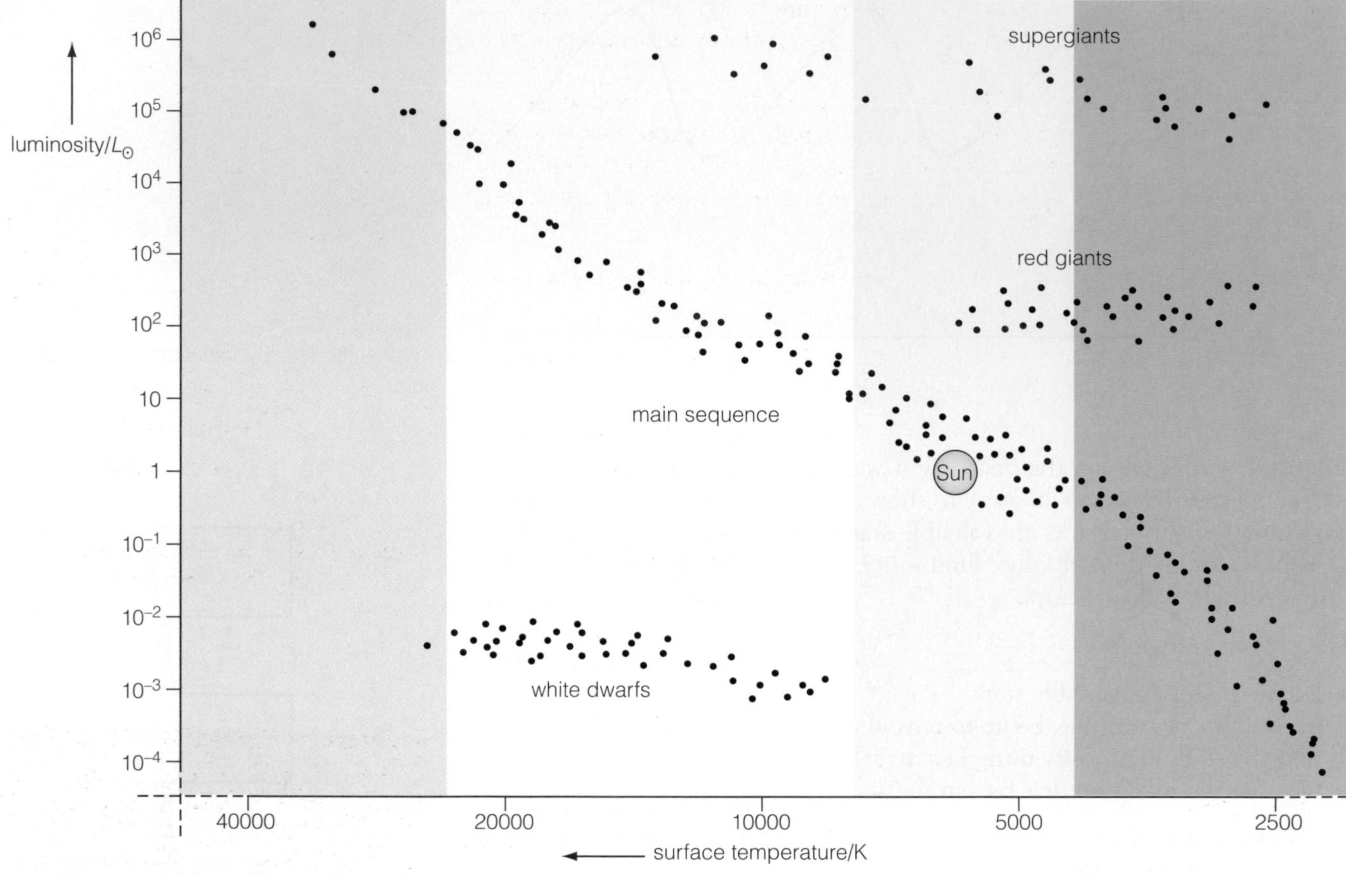

Figure 16.8 ▲
The Hertzsprung–Russell diagram

> **Tip**
> Copying a diagram such as this will help you to remember the detail it holds.

temperatures on the left to low temperatures on the right. The colours give you an indication of what a star will look like. The Sun has a surface temperature of 5800 K.

Worked example

a) Use the H–R diagram in Figure 16.8 to estimate:
   i) the luminosity of a main sequence star with a surface temperature of 5000 K,
   ii) the surface temperature of a main sequence star of luminosity $L_\odot/1000$.

b) Explain why 'main sequence' stars were specified in a).

Answer

a) i) just less than $L_\odot$
   ii) about 2500 K.

b) There are red giants with temperatures of 5000 K and there are white dwarfs with luminosities of $L_\odot/1000$.

> **Tip**
> Remember that an individual star does *not* progress *along* the main sequence band.

The life cycle of a star

The H–R diagram is a snapshot of stars *in our galaxy* at this moment; the vast majority of stars that are now visible lie on the **main sequence**. Once a star is formed (see Section 17.1) it adopts its position on the main sequence and spends most of its life with a fairly constant surface temperature and luminosity (although it does get a little bit brighter during this very long time). More massive stars will have shorter stays on the main sequence, and smaller stars longer stays.

Astrophysicists can predict different stages in the life of an 'average' star like our Sun. They test their predictions by observing the properties of stars of different mass and age that exist in clusters and are therefore all the same distance from Earth. The stages for our Sun are believed to be as follows (see also Figure 16.11 below):

Stage 1: It was formed or 'born' from a cloud of hydrogen and helium (see page 198).

Stage 2: It joined the main sequence about 5 billion years ago and will leave the main sequence about 5 billion years from now.

Stage 3: After leaving the main sequence it will expand and become a red giant (see Figure 16.9) and, after losing about 50% of its mass, will then become a white dwarf.

Stage 4: It will then slowly cool for billions years and effectively 'die'.

Stage 1 takes about 10 million years. The predicted lifetime of our Sun on the main sequence, throughout which time it pours energy to the Earth at over $1\,\text{kW}\,\text{m}^{-2}$, is some 10 billion years! The processes in stage 3 will take place over a billion or so years.

For stars on the main sequence that have a mass $> M_\odot$ ($M_\odot$ is the mass of the Sun), there are two possible fates.

- Stars between $1.4M_\odot$ and $3.0M_\odot$ end up as spinning **neutron stars** or 'pulsars' – stars that were discovered in 1967 by Jocelyn Bell (Figure 16.10), then a graduate student at Cambridge.
- Stars heavier than $3.0M_\odot$ finish up as **black holes**.

Figure 16.11 illustrates the lives of stars of different masses. Note that the time spent on the main sequence is not drawn to scale.

After all stars leave the main sequence, nuclear reactions in their cores produce new elements, with atomic masses up to carbon (12). Further **nucleosynthesis** of heavier elements, up to iron, takes place in the cores of stars of mass $1.4M_\odot$ to $3.0M_\odot$. Elements beyond iron up to uranium are produced in supernova explosions. All these elements are spread through the universe, so we can honestly say that we are all made of star dust!

Figure 16.9 ▲
Relative sizes of an average main sequence star and a red giant

Figure 16.10 ▲
Jocelyn Bell Burnell

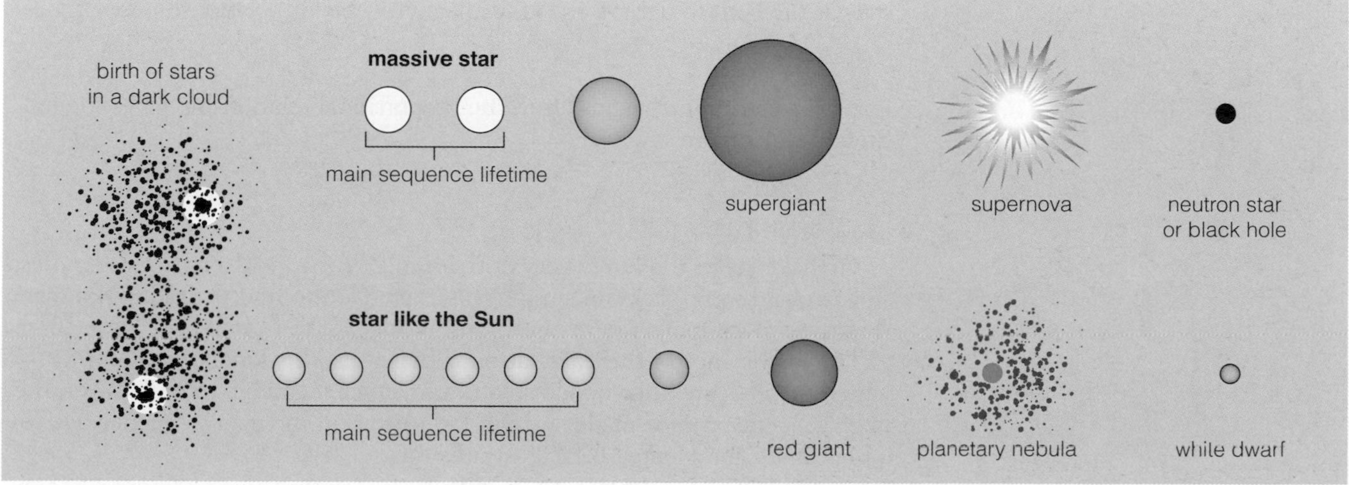

Figure 16.11 ▲
The lives of stars

Exercise

List useful websites where you can find information about black holes.

16.5 Light from the stars

Figure 16.8 (page 190) shows the general picture: very hot stars are blue and cool stars are red. The Sun is a yellow star and those a bit hotter than the Sun look white. There is, however, a quantitative link between the peak wavelength λ_m in the radiation spectrum emitted by a star and its surface temperature T. This is **Wien's law**:

$$\lambda_m T = 2.90 \times 10^{-3}\,\text{m K}$$

To 4 sig. fig. the constant is 2.898×10^{-3} m K (beware – the unit is metre-kelvin, *not* millikelvin, mK) but few constants are given to more than 3 sig. fig.

The law tells us that λ_m and T are inversely proportional and that *you*, at about 310 K, emit electromagnetic waves with a peak in the infrared!

Worked example

Explain, with an example, how a knowledge of λ_m for a given star can lead to a value for its luminosity L. State any assumption you make.

Answer

As λ_m is known, the surface temperature can be calculated from Wien's law. So, if $\lambda_m = 270$ nm, $T = (2.90 \times 10^{-3}\,\text{m K})/(270 \times 10^{-9}\,\text{m}) = 10\,700$ K.

On a Hertzsprung–Russell diagram, a star with a temperature of 10 700 K will have a luminosity about 100 times that of the Sun, i.e. about 4×10^{28} W.

The assumption is that the star lies on the main sequence.

Tip

Any 'calculation' that involves reading off values from T to L or L to T on an H–R diagram will only give results to 1 sig. fig., as the main sequence is a narrow band not a line.

The Stefan–Boltzmann relationship

You will be familiar with the electromagnetic spectrum from γ-rays and X-rays to microwaves and radio waves. A star emits a continuous spectrum for which the total power output, its luminosity L, is proportional not to its temperature T, but to T^4! Think what this means: as you move from 6000 K to 12 000 K, the luminosity of a star would increase by a factor $2^4 = 16$ times.

This does not match up with the $L-T$ relationship shown for main sequence stars on an H–R diagram – try it. This is because the high-luminosity stars on the H–R diagram are larger than stars like the Sun, and the Sun is larger than stars at the bottom right of the H–R diagram. The full relationship is:

$$L \propto 4\pi r^2 T^4$$

Putting a constant σ (sigma) into the proportional relationship gives the full **Stefan–Boltzmann law**:

$$L = 4\pi\sigma r^2 T^4$$

where $\sigma = 5.67 \times 10^{-8}\,\text{W m}^{-2}\,\text{K}^{-4}$.

We have given the luminosity of the Sun, 3.9×10^{26} W, but not the radius, so you could work backwards using this expression to find the radius, knowing that the surface temperature of the Sun is 5800 K. You should get 7.0×10^8 m.

The importance of this calculation is that it can be done for *any* star for which L and T are known, and so an estimate of the radius of any star on the main sequence can be made. A high-radius star on the main sequence will be a high-mass star. Figure 16.12 illustrates this.

Worked example

A main sequence star like the Sun becomes a red giant when it moves off the main sequence. Explain why the red giant star is more luminous than the Sun.

Answer

The radius of a red giant is much greater than that of the Sun, perhaps 100 times greater, so r^2 increases by $100^2 = 10\,000$.

But the red giant is cooler, perhaps half the temperature of the Sun, so T^4 changes by a factor of $(\frac{1}{2})^4 = \frac{1}{16}$.

Using $L \propto 4\pi r^2 T^4$, the overall luminosity of the red giant is $10\,000/16 = 625$ times as luminous as the Sun.

Tip

Beware of confusing two expressions for luminosity L. From the inverse-square law you can get $L = 4\pi d^2 F$ and from the Stefan–Boltzmann law $L = 4\pi\sigma r^2 T^4$. They 'look' a bit the same, but remember that d is the distance **to** a star and r is the radius **of** a star.

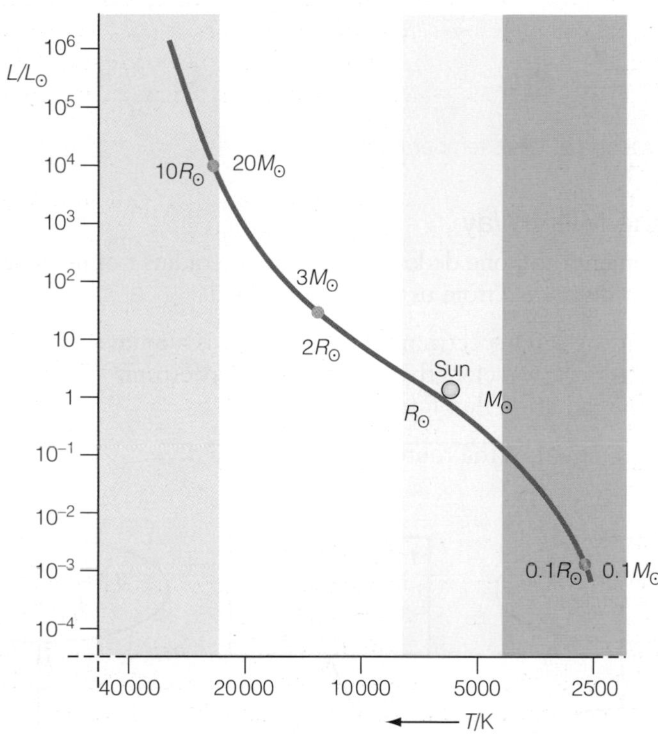

Figure 16.12 ▲
Masses and radii of some main sequence stars

Both Wien's law and the Stefan–Boltzmann law strictly apply only to what are known as **black body radiators**. In fact all stars behave as 'black bodies' – meaning they would absorb any electromagnetic radiation falling on them, but this is not really an issue in the physics of stars.

The three curves in Figure 16.13 show both proportional relationships:

- the $\lambda_m \propto 1/T$ of Wien's law
- the $L \propto T^4$ of the Stefan–Boltzmann law.

The first is quite easy to check. The second says that the total power output is proportional to T^4 and here the total power or luminosity is represented by the area under the curve. As the ratios of 4^4, 5^4 and 6^4 are 256, 625 and 1296, the 6000 K curve should cover about twice the area of the 5000 K curve, and the 5000 K curve should cover an area a bit more than twice the 4000 K curve. They do, more or less.

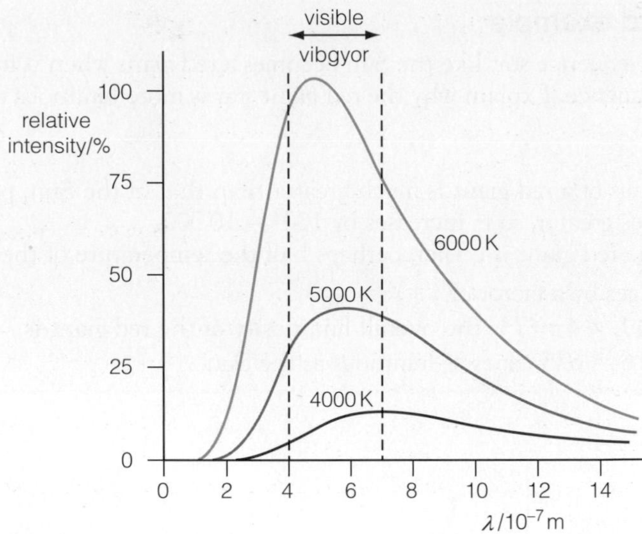

Figure 16.13 ▲
Black body spectra at three temperatures

Stars in the Milky Way

Two measurements and one deduction enable the radius r of main sequence stars *and* their distance d from us to be determined.

1 Ensure that the star's spectrum tells you that it is a main sequence star.
2 Measure the peak wavelength λ_m in the star's spectrum.
3 Measure the radiation flux F from the star.

Figure 16.14 summarises the route to finding r and d.

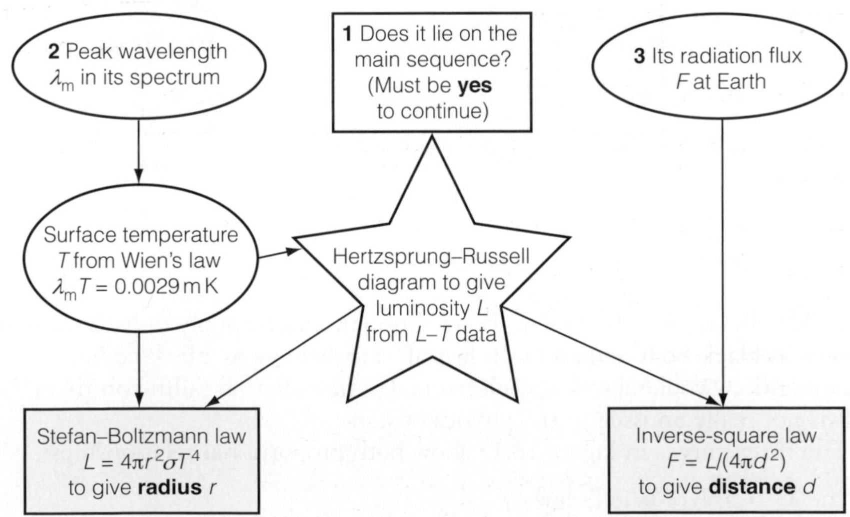

Figure 16.14 ▲
How to find the radius and the distance from Earth of a main sequence star

16.6 Why pay to study the stars?

In the UK all tax payers are contributing to research that has produced the details described in this chapter. Are they getting value for money? They also pay for research into the origins of the universe (Chapter 17), another area of what is called 'pure' research. When you repay your student loan, will you be happy for the government to spend some of this money on pure science?

It would be very easy to decide that the money is wasted, but perhaps the answer is to 'wait and see'. Research in astrophysics may surprise us all and turn up some really useful prizes. After all, when the laser was first developed, scientists themselves saw it as a solution searching for a problem!

Another response is to argue that any civilised society will and should spend money on grand schemes that do not appear to 'pay'. In the Middle Ages, for example, we built great temples – cathedrals and mosques – for religious purposes. The spirituality behind this and the natural curiosity that drives space exploration and astrophysical research are, it can be argued, what makes us human.

There is, of course, no clear and universally agreed answer to the question heading this short section. But it may be worth arguing about.

REVIEW QUESTIONS

1 4.1×10^7 light years is equivalent to:

A 1.3×10^{15} m B 1.6×10^{22} m
C 3.9×10^{23} m D 7.8×10^{24} m

2 Nearby stars might exhibit trigonometric parallax because:

A the Sun (our star) is moving around the Milky Way

B light from distant stars reaches Earth at different angles in March and September

C light from nearby stars reaches Earth at the same angle in September and March

D the Earth is moving around the Sun (our star)

3 The solar radiation flux is reduced by about a factor X between reaching the Earth's atmosphere and reaching the Earth's surface. A value for X to 2 sig. fig. might be:

A 0.95 B 0.85 C 0.75 D 0.65

4 A 'standard candle' in astrophysics is:

A a name for the power output or luminosity of a star

B an alternative name for a Cepheid variable star

C an alternative name for a supernova explosion

D a name for a star of predictable luminosity

5 Which of the following graphs shows the relationship between the radiation flux F at a distance d from a star?

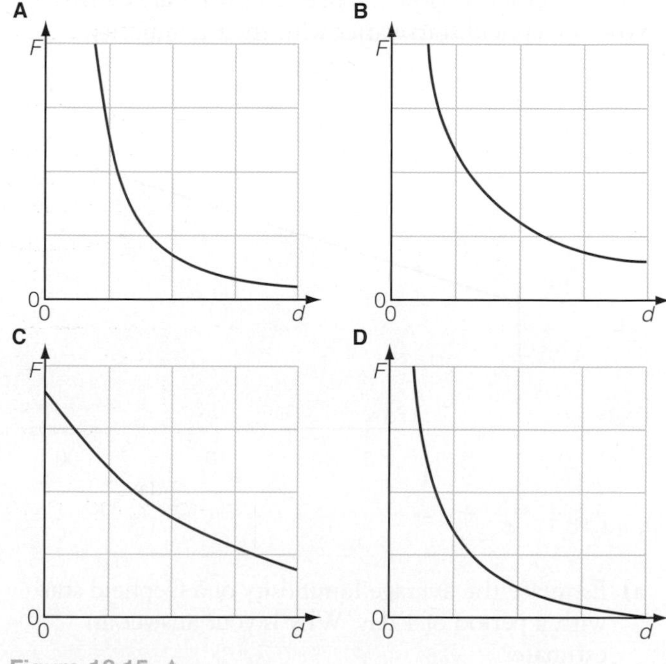

Figure 16.15 ▲

6 Convert 0.01 arcsec (seconds of arc) to degrees and to radians.

7 Figure 16.16 shows how the distance to a nearby star X can be determined using trigonometric parallax. E_1 and E_2 are the positions of the Earth in its orbit around the Sun S in March and September, i.e. six months apart.

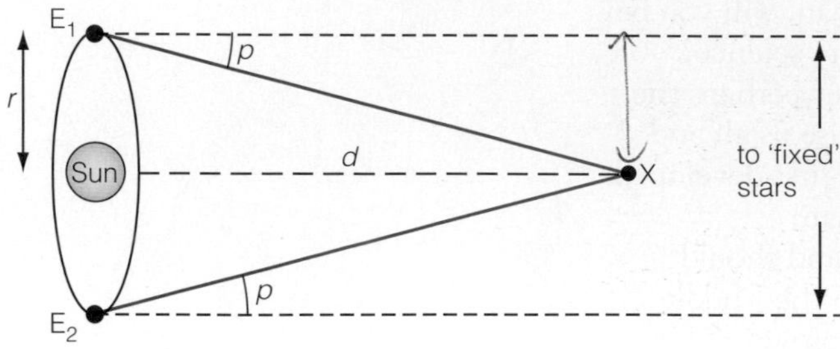

Figure 16.16 ▲

a) How are the radius r of the Earth's orbit round the Sun, the measured parallax angle p and the distance from the Sun to the star d related?

b) Taking r to be 1.5×10^{11} m, calculate

i) the distance to a star for which the parallax angle is found to be 4.5×10^{-5} degrees (0.16 seconds of arc),

ii) the parallax angle in degrees and arcsec for a star that is 240 light years away.

8 Establish the base SI unit for radiation flux.

9 The bright star Rigel has a luminosity of 3.8×10^{31} W. The intensity of radiation from Rigel measured on Earth – its radiation flux – is $5.4 \times 10^{-8}\,\mathrm{W\,m^{-2}}$. Calculate the distance from Earth to Rigel.

10 The graph shows how the period of what are called type II Cepheid stars varies with their luminosity.

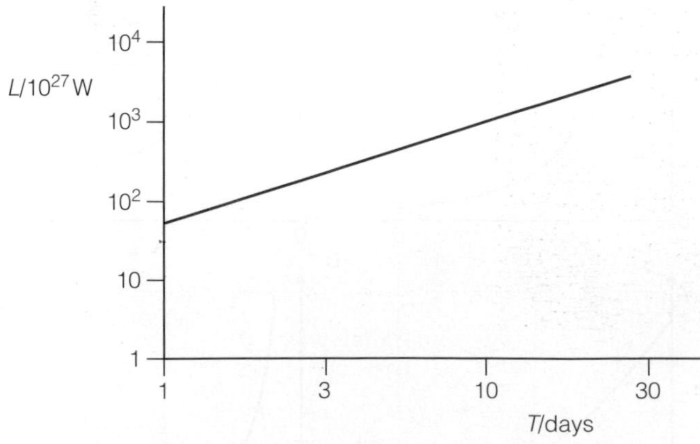

Figure 16.17 ▲

a) Estimate the average luminosity of a Cepheid star with a period of 1 day. Why is your answer an estimate?

b) The star in part a) is known to lie 220 light years from Earth. Calculate its average radiation flux as measured on Earth.

11 A neutron star has a density of $4 \times 10^{17}\,\mathrm{kg\,m^{-3}}$. What would be the approximate mass of a piece of neutron star the size of a grain of rice?

12 The path SXY in Figure 16.18 describes the movement of the Sun from the time of its arrival on the main sequence PQ to its becoming a white dwarf at Y.

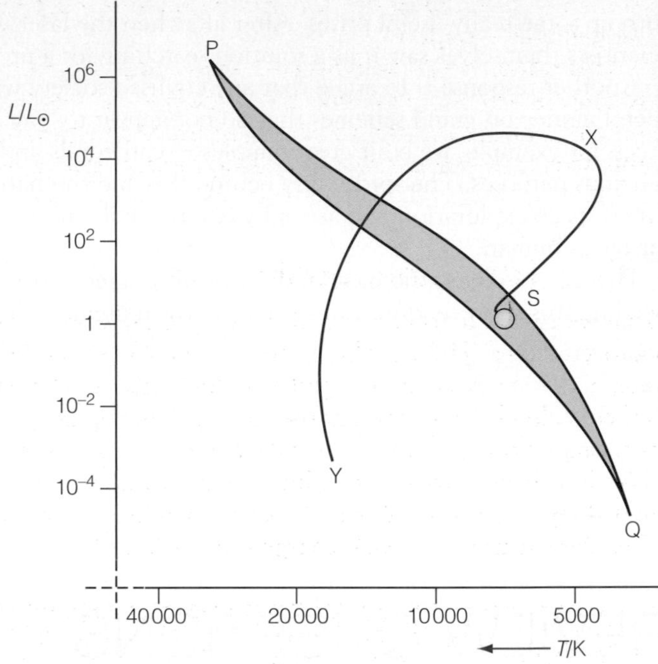

Figure 16.18 ▲

Use the diagram to write a description of the Sun's temperature and luminosity on its journey from S to Y. (Do not attempt to put a timescale on the journey.)

13 Sketch a graph describing the relation between the peak wavelength λ_m in a star's spectrum and its surface temperature T, for values of T from 0 to 10 000 K.

Give the value for λ_m in nm when $T = 5.0$ kK.

In questions 14 and 15 take $L_\odot$ to be 3.9×10^{26} W.

14 A white dwarf star has a luminosity of $0.002L_\odot$ and a surface temperature of 20 kK. Calculate its radius.

15 A main sequence star is observed to have a maximum wavelength λ_m at 440 nm in its spectrum. On Earth its radiation flux F is measured to be $5.8 \times 10^{-10}\,\mathrm{W\,m^{-2}}$.

a) Calculate its distance d from Earth in light years.

b) Explain the least reliable step in your calculation.

17 Cosmology

The 20th century saw some astonishing developments in physics. In 1900 no-one knew how the Sun kept on pumping out energy – the radiation that supports life on Earth. By 2000 physicists could explain fully not only these energy processes in the Sun, but the origin, life and death of stars of all sizes. They were also investigating, at laboratories like that at CERN, the earliest moments of what became known as the 'Big Bang', when our universe – perhaps one of many multiverses – began.

This chapter will develop your knowledge of energy in nuclear reactions – including fusion and fission – and will look deeper into the universe and consider its ultimate fate. Many questions – especially the 'why' questions – will, however, remain unanswered.

17.1 How stars begin

What we call 'outer space' contains a very tenuous gas consisting mainly of hydrogen atoms. There is also a little helium and some tiny dust particles. In some places the gas clumps together to form clouds. The particles and atoms in such a cloud attract one another gravitationally, the rate of contraction increasing as the size of the cloud decreases and its density increases.

We now believe that this 'ordinary' matter (baryons and mesons made from quarks) represents only about 5% of the total mass of the universe. The rest is called **dark matter** and **dark energy** ($\Delta m = \Delta E/c^2$).

Tip

Discovering more about dark matter and dark energy shows how scientists are driven to explore the unknown. As Matt Ridley puts it in his book *Genome*: 'The fuel on which science runs is ignorance'.

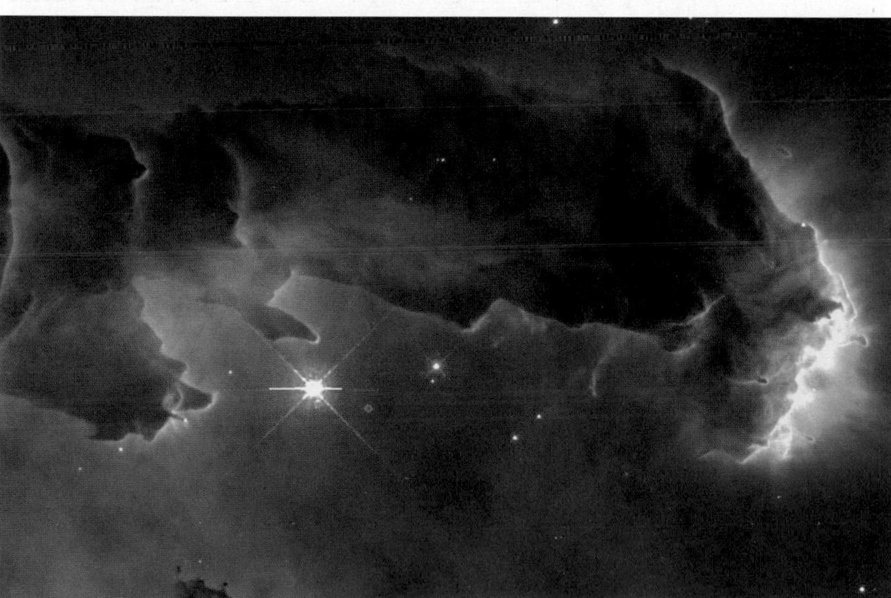

Figure 17.1 ▲
A cloud of matter in deep space, about one-third of a light year across.

Figure 17.1 shows part of a huge cloud, on the right hand edge of which can be seen 'fingers' of matter each about the size of our solar system! Each of these dense blobs will collapse to form a star. As it collapses it loses gravitational potential energy, energy that is transferred to random kinetic energy of the particles. The particles collide more and more violently. The gas gets hotter until ionised hydrogen atoms – protons – undergo **nuclear fusion** at very high temperatures.

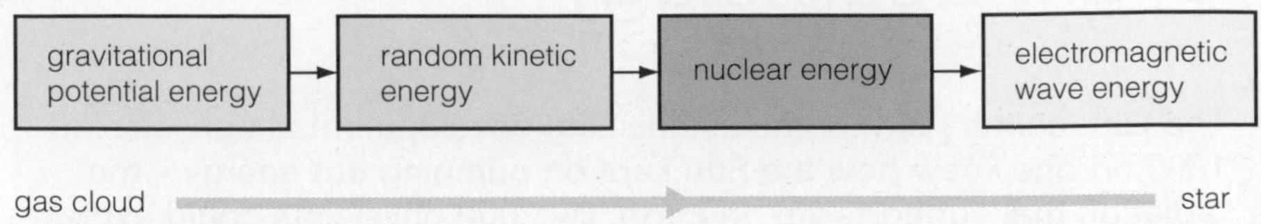

Figure 17.2 ▲
Stages of star formation

The fusion process inside our Sun involves a series of steps that convert four hydrogen nuclei ($^1_1$H) to make one helium nucleus ($^4_2$He). Figure 17.3 shows the steps involved, and the Worked example below develops the nuclear equations.

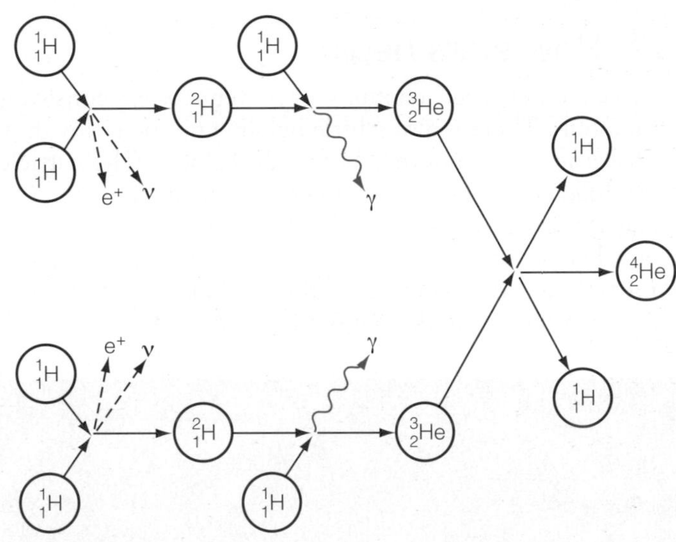

Figure 17.3 ▲
Nuclear reactions in the Sun

Worked example

Write the nuclear equations of the reactions shown in Figure 17.3.

Answer

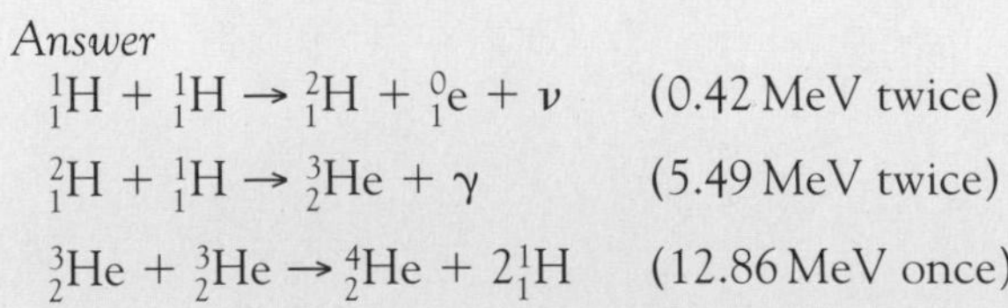

$^1_1\text{H} + ^1_1\text{H} \rightarrow ^2_1\text{H} + ^0_1\text{e} + \nu$ (0.42 MeV twice)

$^2_1\text{H} + ^1_1\text{H} \rightarrow ^3_2\text{He} + \gamma$ (5.49 MeV twice)

$^3_2\text{He} + ^3_2\text{He} \rightarrow ^4_2\text{He} + 2^1_1\text{H}$ (12.86 MeV once)

The net result of this process is:

$$4^1_1\text{H} \rightarrow ^4_2\text{He} + 2^0_1\text{e} + 2\nu + 2\gamma \quad (24.68\,\text{MeV})$$

The two positrons produced annihilate with two electrons from the surrounding plasma, forming four more gamma photons (1.02 MeV twice) – see page 84. The overall result of this **nuclear synthesis** (or nucleosynthesis), which occurs in the Sun and in all main sequence stars, is that hydrogen atoms fuse to form helium. At the same time lots of (gamma) photons and neutrinos are produced. The neutrinos pour rapidly out of the Sun; tens of millions pass through you every second! The photons take thousands of years to 'fight' their way to the surface of the Sun, but then escape into space as visible and near-visible photons at the speed of light.

Figure 17.4 ▲
Neutrinos are unreactive, so pass through matter, even you!

17.2 Nuclear binding energy

To learn more about energy conservation in these fusion processes, we need to use the relationship $\Delta E = c^2 \Delta m$ and/or the equivalences:
$1\,u = 1.66 \times 10^{-27}\,kg \equiv 930\,MeV$, from Table 9.1 (page 83).

The masses of a proton (m_p), a neutron (m_n) and a helium nucleus (^4_2He) or α-particle (m_α) are:

$$m_p = 1.0073\,u = 1.673 \times 10^{-27}\,kg$$
$$m_n = 1.0087\,u = 1.675 \times 10^{-27}\,kg$$
$$m_\alpha = 4.0015\,u = 6.645 \times 10^{-27}\,kg$$

so you can see that m_α is *not* equal to $2m_p + 2m_n$, but less. The difference $(2m_p + 2m_n - m_\alpha) = \Delta m = 0.0305\,u$ or $0.051 \times 10^{-27}\,kg$.

This mass loss or **mass deficit** must be something to do with energy. The energy equivalent of 0.0305 u is:

$$0.0305\,u \times 930\,MeV\,u^{-1} = 28\,MeV$$

(Or you can use $\Delta E = c^2\Delta m$ and convert joules to get eV.)

This is the energy that needs to be given to an α-particle, or helium nucleus, to break it up into its four components: two protons and two neutrons. We can write this as:

$$^4_2He + \Delta E \rightarrow 2\,^1_1p + 2\,^1_0n$$

Running this equation backwards, this means that if you can join two protons and two neutrons together, you can create an α-particle and have **lots of energy** to spare.

As all nuclei are built from protons and neutrons, you can similarly add up the total mass of the building blocks and compare this total with the mass of the resulting nucleus. In every case the nucleus is lighter than the constituent protons and neutrons. Figure 17.5 illustrates this idea for carbon-12.

The missing mass has become nuclear energy or **nuclear binding energy**, and, as in the example above for an α-particle, you need that much energy if you want to tear the nucleus apart. For carbon-12 the mass deficit, is $1.61 \times 10^{-28}\,kg$, so the binding energy is:

$$(1.61 \times 10^{-28}\,kg)(3.00 \times 10^8\,m\,s^{-1})^2 = 1.45 \times 10^{-11}\,J \text{ (about 90 MeV)}$$

Table 17.1 lists some atomic rest masses.

Worked example

Use values from above and from Table 17.1 to calculate:

a) the binding energy B of the ^{56}Fe nucleus,

b) the binding energy per nucleon B/A for ^{56}Fe.

Answer

a) Iron-56 has 26 protons and 30 neutrons. Using atomic masses,

$$B = 26 \times 1.0078\,u + 30 \times 1.0087\,u - 55.9349\,u$$
$$= 0.5289\,u \text{ or } 492\,MeV$$

b) B/A for this isotope of iron is therefore $492\,MeV \div 56 = 8.79\,MeV$ per nucleon.

Note

Remember that an α-particle is made up of two protons and two neutrons.

Figure 17.5 ▲
Where has the mass gone?

Nucleus	Symbol	Mass
hydrogen-1	^1_1H	1.0078 u
helium-4	^4_2He	4.0026 u
carbon-12	$^{12}_6C$	12.0000 u
iron-56	$^{56}_{26}Fe$	55.9349 u
uranium-235	$^{235}_{92}U$	235.0439 u
uranium-238	$^{238}_{92}U$	238.0508 u

Table 17.1 ▲

Tip

Table 17.1 gives the atomic mass of $^{56}_{26}Fe$, so the mass of ^1_1H must be used here.

Figure 17.6 shows how the binding energy per nucleon, B/A, varies with nucleon number, A, for different elements. The main feature of the plot is that B/A peaks at or near iron-56, and this means that this isotope of iron is the most stable nucleus, i.e. the one requiring most energy per nucleon to tear it apart.

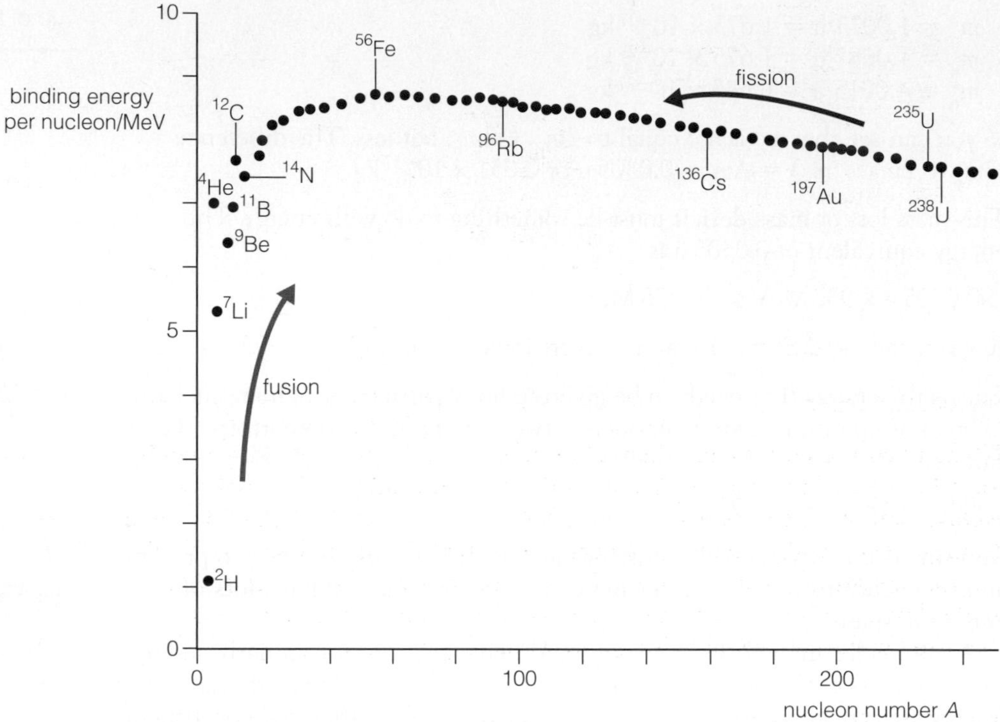

Figure 17.6 ▲
Variation of binding energy per nucleon

Figure 17.6 has much more information to offer:

1 If a nucleus with a very high mass number (A = 230 to 250) can be made to break up into smaller nuclei, there will be a large release of energy.
2 If four very light nuclei (e.g. the $4\,^{1}_{1}H$ in Section 17.1) can be made to join together there will be a very large release of energy.

Both these processes are possible; they are called **fission** and **fusion** respectively.

17.3 Uranium fission

Nuclei that have high A are rich in neutrons. When an extra neutron is absorbed, they can immediately break up into two fragments. A common result of such fission is that, as well as the two fragments, two or three neutrons are released. For example:

$$^{235}_{92}U + ^{1}_{0}n \rightarrow \text{fission fragments} + 2 \text{ or } 3 \text{ neutrons} + \text{energy}$$

One likely outcome of the fission of uranium-235, in which the two fission fragments together have a kinetic energy of 168 MeV, is:

$$^{235}_{92}U + ^{1}_{0}n \rightarrow ^{144}_{56}Ba + ^{89}_{36}Kr + 3^{1}_{0}n$$

Figure 17.7 illustrates this.

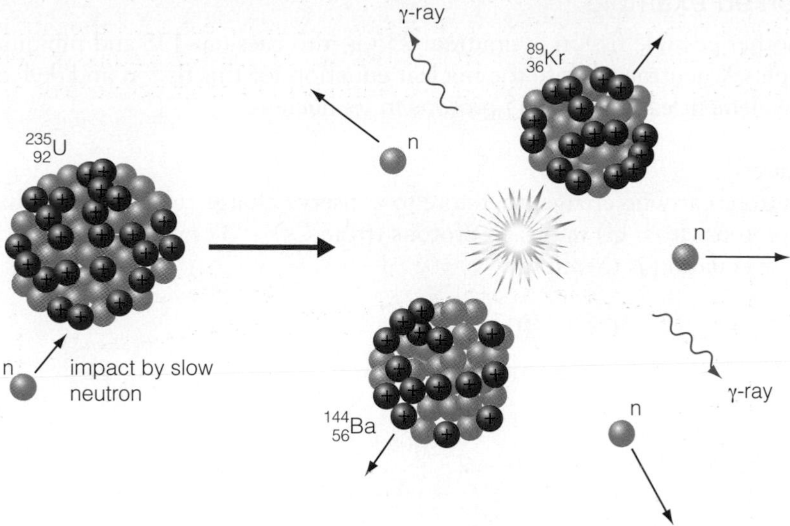

Figure 17.7 ▲
Uranium-235 fission

When a neutron enters a uranium-235 nucleus, and it splits into a barium-144 and a krypton-89 nucleus, the graph in Figure 17.6 would suggest that the average B/A of the fragments is about 1 MeV greater than the B/A of uranium. A single fission will therefore produce over 200 MeV – more than 2×10^8 eV. You can see that with an average energy of only 10–20 eV available from a single *chemical* reaction, the energy available from uranium fission is vast compared with that available from burning coal, gas or oil.

Worked example

Assuming 200 MeV per fission, calculate the number of fission events occurring each second in a nuclear reactor whose thermal power output is 2400 MW. Comment on your answer.

Answer

$$200\,\text{MeV} \equiv (200 \times 10^6\,\text{eV})(1.6 \times 10^{-19}\,\text{J}\,\text{eV}^{-1}) = 3.2 \times 10^{-11}\,\text{J}$$

So from each fission there is 3.2×10^{-11} J of energy.

$$2400\,\text{MW} \equiv 2.4 \times 10^9\,\text{J}\,\text{s}^{-1}$$

∴ The number of fission events per second is

$$\frac{2.4 \times 10^9\,\text{J}\,\text{s}^{-1}}{3.2 \times 10^{-11}\,\text{J}} = 7.5 \times 10^{19}\,\text{s}^{-1}$$

As the power station will be less than 50% efficient, the real number of fission events will be at least twice this, perhaps $20 \times 10^{19}\,\text{s}^{-1}$.

This looks to be a very large number, but 1 kg of uranium will contain more than 10^{24} atoms, and a nuclear reactor will contain several tonnes of uranium – hence enough fuel for a year. By comparison, a 2400 MW coal- or gas-burning power station needs nearly 10 tonnes of coal or gas *per day*!

Worked example

Another possible fission of uranium-235 is into caesium-138 and rubidium-96 plus X neutrons. Write the nuclear equation for this fission and deduce X. The element caesium has 55 protons in its nucleus.

Answer

Neutrons carry no charge, therefore to conserve charge rubidium must have 92 protons (from U) minus 55 protons (from Cs) = 37 protons in its nucleus.

The equation is therefore

$$^{235}_{92}U + ^{1}_{0}n \rightarrow ^{138}_{55}Cs + ^{96}_{37}Rb + X\ ^{1}_{0}n$$

and hence

$$X = 235 + 1 - 138 - 96 = 2$$

Splitting the atom

The first success in the race to 'split the atom' took place in Cambridge in 1932. Two researchers, John Cockcroft and Ernest Walton, working under the direction of Lord Rutherford (all three are shown in Figure 17.8), fired a proton of a few hundred keV to strike a lithium nucleus ($^{7}_{3}Li$). The remarkable and unexpected result was that the lithium nucleus 'disappeared' and two α-particles were detected. The energy of the two resulting α-particles was estimated by the length of their tracks in a cloud chamber.

This process does not produce energy, indeed it requires the kinetic energy of the proton to get it to work, but it led the way to the fission we use today.

Figure 17.8 ▲
Walton, Rutherford and Cockcroft

Fission in war and peace

Two neutrons result from each fission in the U → Cs + Rb reaction, and three neutrons result from each fission in the U → Ba + Kr reaction. There are other ways for the uranium-235 to break up after swallowing a neutron, but the average number of neutrons per fission is about 2.5. These neutrons can then cause other nuclei to break up, which can trigger the fission of an even larger number of nuclei, and so on. This **chain reaction** is illustrated in Figure 17.9.

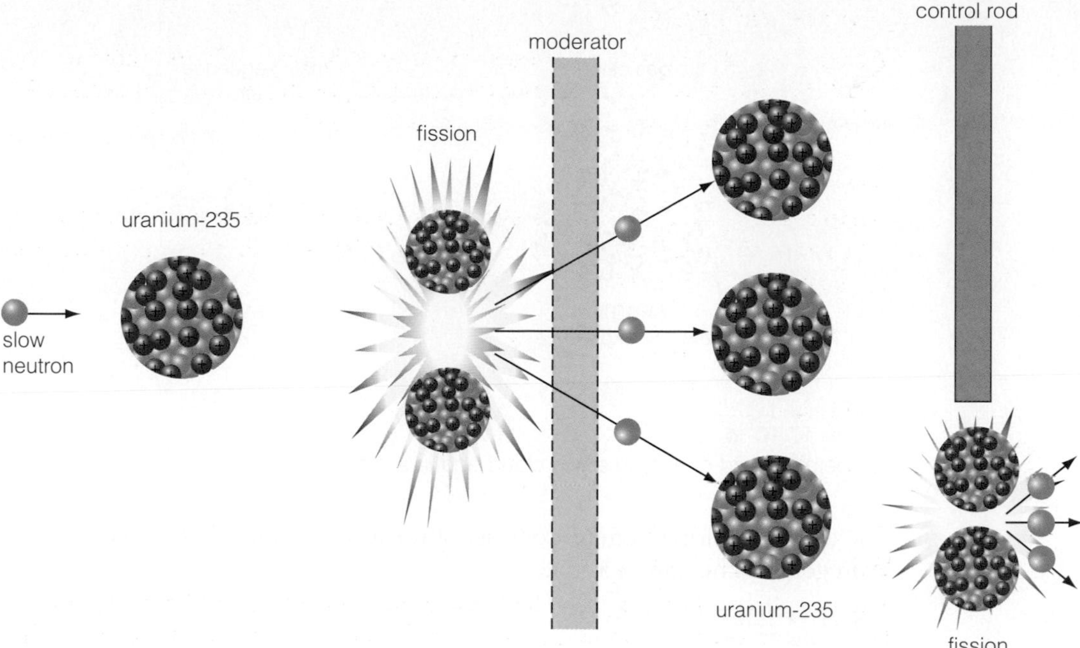

Figure 17.9 ▲
Useful chain reaction

The neutrons resulting from such fissions have a lot of energy. It is necessary to slow them down before they can induce further fissions, and this is shown by the presence in Figure 17.9 of a 'moderator' such as graphite.

When a chain reaction such as this is uncontrolled, you have an explosive situation – a **nuclear bomb** (or atomic bomb as it is sometimes called). Fortunately the world has only twice seen the use of an atomic bomb in war, both in Japan in 1945.

If, using control rods to absorb some of the neutrons, on average only one of these neutrons induces a new fission, the outcome is a useful, controlled **nuclear reactor** for use, for example, in an electricity generating power station.

There are two main problems with nuclear power generation:

- Firstly, mined uranium is 99.28% U-238 and only 0.72% U-235, and the U-238 absorbs neutrons but does not undergo fission. This means that the percentage of U-235 must be increased for use in power generation – a difficult 'enrichment' process that cannot be achieved chemically.
- Secondly, the U-238 reaction is ${}^{238}_{92}\text{U} + {}^{1}_{0}\text{n} \rightarrow {}^{239}_{92}\text{U}$, and the ${}^{239}_{92}\text{U}$ then decays naturally to ${}^{239}_{94}\text{Pu}$ after two beta decays. This isotope of plutonium has a half-life of 24 000 years, and poses a severe problem when it comes to disposing of the radioactive waste – spent fuel – from nuclear power stations.

17.4 Stellar fusion

The fusion process that generates energy in stars was described in the first section of this chapter. In order for fusion to occur, the temperature T and density ρ of the hydrogen atoms needs to be extremely high. This is because two positively charged protons will repel each other with a force that increases as they get closer. You have met this Coulomb force in Chapter 5 (page 36). Figure 17.10 gives some 'order of magnitude' values for T and ρ for fusion to occur. A star is a continuously exploding hydrogen bomb!

Definition

The Coulomb force is

$F = \frac{kQ_1Q_2}{r^2} = \frac{ke^2}{r^2}$ for two protons

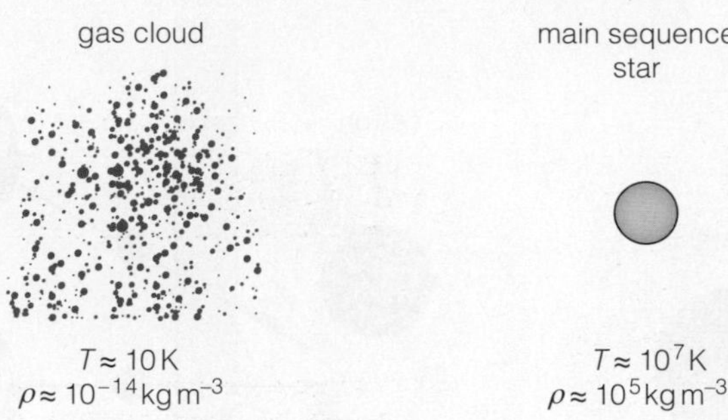

Figure 17.10 ▲
Temperature and density in a gas cloud and in a main sequence star

The energy resulting from just one set of nuclear reactions that convert hydrogen into helium in a star:

$$4{}^1_1H \rightarrow {}^4_2He + 2\nu + 6\gamma$$

is about 27 MeV. This is not as great as the result of one uranium fission (about 200 MeV), but in the Sun there are, to an order of magnitude, 10^{38} of these reactions occurring in the core *every second*!

Worked example

Show that $27 \times 10^{38}\,MeV\,s^{-1}$ is approximately equal to $L_\odot$, the output power of the Sun.

Answer

$$27 \times 10^{38}\,MeV\,s^{-1} = (27 \times 10^{38})(1.6 \times 10^{-13}\,J\,s^{-1}) \approx 4 \times 10^{26}\,W$$

which agrees with the power output of the Sun ($L_\odot = 3.9 \times 10^{26}\,W$).

Fusion on Earth

The hot dense material involved in stellar fusion is held together in space by gravitational forces. On Earth, magnetic fields are used to 'hold' the hot plasma consisting of ionised hydrogen atoms away from the containing walls – see Figure 11.6 on page 109. But controlled fusion has so far only been achieved for very short periods, taking more energy to heat the plasma than can be extracted. The possibility of useful nuclear fusion reactors is still a long way off. Hydrogen or 'thermonuclear' bombs, by contrast, have been with us for over 50 years, but thankfully have never been used in warfare.

17.5 The expanding universe

Stars have a gaseous 'atmosphere' around them that contains atoms of the same elements as the star itself. The atoms under high pressure in the outer layers of the star emit a **continuous spectrum** of electromagnetic radiation (remember, stars act as black bodies). Atoms in the low-pressure atmosphere selectively absorb photons at wavelengths that excite their electrons to a higher energy level. You will have met the idea of atomic energy levels in your AS course. When the electrons drop back to the lower energy level the resulting photons are emitted as an **emission spectrum** *in all directions*, but the emission spectrum is not observed by us. What is seen is the continuous spectrum crossed by dark lines that match the atoms' emission spectrum: it is called an **absorption spectrum**. Figure 17.11 illustrates this.

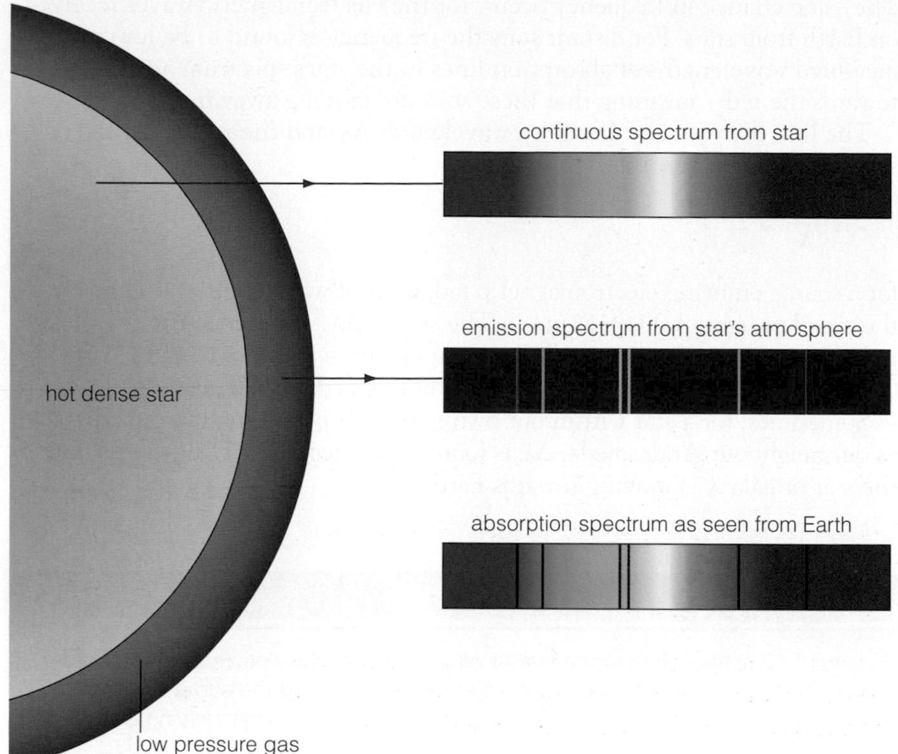

Figure 17.11 ◀
What we see from Earth

The Doppler shift

In your AS course you will have met the idea that the frequency of sound from a source moving towards or away from you changes. This is the Doppler effect. Figure 17.12 shows how the wavefronts become squashed or spread out.

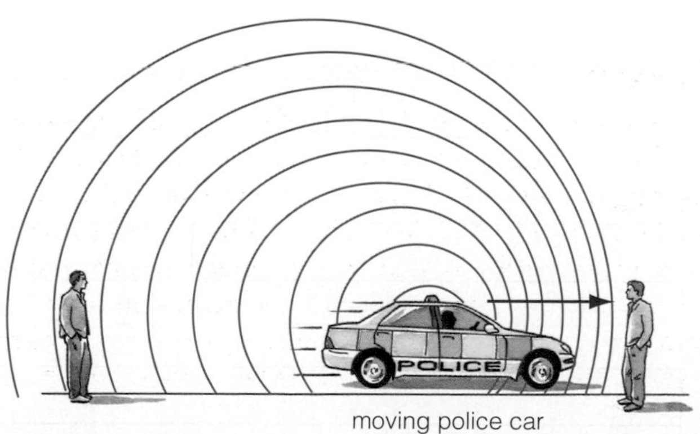

Figure 17.12 ◀
Doppler change of wavelength

Worked example

Describe how an understanding of the Doppler effect can explain what you hear when a police car passes you at high speed with its siren blaring.

Answer

- As the car approaches you, the frequency of the note you hear from the siren is higher than the frequency f_0 heard by the police in the car.
- As the car passes and moves away from you this frequency becomes lower than f_0.
- Therefore the effect is to hear a note that drops in frequency as the police car passes you.

Tip

A full answer to this question would be supported by a diagram similar to Figure 17.12.

The same change in frequency occurs for the electromagnetic waves received on Earth from stars. For distant stars the frequency is found to be lower – the measured wavelengths of absorption lines in the star's spectrum are moved towards the red – meaning that these stars are moving away from us.

The link between the *change* in wavelength $\Delta\lambda$ and the relative speed of source and observer v is:

$$z = \frac{\Delta\lambda}{\lambda} \approx \frac{v}{c}$$

for a source emitting electromagnetic radiation of wavelength λ and where c is the speed of light, $3.00 \times 10^8\,\mathrm{m\,s^{-1}}$. The size of $\Delta\lambda$, and hence the size of the **red shift** z, is usually very small. For example, for a speed of $60\,\mathrm{km\,s^{-1}}$, $z = 0.0002$ and $\Delta\lambda = 0.1\,\mathrm{nm}$ for an absorption line at $\lambda = 500\,\mathrm{nm}$.

Sometimes, for a star within our own galaxy, or occasionally a galaxy such as our neighbour Andromeda, $\Delta\lambda$ is found to be negative. This tells us that the star or galaxy is moving towards Earth.

Exercise

Figure 17.13 shows diagrammatically four well known absorption spectra, B–E. Above them is a laboratory emission spectrum, A, showing the same two lines. The absorption spectra are from four distant galaxies, which are moving away from the Earth.

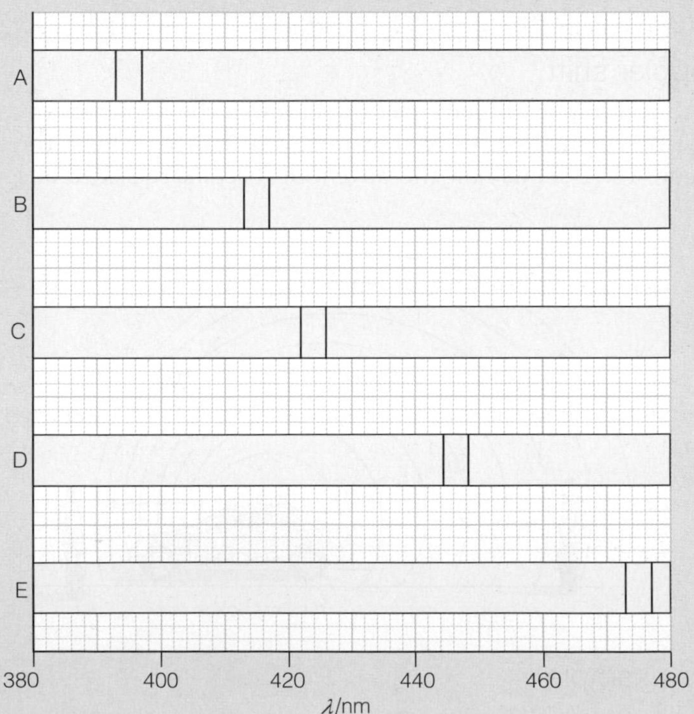

Figure 17.13 ▲

Copy and complete Table 17.2 to show:

1 The wavelengths λ of the left-hand line in each spectrum.
2 The change $\Delta\lambda$ in nanometres of the left-hand line in each absorption spectrum.
3 The red shift $z = \Delta\lambda/\lambda$ for the left-hand line in each absorption spectrum.
4 The speed in $\mathrm{km\,s^{-1}}$ to 2 sig. fig. at which each galaxy is moving away from Earth.

(C has been done for you.)

	A	B	C	D	E
λ/nm	393		422		
$\Delta\lambda$	–		29		
$z = \Delta\lambda/\lambda$	–		0.0738		
v/km s^{-1}	–		22 000		

Table 17.2 ▲

The distances *d* in metres to these galaxies are given in Table 17.3.

	A	B	C	D	E
$d/10^{24}$ m	–	6.0	9.7	17	27

Table 17.3 ▲

Plot a graph of *v* (up) against *d* (along) and calculate the gradient of this straight-line graph (it should come out at about $2 \times 10^{-18}\,s^{-1}$).

The Hubble constant

Uncertainty in the measured values of *d* in the above Exercise means that the gradient of your graph is not reliable to better than ±10%. However, it is clear that *v* is proportional to *d*. This fact is known as **Hubble's law** after Edwin Hubble, an American astronomer working in the 1920s and 1930s.

The constant linking *v* and *d* is called the Hubble constant. It has a value $H_0 \approx 2.3 \times 10^{-18}\,s^{-1}$.

$$v = H_0 d \qquad \text{and} \qquad z = \frac{v}{c} = \frac{H_0 d}{c}$$

Figure 16.7 on page 189 summarises methods for measuring *d*. (Note that for these to work as *d* gets bigger and bigger there must be overlap between adjacent methods.) We now have a method of finding *d* for any distant galaxy, provided we can measure its red shift, the value of *z*. We will need to assume that the straight line in the graph in Figure 17.14 (page 208) remains linear as the red shift becomes bigger and bigger, i.e. that Hubble's law holds universally.

Worked example

How far away is a galaxy that shows a red shift of $z = 0.20$?

Answer

$$z = \frac{v}{c} = \frac{H_0 d}{c} \Rightarrow d = \frac{zc}{H_0}$$

$$\therefore d = \frac{(0.20)(3.0 \times 10^8\,m\,s^{-1})}{2.3 \times 10^{-18}\,s^{-1}}$$

$$= 2.6 \times 10^{25}\,m$$

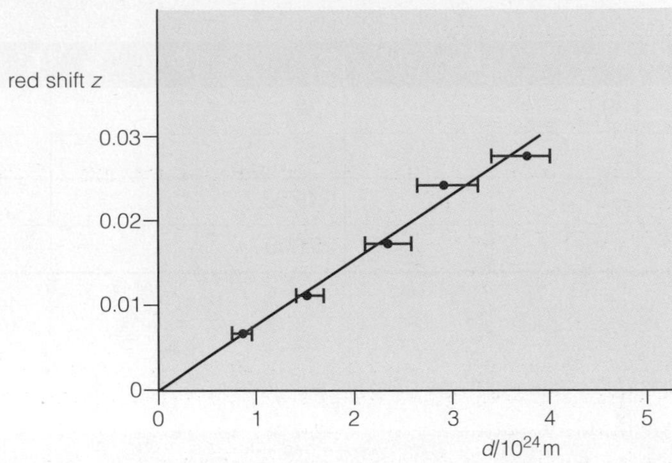

Figure 17.14 ▲

Physicists believe that the universe started with a **Big Bang** about 14 billion years ago, that is 14 thousand million years ago. Evidence for such a Big Bang depends on the observed expansion of the universe as established by Hubble's law, on predictions such as the existence of a 'cosmic background' microwave radiation and of the proportion of helium to hydrogen in the early universe (about one He atom for every four H atoms).

When cosmologists (people who study the origin and the nature of our universe) talk of the universe expanding, they do not mean that matter flew outwards into space in all directions. They prefer to think of space itself expanding. Think of all the galaxies painted at random on the surface of a part blown-up balloon – Figure 17.15. Now think of time passing as the balloon is gradually blown up more and more. You can try it for yourself.

Figure 17.15 ▲

As the figure shows, the distance between any two galaxies increases. A red shift would be measured by an observer in *any* galaxy and the value of the Hubble constant would be the same everywhere. You may think that the surface of a balloon does not seem like a three-dimensional universe. In fact the model is quite good although, like all analogies, limited in usefulness. You should try to keep in mind both a Big Bang and expanding space when thinking about our universe.

Exercise

Put 'Big Bang' or 'cosmic background microwave radiation' or 'cosmological red shift' into a search engine if you want to learn more about the universe from its start until now.

Worked example

Rearranging Hubble's law gives

$$\frac{1}{H_0} = \frac{\text{distance to a galaxy}}{\text{its speed of recession}}$$

It is suggested that this reciprocal of the Hubble constant tells us how long has elapsed since the Big Bang.

a) Discuss the ideas behind this suggestion.

b) Calculate $1/H_0$ and comment on its value.

Answer

a) Suppose the universe was created a time t_u ago. For a galaxy that has been moving away from us at a steady rate v for a time t_u, its distance d from us will now be vt_u.

Hubble's law tells us that $v = H_0d$, so substituting for d gives us $v = H_0vt_u$ or simply $H_0t_u = 1$ and hence that the age of the universe is $t_u = 1/H_0$.

b) $$\frac{1}{H_0} = \frac{1}{2.3 \times 10^{-18}\,\text{s}^{-1}} = 4.35 \times 10^{17}\,\text{s}$$

$$= 1.38 \times 10^{10}\,\text{years or 14 billion years to 2 sig. fig.}$$

In practice gravitational forces will mean that the present rate of expansion is less than that in the past, and hence $t_u < 1/H_0$.

17.6 How will the universe end?

What the fate of the universe will be depends on the average density of matter it now contains. Figure 17.16 shows some possible pasts and futures. With a high density the universe implodes into a Big Crunch (don't panic – this would be billions of years away!); with a small density it will go on expanding forever. The dashed line represents what might have happened had gravity not interfered with the expansion. Notice that the reciprocal of the Hubble constant is bigger than the age of the universe suggested by both 'open' and 'closed' scenarios.

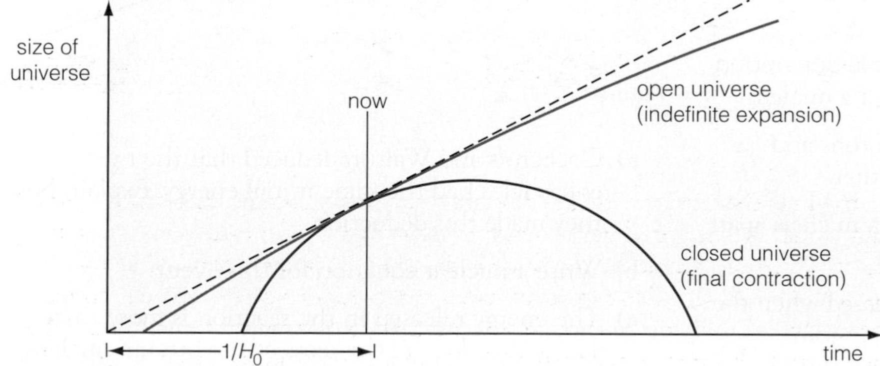

Figure 17.16 ▲
Towards a Big Crunch?

When cosmologists try to estimate the average density of matter in the universe, they discover problems. Spiral galaxies are spinning as if they contain much 'more stuff with mass' than the luminosity of the galaxy would predict. The expansion of the universe appears to be slower than predicted by the gravitational effect of observable mass. Cosmologists estimate that only about 5% of the mass of the universe consists of atoms as we know them. What is the other 90%? We don't know, so we call it **dark matter** and **dark energy**. Unobserved neutrinos might contribute to some of this mass. There are lots of them about – see Figure 17.4 on page 198.

One recent intriguing suggestion is that dark matter and dark energy plus ordinary matter together make the average density of the universe just right to end up as neither an open nor a closed universe. Such a 'flat' universe, one that stops expanding after an infinite time, satisfies many cosmologists' gut feelings. Perhaps you will live to see these mysteries resolved.

REVIEW QUESTIONS

1 In the fission reaction

$$^{235}_{92}U + ^{1}_{0}n \rightarrow ^{144}_{56}Ba + ^{89}_{36}Kr + 3^{1}_{0}n$$

the number of neutrons in Ba-56 and Kr-89 are respectively:

A 88 and 53 B 89 and 53
C 143 and 88 D 144 and 89

2 A galaxy shows a red shift of 4.0%.

a) Its speed of recession is:

A $7.5 \times 10^{9}\,m\,s^{-1}$ B $7.5 \times 10^{7}\,m\,s^{-1}$
C $1.2 \times 10^{8}\,m\,s^{-1}$ D $1.2 \times 10^{7}\,m\,s^{-1}$

b) Its distance from Earth is:

A $5.2 \times 10^{24}\,m$ B $5.2 \times 10^{22}\,m$
C $5.5 \times 10^{9}\,ly$ D $5.5 \times 10^{7}\,ly$

3 During fission a uranium nucleus splits into two parts of roughly equal mass. The ratio of the radius r of each fission fragment to the radius R of the uranium nucleus R is about:

A 0.8 B 0.7 C 0.6 D 0.5

4 Which of the following is **not** a reasonable description of what is meant by the binding energy of a nucleus?

A It is the energy released when the protons and neutrons forming a nucleus join together.

B It is the energy it would take to tear a nucleus apart into its protons and neutrons.

C It is the electric potential energy released when the protons and neutrons forming a nucleus join together.

D It is the work done in separating the nucleus into its constituent protons and neutrons.

5 The nuclei of the first three isotopes of hydrogen are:

$^{1}_{1}H$ with a mass of 1.007 28 u

$^{2}_{1}H$ with a mass of 2.013 55 u

$^{3}_{1}H$ with a mass of 3.015 50 u

a) Describe the structure of these three nuclei.

b) Calculate the mass deficit in the nuclear reaction

$$^{2}_{1}H + ^{2}_{1}H \rightarrow ^{3}_{1}H + ^{1}_{1}H$$

c) How much energy, in MeV, is released in such a reaction?

6 How would you explain to a friend who was not studying physics what is a meant by 'nuclear energy'?

7 Estimate the size of the little 'fingers' in the contracting cloud of Figure 17.1 (page 197).

8 Figure 17.17 represents a bubble chamber photograph showing the first successful 'splitting the atom' experiment in 1932 (see page 202).

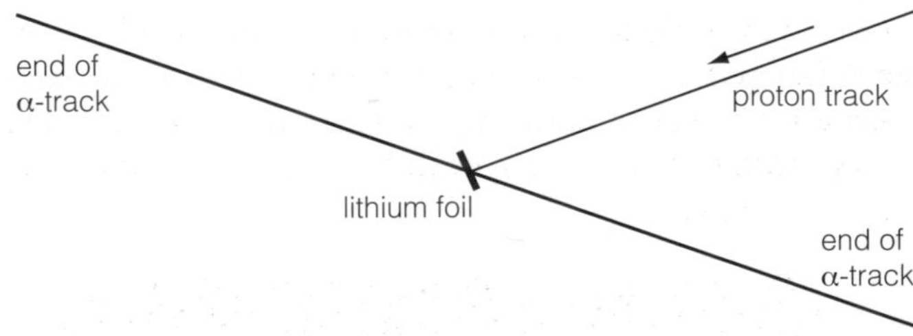

Figure 17.17 ▲

a) Cockcroft and Walton deduced that the two α-particles had the same initial energy. Explain how they made this deduction.

b) Write a nuclear equation for this event.

c) The energy released in the reaction is more than 17 MeV. Calculate the mass of the lithium nucleus, given that the rest masses of a proton and an α-particle are 1.0073 u and 4.0015 u respectively.

9 Figure 17.18 illustrates the Doppler effect. Explain what the figure is telling you about the waves from moving sources.

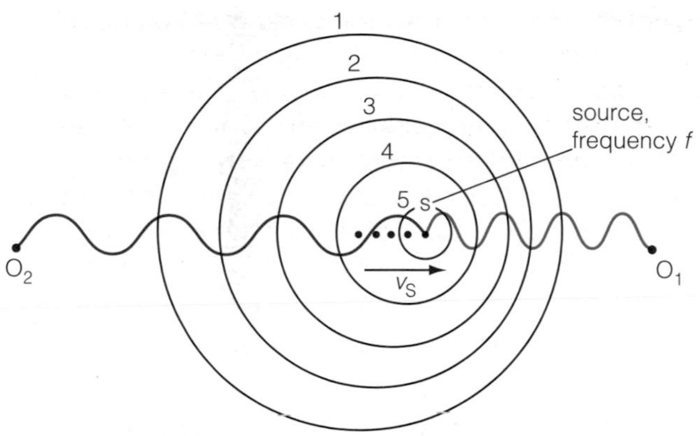

Figure 17.18 ▲

10 The Sun spins about its own axis with a period of rotation of 2.2×10^6 s. Its mean diameter is 1.4×10^9 m.

a) Calculate the change in wavelength for light of wavelength 520 nm coming from the edge of the Sun's equator that is moving away from us.

b) Discuss whether or not a sharp line in the Sun's absorption spectrum, when it is examined *very* closely, is found to be split into two lines.

11 The spectrum of the star Arcturus was photographed six months apart. M (March) and S (September) in Figure 17.19 are drawings of some of the most prominent lines in the two absorption spectra. Above and below are laboratory emission spectra drawn to the same scale. The spectra cover wavelengths from 425 nm to 430 nm from left to right. In Figure 17.19 the spectra are 74 mm wide.

Figure 17.19 ▲

Careful measurements of the positions of the spectral lines drawn in the figure show a red shift in M of 0.38 mm and a blue shift in S of 0.67 mm.

a) Calculate the velocities with which Arcturus appears to be approaching and receding from Earth.

b) What can you deduce from your results?

12 Explain how cosmologists determine the distance to galaxies that show very large red shifts, e.g. values of $z > 0.2$.

13 The average density of matter in the universe ρ_0 can be calculated from the relationship

$$\rho_0 = \frac{3H_0^2}{8\pi G}$$

where H_0 is the Hubble constant and G is the gravitational constant.

a) Calculate a value for ρ_0, add approximate $\pm$ values, and show that ρ_0 has the unit kg m^{-3}.

b) How many hydrogen atoms would you find on average in each cubic metre of the present universe?

c) Explain why cosmologists want a precise knowledge of the Hubble constant.

14 Figure 17.20 shows a typical galaxy like the Milky Way, seen from 'the side'. The speed v of star S as it moves round the centre of the galaxy can be measured, as can the distance r.

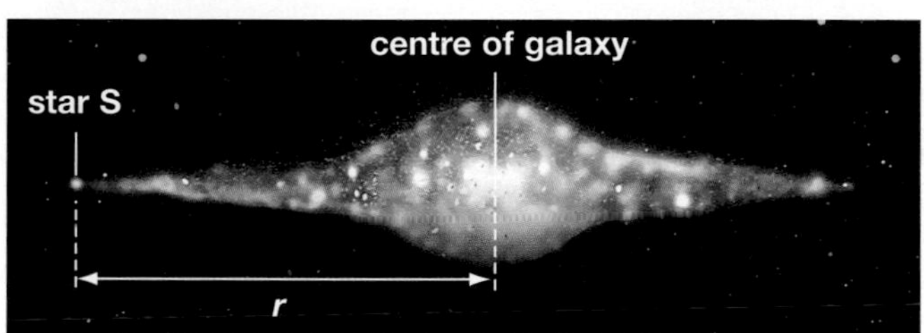

Figure 17.20 ▲

a) Show that the mass m_G of the galaxy is equal to v^2r/G.

b) This turns out to be much bigger than the mass expected by studying the luminosity of the galaxy. How do cosmologists explain this difference?

18 A guide to practical work

Physics is a very practical subject and, as in your AS Physics course, experimental work should form a significant part of your A2 studies. You should build on and develop the practical skills that you acquired last year and apply them to the new areas of physics in the A2 course. There are a number of experiments mentioned in the Edexcel specification that you should have undertaken and which you may be asked to describe in the examination. Through performing these experiments you should get a better understanding of the theoretical aspects as well as improving your practical skills. That is why this book is illustrated throughout by experiments, usually with a set of data for you to work through. However, nothing is like the real thing, so you should be carrying out as many as possible of these experiments for yourself in the laboratory.

As part of your practical assessment, you will be tested on:

- **planning**
- **implementation and measurements**
- **analysis.**

These elements were also the basis of your practical assessment at AS level. They were discussed in detail, with several examples and exercises, in Chapter 2 of the AS book, to which you should refer. The emphasis in this chapter is on the further requirements for A2 assessment.

18.1 Planning

You will be expected to:

- draw an appropriately labelled diagram of the apparatus to be used
- state how to measure one quantity using the most appropriate instrument
- explain the choice of the measuring instrument with reference to the scale of the instrument as appropriate and/or the number of measurements to be taken
- state how to measure a second quantity using the most appropriate instrument
- explain the choice of the second measuring instrument with reference to the scale of the instrument as appropriate and/or the number of measurements to be taken
- demonstrate knowledge of correct measuring techniques
- identify and state how to control all other relevant quantities to make it a fair test.

Worked example

A student makes some measurements of the time period T of the oscillations of a wooden metre rule in order to find a value for the Young modulus of the wood, using the arrangement shown in Figure 18.1.

She finds in a textbook that the Young modulus E for the wood is given by the formula

$$E = \frac{16\pi^2 ML^3}{wt^3T^2}$$

where M is the mass suspended from the rule at a horizontal distance L from where the rule is clamped, and w and t are the width and thickness of the rule respectively.

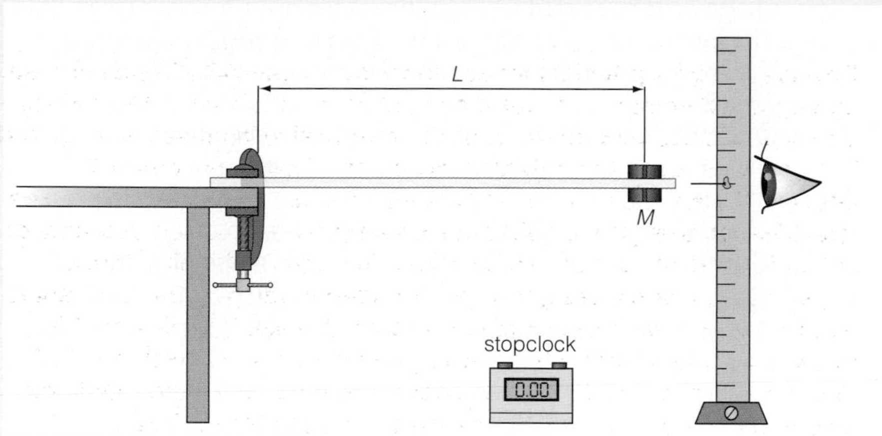

Figure 18.1 ▲

She plans to use:

- **two 100 g slotted masses** attached with elastic bands to either side of the rule, near the end, to make L as long as possible – with the slots in the masses perpendicular to the rule so that she can read the scale to determine the position of the centre of mass;
- **a digital stopwatch** to time the oscillations – one that can record to a precision of 0.01 s, which she considers to be more than adequate for timings that are likely to be of the order of 10 s;
- **vernier callipers** to measure the width and thickness of the rule, at different points along its length; the callipers can read to a precision of 0.1 mm;
- a pin secured to the metre rule with a small piece of Blu-Tack acting as a fiducial mark to help her judge the centre of the oscillations.

She records the following results:

M = 200 g
L = 937 − 40 = 897 mm
w = 28.5, 28.0, 28.6, 28.9 mm; average 28.5 mm
t = 6.6, 6.7, 6.6, 6.6 mm; average 6.6 mm
$20T$ = 8.53, 8.54, 8.67 s; average 8.58 s

1 Show that this data gives a value of the order of 10^{10} Pa for the Young modulus of the wood.
2 Explain the number of significant figures to which the final value should be given.
3 Discuss how the student could improve her experiment to try to get a more reliable value for the Young modulus.

Answer

1 Care needs to be exercised with units – all of the measurements need to be converted to SI units:

$M = 0.200\,\text{kg}$ $\quad t = 6.6 \times 10^{-3}\,\text{m}$
$L = 0.897\,\text{m}$ $\quad T = 8.58 \div 20 = 0.429\,\text{s}$
$w = 28.5 \times 10^{-3}\,\text{m}$

$$\Rightarrow E = \frac{16\pi^2 ML^3}{wt^3T^2} = \frac{16\pi^2 \times 0.200\,\text{kg} \times (0.897\,\text{m})^3}{28.5 \times 10^{-3} \times (6.6 \times 10^{-3}\,\text{m})^3 \times (0.429\,\text{s})^2}$$

$$= 1.5 \times 10^{10}\,\text{Pa} \sim 10^{10}\,\text{Pa}$$

2 The value should be quoted to 2 significant figures, 1.5×10^{10} Pa, because this is the number of significant figures of the least precise measurement – the thickness of the rule, 6.6 mm, has been measured to only 2 significant figures.
3 A more reliable value for the Young modulus could be obtained by using a suitable graphical method. There are two possibilities here, either:

- keep L constant and vary the mass M, plotting a graph of T^2 against M; or
- keep M constant and vary the length L, plotting a graph of T^2 against L^3.

This would give an average of a number of readings and reduce both random and systematic errors.

You will also be expected to:

- comment on all safety aspects of the experiment
- identify the main sources of uncertainty and/or systematic error.

Worked example

For safety reasons, a teacher demonstrates the absorption of gamma rays emitted from a sealed cobalt-60 source. He emphasises the importance of **safety procedures** when carrying out such an experiment, because of the dangers of radioactivity. The source is kept in a lead-lined box in a special locked metal cupboard in a separate store room, labelled with the radioactivity symbol. He is always careful to handle the source with tongs and keep the watching students a reasonable distance away. As soon as the experiment is finished he locks the source away to minimise the exposure time.

It is proposed to absorb the gamma rays by means of some circular lead discs of thickness ranging from about 2 mm to 7 mm. The students are set the task of measuring the thickness. They decide to use vernier callipers, measuring to a precision of 0.1 mm, and they record the values in Table 18.1.

Disc number	1	2	3	4
Thickness/mm	1.7	3.3	5.0	6.6

Table 18.1 ▲

Before bringing the source into the laboratory, the teacher asks the students to measure the background count. They monitor the background for 5 minutes and get 73 counts. They know that radioactivity is a random process and so they repeat for a further 5 minutes, getting 67 counts this time.

The teacher then sets up the apparatus as shown in Figure 18.2. He emphasises that the distance between the γ-ray source and the G-M tube must be kept constant.

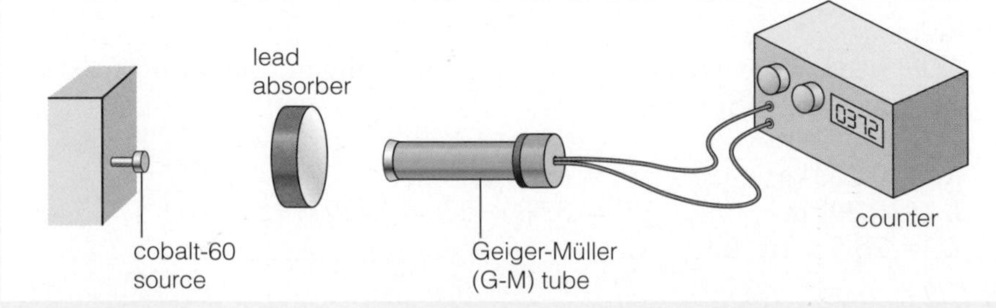

Figure 18.2 ▲

The following data were then recorded.

Total thickness of lead/mm	0.0	1.7	3.3	5.0	6.6	8.3	9.9	11.6	13.3	14.9	16.6
Counts/min	372	342	304	283	259	234	216	194	178	162	150
Corrected counts/min	358	328									

Table 18.2 ▲

1 Complete the table by calculating the rest of the count rates, corrected for background radiation.
2 Plot a graph of the corrected count rate, N, against the thickness of absorber, x.
3 Discuss what advantage, if any, would have been achieved if the students had used a micrometer measuring to a precision of 0.01 mm for determining the thickness of the discs.
4 Explain how a thickness of 14.9 mm was obtained.

Answer

1

Total thickness of lead/mm	0.0	1.7	3.3	5.0	6.6	8.3	9.9	11.6	13.3	14.9	16.6
Counts/min	372	342	304	283	259	234	216	194	178	162	150
Corrected counts/min	358	328	**290**	**269**	**245**	**220**	**202**	**180**	**164**	**148**	**136**

Table 18.3 ▲

2

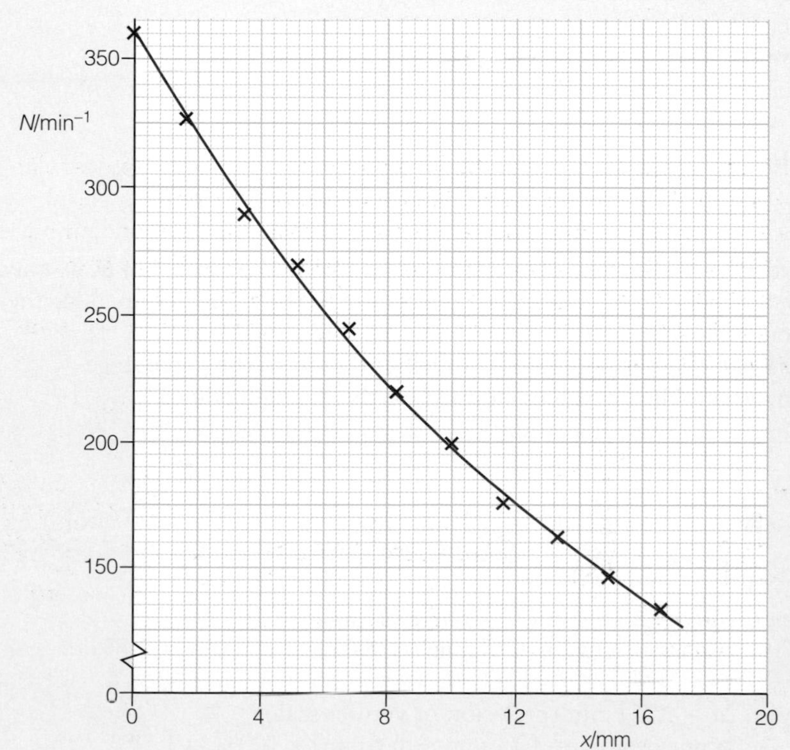

Figure 18.3 ▲

3 As the thickness of the discs was of the order of a few millimetres, it might seem advantageous to have used the micrometer, particularly as the discs are combined to get the larger thicknesses. For example, the percentage uncertainty in the thickness of disc 1 is (0.1 mm ÷ 1.7 mm) × 100 ≈ 6%. However, because of the wide range of thicknesses, it was not possible to plot the thickness to a precision greater than 0.1 mm, so the choice of vernier callipers was sensible.

4 A thickness of 14.9 mm could have been obtained by using discs 2, 3 and 4 together, giving 3.3 mm + 5.0 mm + 6.6 mm = 14.9 mm.

18.2 Implementation and measurements

You will be expected to:

- record all measurements with appropriate precision, using a table where appropriate
- use correct units throughout
- show an appreciation of uncertainty
- obtain an appropriate number of measurements
- obtain measurements over an appropriate range.

The radioactivity experiment in the Worked example above illustrates these points, with ten different thicknesses over a wide range (a factor of about 10) recorded to a sensible precision, with units, in a table. The uncertainty in the values of the thicknesses was discussed and the need to repeat the background count was explained. The experiment to find the Young modulus (page 212) emphasised the care that needs to be taken with units.

18.3 Analysis

This is where we see a significant difference between the AS and A2 assessment criteria. In particular, at A2 you will be expected to:

- discuss more than one source of error qualitatively
- calculate errors quantitatively
- compound errors correctly
- discuss realistic modifications to reduce error/improve experiment.

Worked example

Using the data from the Young modulus experiment (page 212):

1 Estimate the percentage uncertainty in the measurements of L, w, t and T, and hence the percentage uncertainty in the value obtained for the Young modulus.

2 Discuss which measurement contributes most to the percentage uncertainty in the value obtained for the Young modulus.

3 Suggest how this uncertainty could have been reduced.

Answer

1 For L: uncertainty (ΔL) = ±2 mm (1 mm at either end of the rule)
% uncertainty = (2 mm ÷ 897 mm) × 100% = 0.2%

For w: Δw = ±0.5 mm ($\frac{1}{2}$ × range of values)
% uncertainty = (0.5 mm ÷ 28.5 mm) × 100% = 1.8%

For t: Δt = ±0.1 mm (precision of vernier scale)
% uncertainty = (0.1 mm ÷ 6.6 mm) × 100% = 1.5%

For T: $\Delta(20T) = 0.1\,\text{s}$ (reaction time$>\frac{1}{2}\times$ range)
% uncertainty in $20T = (0.1\,\text{s} \div 8.58\,\text{s}) \times 100\% = 1.2\%$

For $E = \dfrac{16\pi^2 ML^3}{wt^3T^2}$

% uncertainty in: $L^3 = 3 \times 0.2\% = 0.6\%$
$w = 1.8\%$
$t^3 = 3 \times 1.5\% = 4.5\%$
$T^2 = 2 \times 1.2\% = 2.4\%$

% uncertainty in $E = (0.6 + 1.8 + 4.5 + 2.4)\% \approx 9\%$

2 From the above calculations it can be seen that the thickness t of the rule contributes most to the percentage uncertainty in the value obtained for the Young modulus, because the percentage uncertainty in t^3 is 4.5%, approximately twice as much as any of the other quantities.
3 If t had been measured with a micrometer (0.01 mm precision) or a digital vernier (0.02 mm precision), this uncertainty would have been considerably reduced. As discussed earlier (page 214), using a graphical method would appreciably reduce the uncertainty in T, which is the next largest source of error.

Note

This is also the **percentage** uncertainty in T. A common mistake that candidates make is to take the uncertainty in the timing ($20T$) to be the uncertainty in T, forgetting about the factor of 20.

Tip

Remember:

- if a quantity appears raised to a power in an equation, the % uncertainty it contributes is the power × the % uncertainty in the quantity
- the % uncertainty of a product or quotient is equal to the sum of the % uncertainties of the individual quantities.

You will also be expected to:

- process and display data appropriately to obtain a straight line where possible, for example, using a log/log graph
- suggest relevant further work.

Although the specification states a 'log/log' graph, you will be expected to use both a log/log graph to test a power relationship and a ln/linear graph to test an exponential.

Tip

To test for an exponential you *must* use ln. To test a power relationship you can use either log/log or ln/ln. As ln works for both cases, it is a good idea to always use ln.

Worked example

The students think that the graph they obtained for the radioactivity experiment (Figure 18.3, page 215) looks as though it could be an exponential function of the form

$$N = N_0 e^{-\mu x}$$

where μ is a constant (a property of the absorbing material). To test this suggestion they decide to plot a graph of $\ln N$ against x.

1 Complete the table below to show the values of $\ln N$.

Thickness *x*/mm	0.0	1.7	3.3	5.0	6.6	8.3	9.9	11.6	13.3	14.9	16.6
Corrected count rate *N*/min	358	328	290	269	245	220	202	180	164	148	136
ln (*N*/min)	5.88	5.79									

Table 18.4 ▲

2 Explain why a graph of $\ln N$ against x should be a straight line.
3 Plot a graph of ln (N/min) against x/mm.
4 Discuss the extent to which your graph verifies the proposed equation.
5 Use the gradient of your graph to find a value for the absorption coefficient μ.

Answer

1

Thickness x/mm	0.0	1.7	3.3	5.0	6.6	8.3	9.9	11.6	13.3	14.9	16.6
Corrected count rate N/min	358	328	290	269	245	220	202	180	164	148	136
ln (N/min)	5.88	5.79	**5.67**	**5.59**	**5.50**	**5.39**	**5.31**	**5.19**	**5.10**	**5.00**	**4.91**

Table 18.5 ▲

2 If we take logarithms to base 'e' of both sides of the equation $N = N_0 e^{-\mu x}$ we get

$\ln N = -\mu x + \ln N_0$

which is of the form

$y = mx + c$

The gradient of the straight line graph will be $-\mu$. The intercept when $x = 0$ will be $\ln N_0$. (Note the intercept is $\ln N_0$ and **not** N_0 – a common mistake that candidates make.)

3

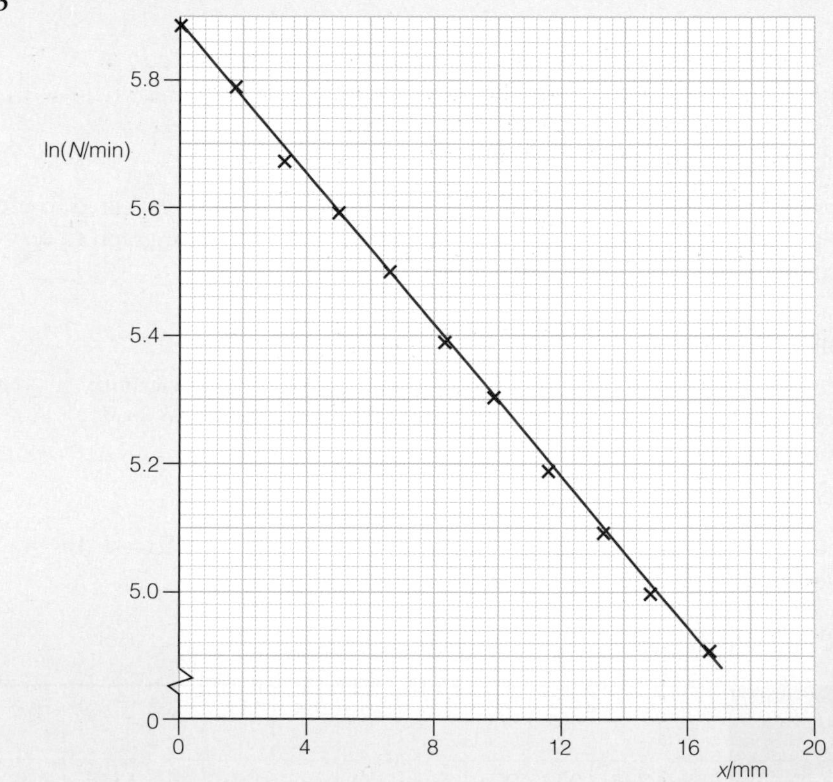

Figure 18.4 ▲

4 Allowing for a bit of scatter, which is to be expected because of the random nature of radioactive decay, the graph is a straight line of negative gradient that has an intercept equal to $\ln N_0$. The results therefore confirm that the absorption of the gamma radiation varies exponentially with the thickness of the absorber, as suggested by the equation.

5 The gradient of the graph, found by drawing a large triangle, is about -0.059, giving a value for μ of $0.059\,mm^{-1}$. (Note that the units are mm^{-1} because the value of a logarithm, ln (N/min), is simply a number.)

Worked example

One of the students decides to find out more information about the absorption of gamma radiation and discovers that knowledge of the absorption coefficient μ can be used to find the energy of the gamma rays incident on the absorber. A property of an absorber, called its 'mass absorption coefficient', is given by μ/ρ, where ρ is the density of the material of the absorber. The mass absorption coefficient varies with the energy of the incident gamma radiation. The student finds the information in Table 18.6 in a radiation data book.

Energy/MeV	0.80	1.00	1.20	1.40	1.60
Mass absorption coefficient/$10^{-3}\,m^2\,kg^{-1}$	8.75	7.04	6.13	5.42	5.01

Table 18.6 ▲

The student then determines the density of the lead absorbers and finds its value to be $1.06 \times 10^4\,kg\,m^{-3}$.

1 Suggest how the density of the lead absorbers could be determined.
2 Calculate the value of the mass absorption coefficient for the lead absorbers, using a value for μ of $0.0590\,mm^{-1}$. (Hint: be careful with units.)
3 Plot a suitable graph to determine the energy of the gamma rays.

Answer

1 The thickness t of the circular lead discs has already been measured. If the diameter d of each absorber is measured with a vernier, the volume v of each disc can be calculated from $v = \pi d^2 t/4$. If all four discs are weighed together to find their total mass M, the density can be found from $\rho = M/V$, where V is the total volume of all four discs.
2 The first thing we have to do is to put the value for $\mu = 0.0590\,mm^{-1}$ into base SI units, that is $\mu = 59.0\,m^{-1}$. Then:

$$\text{mass absorption coefficient} = \frac{\mu}{\rho} = \frac{59.0\,m^{-1}}{1.06 \times 10^4\,kg\,m^{-3}}$$
$$= 5.57 \times 10^{-3}\,m^2\,kg^{-1}$$

3

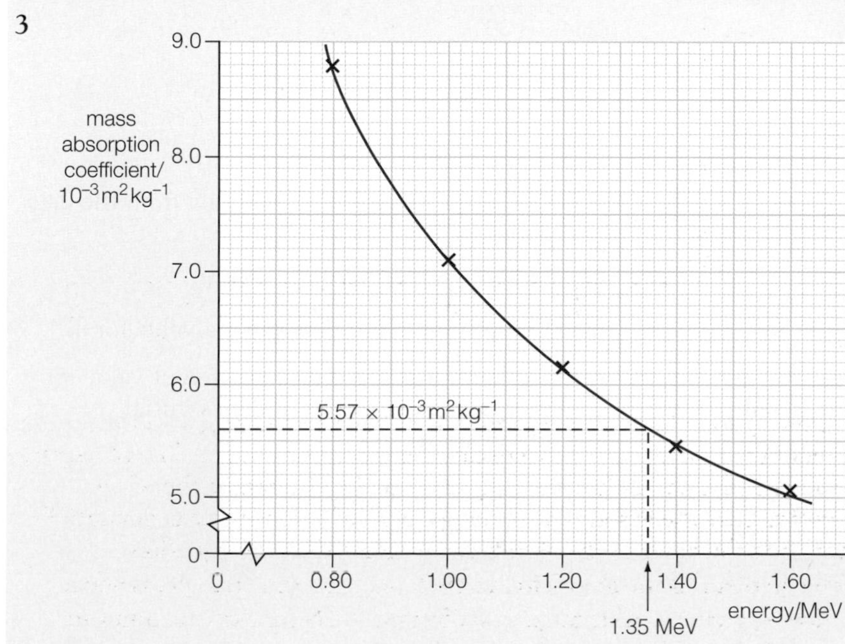

Figure 18.5 ▲

From the graph we can read off that the energy of the gamma radiation is 1.35 MeV. The nature of the experiment is such that we can only just about justify 3 significant figures, so a final answer of 1.35 ± 0.05 MeV would be sensible.

If you look up the details of cobalt-60, you will find that it decays by β^--emission of maximum energy 0.31 MeV, and then the excited nucleus decays in two stages, emitting γ-rays of energy 1.33 MeV and 1.12 MeV.

REVIEW QUESTION

1 A student plans to investigate how the frequency of vibration of air in a conical flask depends on the volume of air. She blows directly into the neck of the flask and listens to the sound of the air vibrating. She then pours water into the flask until it is approximately half filled and blows into the flask as before. She notices that the pitch (frequency) of the vibrating air is higher when the flask is half full of water.

She thinks that there might be a relationship between the natural frequency of vibration f of the air in the flask and the volume V of air in the flask of the form $f \propto V^n$, where n is a constant.

She sets up the arrangement shown in Figure 18.6.

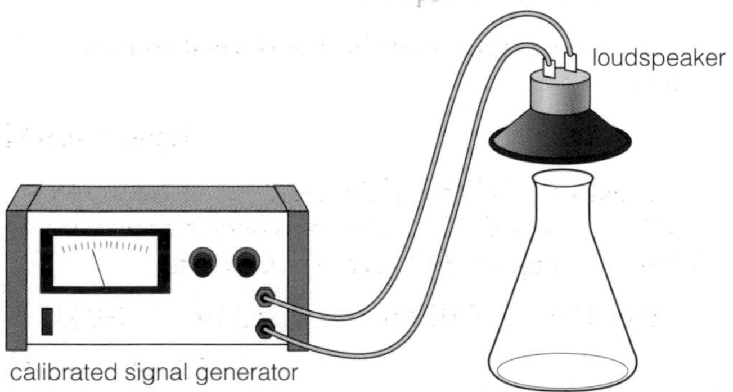

Figure 18.6 ▲

a) The student increases the frequency of the signal generator until the air in the flask vibrates very loudly.

i) Explain why this happens at the natural frequency of vibration of the air.

ii) Describe how she could vary, and measure, the volume of air in the flask.

iii) Explain how plotting a graph of $\ln f$ against $\ln V$ would enable her to test whether $f \propto V^n$ and would enable her to find a value for n.

b) The student obtained the following data:

V/cm^3	f/Hz
554	219
454	242
354	274
254	324
204	361
154	415

Table 18.7 ▲

i) Calculate values of $\ln V$ and $\ln f$ from the data above.

ii) Plot a graph of $\ln f$ against $\ln V$.

iii) Use your graph to determine a value for n.

iv) Explain **qualitatively** whether your value for n is consistent with the student's initial observations.

Unit 5 test

Time allowed: 1 hour 35 minutes
Answer **all of** the questions.
There is a data sheet on page 225.

For Questions 1–10, select one answer from A–D.

1 A kettle rated at 3 kW contains 1.5 litres ($1500\,cm^3$) of water at a temperature of 15 °C. Assuming that the energy transferred to the kettle itself can be ignored, the time taken for the water to come to the boil will be about:

A 1 min B 2 min C 3 min D 4 min

[Total: 1 mark]

2 Air is mainly composed of nitrogen-14 and oxygen-16. For air at room temperature, which of the following statements is true?

A The average speed of the oxygen molecules is greater than that of the nitrogen molecules.

B The nitrogen and oxygen molecules have the same average speed.

C The average kinetic energy of the oxygen molecules is greater than that of the nitrogen molecules.

D The nitrogen and oxygen molecules have the same average kinetic energy.

[Total: 1 mark]

3 Which of the graphs below represents the behaviour of an ideal gas?

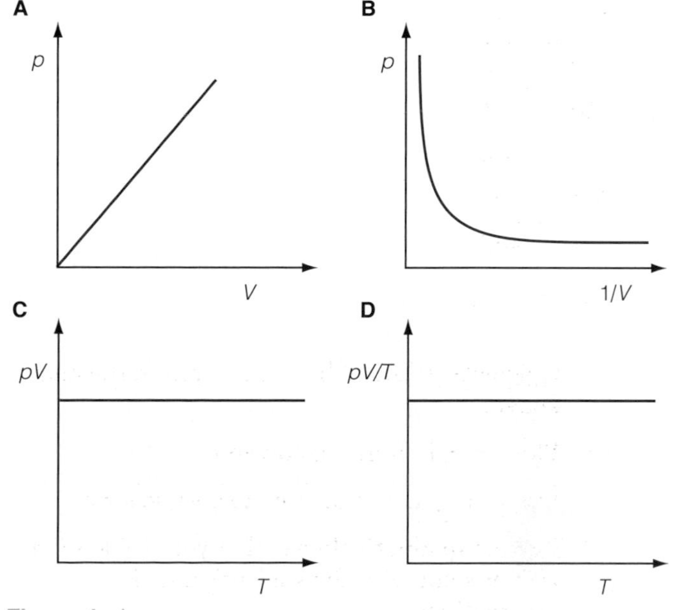

Figure 1 ▲

[Total: 1 mark]

4 The product Gm_E, which has a numerical value 4.0×10^{14}, has SI unit:

A $m^3\,s^{-2}$ B $N\,m^2$ C $m^2\,kg^{-4}\,s^{-2}$ D $kg^2\,m\,s^{-2}$

[Total: 1 mark]

5 Polonium-210 decays to lead-206 by emission of which type of radiation?

A α B β^- C β^+ D γ

[Total: 1 mark]

6 Bromine-85 decays by β^--emission to form an isotope of krypton. This is:

A krypton-83 B krypton-84
C krypton-85 D krypton-86

[Total: 1 mark]

Questions 7 and 8 relate to a mass suspended on a vertical spring.

7 The mass is pulled down a distance of 50 mm and released. It makes 10 simple harmonic oscillations in 14.1 s. The maximum acceleration of the mass is:

A $0.03g$ B $0.05g$ C $0.10g$ D $0.14g$

[Total: 1 mark]

8 At which point, or points, of the oscillation does the mass have maximum acceleration?

A middle B bottom only
C top only D top and bottom

[Total: 1 mark]

9 The radiation flux at the surface of a star is equal to:

A the luminosity of the star divided by the Sun's luminosity

B the luminosity of the star divided by its surface area

C the radiation flux from the star at Earth times the distance to the star squared

D the power output from the star times its surface area.

[Total: 1 mark]

10 The masses of ^1_1H and of the neutron are respectively 1.0078 u and 1.0087 u. The mass deficit or nuclear binding energy of $^{12}_6C$, mass 12.0000 u, is:

A 0.0936 u B 0.0990 u C 0.0522 u D 0.0438 u

[Total: 1 mark]

11 A student sees on the worldwide weather forecast that the temperature in Hong Kong is 36 °C on a day when the temperature in London is 18 °C. He says to his friend that it is 'twice as hot in Hong Kong as it is in London'. His friend, a physicist, says 'it's actually only 6% hotter in Hong Kong'. Explain which of the friends is correct.

[Total: 4 marks]

12 From a hypothesis put forward by Avogadro in 1811, it can be deduced that a mole of an ideal gas at standard temperature (0 °C) and pressure (101 kPa) occupies a volume of 22 litres (1 litre = 10^3 cm^3).

a) What is meant by an 'ideal gas'? [1]

b) Under what conditions will a real gas behave like an ideal gas? [2]

c) Use the data given above to show that a mole of an ideal gas contains about 6×10^{23} molecules. [3]

[Total: 6 marks]

13 A star observed by the Hipparcos satellite shows an annual movement against the very distant stars of 0.017 arcsec. It is known that this star lies a distance 3.6×10^{18} m from Earth. Deduce the diameter of the Earth's orbit round the Sun.

[Total: 4 marks]

14 a) Sketch a Hertzsprung-Russell diagram showing the main sequence as a curve. State what each axis represents (no values of the variables are expected). [3]

b) Add a position for the Sun on your diagram and indicate where red giant stars would be found. [1]

[Total: 4 marks]

15 In recent years observations of the red shifts of a certain class of standard candles called type Ia supernovas have led astrophysicists to believe that the expansion of the universe is speeding up.

a) Describe the nature of a supernova and explain why supernovas can be used as standard candles. [2]

b) Explain how a 'red shift' is measured and calculate the recession speed v of a galaxy containing a supernova that shows a red shift of 3%. [3]

c) Explain how the inverse-square law applied to radiation from supernovas and the Hubble law applied to their red shift can lead to two different values for the distance from us to the supernova's galaxy. [3]

d) Suggest why physicists expect the expansion of the universe to be slowing down rather than speeding up. [1]

[Total: 9 marks]

16 In 1972 a *natural* uranium fission reactor was discovered in Gambia (Africa). This is believed to have operated 2 billion years ago over a period of several hundred thousand years in a region rich in uranium ores. Uranium-235, with a half-life of 7×10^8 years, is now found to form just 0.7% of mined uranium. Ground water was available as a moderator, but there is now no evidence of the products of the fission having leached any distance from the site of the reactor.

a) Explain how 2 billion years ago the uranium was U-235 rich. [3]

b) Suggest how such a reactor might have worked over such a long period. [2]

c) What tests confirmed that uranium fission had occurred? [2]

d) Suggest how the discovery of this natural reactor might have lessons for us today. [2]

[Total: 9 marks]

17 Consider a small star S that orbits a much more massive star M with a period T of 2.7×10^8 s (between 8 and 9 years). A hydrogen absorption line in the spectrum of S shows a variation from 656.323 nm to 656.241 nm as it moves in its circular orbit around M.

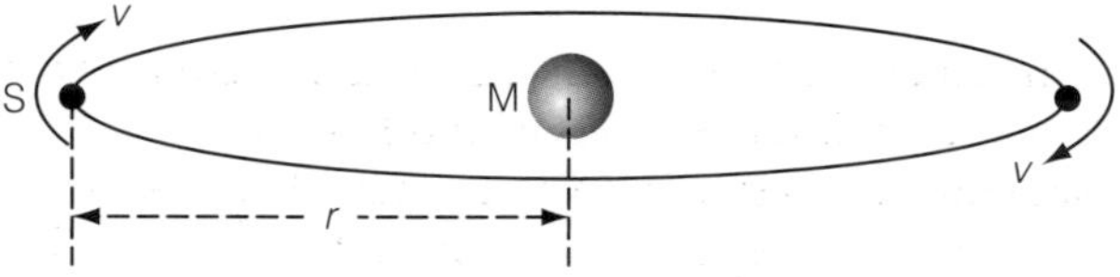

Figure 2 ▲

a) Explain this variation and show that the speed of S in its orbit is about 20 km s^{-1}. [5]

b) Calculate:

i) the radius r of the orbit of S, and hence

ii) deduce the mass of M_M of the massive star. [5]

[Total: 10 marks]

18 The following was taken from a press article about the suspected murder of a former Russian spy in 2006.

The death of Alexander Litvinenko is linked to a massive dose of radioactive polonium-210 found in his body. Polonium-210 decays by emitting high-energy alpha particles which cause terrible internal cell damage, but are difficult to detect outside the body. How the polonium was transported is a mystery, as even a small quantity of it would melt a glass capsule.

a) Explain briefly why alpha particles:

i) 'cause terrible internal cell damage',

ii) 'are difficult to detect outside the body'. [2]

b) Polonium-210 has a half life of 138 days.

i) Calculate the percentage of a given sample remaining after one year.

Sketch a graph of the decay over a 3-year period of a sample that has an initial activity of 2.5×10^8 Bq. [5]

ii) Calculate the decay constant for polonium-210.

Hence calculate the activity of 1.0 μg of polonium-210, given that there are 2.9×10^{15} atoms in 1.0 μg. [3]

c) The energy of the high-energy alpha particles emitted by polonium-210 is 5.3 MeV. Show that 1 g of polonium would generate a power of about 140 W.

Suggest why 'even a small quantity of it would melt a glass capsule'. [4]

[Total: 14 marks]

19 Some physics students go to their local park to investigate the motion of a swing.

One student sits on the swing. Another student pulls her back a distance of 1.2 m and lets go, simultaneously starting a stopwatch. Two students then measure her displacement at the end of each swing. The students sketch a graph of the motion (Figure 3).

a) Draw free-body force diagrams for the student sitting on the swing as she goes through the mid-point of the motion for the first time and as she just reaches the end of the first complete oscillation. [3]

b) i) In sketching the graph, the students have assumed that the motion is simple harmonic. Suggest why this may not be the case. [1]

ii) Use the graph to estimate the velocity of the student as she goes through the midpoint of the motion for the first time. [2]

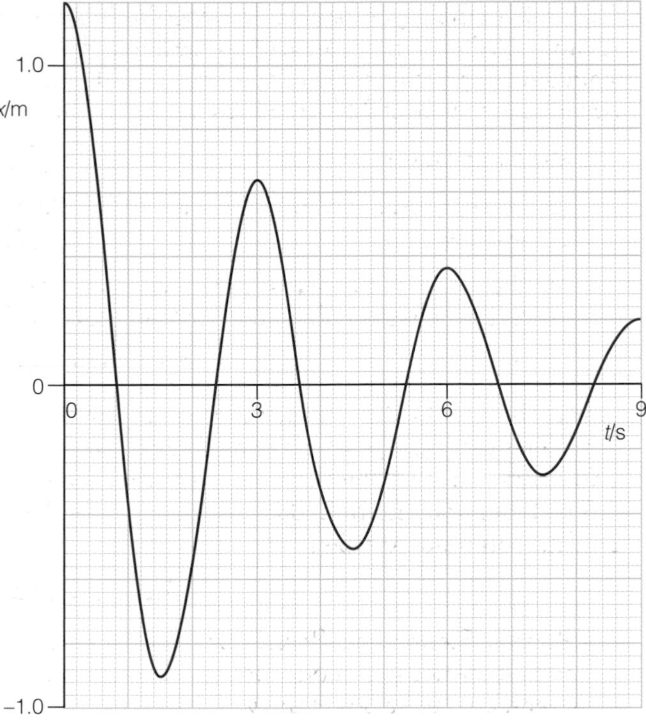

Figure 3 ▲

c) If the motion of the swing was simple harmonic, its velocity would be given by

$$v = -2\pi f A \sin 2\pi f t$$

Use this formula to calculate a theoretical maximum value for the velocity. [2]

d) Explain how the students could use the swing to demonstrate resonance. [2]

[Total: 10 marks]

[TOTAL: 80 marks]

$F = \frac{kQ^1Q^2}{r^2}$

$kg\,m\,s^{-2} = \frac{kA^2s^2}{m^2}$

$kg\,m^3 s^{-2} = kA^2s^2$

$kg\,m^3 s^{-4} A^{-2}$

A2 Data sheet

The values here are given to 3 significant figures.
The unit quoted is the unit usually associated with a definition of the constant.

Name of constant	Symbol, value and unit	Comment
acceleration of free fall	$g = 9.81 \text{ m s}^{-2}$	Close to Earth's surface
Boltzmann constant	$k = 1.38 \times 10^{-23} \text{ J K}^{-1}$	Used in $pV = NkT$
Coulomb's law constant	$k = 8.99 \times 10^{9} \text{ N m}^2 \text{ C}^{-2}$	Defined in $F = kQ_1Q_2/r^2$
density of water	$\rho_w = 1000 \text{ kg m}^{-3}$	
electron charge	$e = -1.60 \times 10^{-19} \text{ C}$	Proton charge = $+1.6 \times 10^{-19}$ C
electron mass	$m_e = 9.11 \times 10^{-31} \text{ kg}$	
electron-volt	$1 \text{ eV} \equiv 1.60 \times 10^{-19} \text{ J}$	Often keV or MeV used
gravitational constant	$G = 6.67 \times 10^{-11} \text{ N m}^2 \text{ kg}^{-2}$	Defined in $F = Gm_1m_2/r^2$
gravitational field strength	$g = 9.81 \text{ N kg}^{-1}$	Close to Earth's surface
Hubble constant	$H_0 = 2.3 \times 10^{-18} \text{ s}^{-1}$	Only known to 2 sig. fig.
Planck constant	$h = 6.63 \times 10^{-34} \text{ J s}$	Used in $E = hf$ and $p = h/\lambda$
proton mass	$m_p = 1.67 \times 10^{-27} \text{ kg}$	
radius of Earth	$r_E = 6.38 \times 10^{6} \text{ m}$	
specific heat capacity of water	$c_w = 4200 \text{ J kg}^{-1} \text{ K}^{-1}$	
speed of light in a vacuum	$c = 3.00 \times 10^{8} \text{ m s}^{-1}$	$c^2 = 9.00 \times 10^{16} \text{ m}^2 \text{ s}^{-2}$
Stefan-Boltzmann constant	$\sigma = 5.67 \times 10^{-8} \text{ W m}^2 \text{ K}^{-4}$	Defined in $L = \sigma AT^4$
unified atomic mass unit	$1 \text{ u} \equiv 1.66 \times 10^{-27} \text{ kg}$	
Wien constant	$2.90 \times 10^{-3} \text{ m K}$	No symbol, but $= \lambda_m T$

Numerical Answers to Review Questions

Note: If you hold numbers in your calculator as you work through a multi-stage calculation, you sometimes get a slightly different answer.

1 Review and revision

1 a) C b) D c) A d) D
e) B f) C g) C

2 a) $v_v = 39\,\mathrm{m\,s^{-1}}$
b) $v_h = 10\,\mathrm{m\,s^{-1}}$

3 $\mathrm{kg\,m^{-1}\,s^{-1}}$

2 Linear momentum

1 a) $2.7\,\mathrm{kg\,m\,s^{-1}}$
b) $18\,000\,\mathrm{kg\,m\,s^{-1}}$ due east
c) $1.2 \times 10^5\,\mathrm{kg\,m\,s^{-1}}$

2 B

3 A

4 C

5 C

6 $\frac{1}{12}$ × mass of carbon nucleus

7 B

8 C

9 D

10 $0.55\,\mathrm{kg\,m\,s^{-1}}$ upwards

11 b) $112\,\mathrm{kg\,m\,s^{-1}}$ 63° W of N

12 Recoil speed $v = 0.20\,\mathrm{m\,s^{-1}}$

13 $104\,\mathrm{m\,s^{-1}}$

14 $73\,\mathrm{m\,s^{-1}}$

15 Yes;
recoil speed = $2 \times 10^{-14}\,\mathrm{m\,s^{-1}}$

16 $590\,\mathrm{kg\,s^{-1}}$

18 98 000 N

19 $118\,\mathrm{kg\,m\,s^{-1}}$ at an angle of 22.5° north of west

3 Momentum and energy

1 D

4 b) 26%

5 a) $8.1\,\mathrm{m\,s^{-1}}$
b) Increase of 320 J

6 A

8 a) Shell before: $3840\,\mathrm{kg\,m\,s^{-1}}$
A after: $1800\,\mathrm{kg\,m\,s^{-1}}$
B after: $2520\,\mathrm{kg\,m\,s^{-1}}$
c) 925 kJ

4 Motion in a circle

1 C

2 A

3 B

4 B

5 490g

6 a) $0.5\,\mathrm{rad\,s^{-1}}$

7 120 kN

8 a) $0.034\,\mathrm{m\,s^{-2}}$

10 $3.4\,\mathrm{m\,s^{-1}}$

12 51 kN

5 Electric fields

1 B

2 B

3 D

7 a) $6.1 \times 10^4\,\mathrm{V\,m^{-1}}$
c) $6.4 \times 10^{-19}\,\mathrm{C}$

8 0.021 m

9 $10^{13}\,\mathrm{m\,s^{-2}}$

10 a) $a = qV/mx$

11 b) $k = 9.1 \times 10^9\,\mathrm{N\,m^2\,C^{-2}}$

12 2000 V

13 350 N

6 Capacitance

1 22 nF

2 A

3 C

4 D

5 a) 6.0 V b) 1.3 C

6 B

7 a) 1.9 V
b) 4.1 V
c) $1.9 \times 10^{-3}\,\mathrm{C}$ or 1.9 mC

9 a) Charge = 0.22 C
Energy stored = 1.1 J
b) i) 0.55 J

10 21 000 V

11 a) Charge on the 22 μF capacitor = 264 μC
Charge on the 47 μF capacitor = 564 μC
b) Capacitor of capacitance 69 μF

12 a) The p.d. across the 20 μF capacitor = 4.0 V
The p.d. across the 10 μF capacitor = 8.0 V
b) Capacitor of capacitance 6.66 μF

14 At $t_{\frac{1}{2}}$ = 28 s, current = 1.6(4) μA
At $2t_{\frac{1}{2}}$ = 56 s, current = 0.78(4) μA

15 $2.2(3) \times 10^{-3}\,\mathrm{F}$ or 2.2 mF

16 24 μC to 2 sig. fig.

7 Magnetic fields

1 B

2 A

3 C

4 a) 22 μT
b) 83 μT

5 980 N

6 B

7 Force on OP = 0.25 N upwards
Force on OR= 0.25 N upwards
Force on OQ is zero

9 0.14 T

11 a) $8.5 \times 10^4\,\mathrm{m\,s^{-1}}$
b) Zero

12 a) $2.4 \times 10^{-13}\,\mathrm{N}$
b) $1.7 \times 10^{-13}\,\mathrm{N}$
c) $1.2 \times 10^{-13}\,\mathrm{N}$

13 Gravitational : electric : magnetic = $1 : 10^9 : 10^{11}$

14 D

17 0.14 T or 140 mT

18 a) Maximum $\Phi = 3.8 \times 10^{-4}\,\mathrm{Wb}$
Minimum $\Phi = 0$
b) $1.3 \times 10^{-3}\,\mathrm{V}$ or 1.3 mV

21 a) 3800 to 2 sig. fig.
b) 0.078 A or 78 mA

8 Electrons and nuclei

1 C

2 B

3 B

4 a) D
b) A

5 D

7 a) 7.3×10^{-13} J
b) 4.6 MeV

9 $E_k = B^2e^2r^2/2m$

11 8.0×10^6 V m^{-1} or 8.0 MN C^{-1}

12 6.6 MeV

13 $\Delta m = 1.78 \times 10^{-29}$ kg
$\Delta m/m = 0.0106$

14 a) At A, approximately 70 mm
b) At A, 1.3×10^{-20} kg m s^{-1}
c) At A, 4.5×10^{-29} kg
d) At A, mass $\approx 50m_e$

9 Particle physics

1 D

2 B

3 D

4 A

5 B

6 C

7 B

8 D

9 C

11 a) 5.3×10^{-26} kg
b) Mass = $5800m_e$

12 $s\bar{u} \rightarrow u\bar{u} + d\bar{u}$
$\{(-\frac{1}{3}e) + (-\frac{2}{3}e)\} \rightarrow \{(+\frac{2}{3}e) + (-\frac{2}{3}e)\} + \{(-\frac{1}{3}e) + (-\frac{2}{3}e)\}$

13 a) Upwards out of the paper

15 1.9×10^{-34} m

10 Specific heat capacity

1 a) B b) D c) B d) D

2 a) 14 K b) 1813 K c) −210 °C
d) 115 °C

3 a) 21 MJ

4 11 kJ

5 b) 98 W

6 b) 27 °C

7 b) i) 9.5% ii) 10.7 kJ
iii) 1.07 kJ

8 b) 6p c) 120 W

11 Internal energy and absolute zero

1 C

2 D

3 C

4 a) 234 K b) 357 °C
c) −117 °C d) 352 K
e) −219 °C f) 90 K g) 1356 K
h) 2580 °C

8 b) Yes (52%)

12 Gas laws and kinetic theory

1 a) C b) D

2 a) C b) B

3 C

4 a) Both sides kg m^{-1} s^{-2}
b) 3.04 kPa

5 a) Both sides kg m^2 s^{-2}
b) Mean square speed of molecules
c) 2.7×10^5 m^2 s^{-2}
d) i) 2.5×10^5 m^2 s^{-2}
ii) 1.8×10^6 m^2 s^{-2}

6 e) 6.0×10^{27} molecules

7 a) 105 kPa
b) 2.4×10^{21} molecules
c) 0.16 g

8 b) 0.67 atmospheres above atmospheric pressure, so yes

9 b) pV at A = pV at B = pV at C = 120 J
c) i) 200 kPa ii) 4.0×10^{-4} m^3
iii) 2.5×10^{-4} m^3

13 Nuclear decay

1 D

2 A

3 C

4 D

5 B

6 C

8 d) ii) 95 protons, 146 neutrons
e) iii) 4.0×10^{13} nuclei
iv) 1.9×10^3 Bq

9 c) 3.2×10^{10} nuclei
d) 7.0×10^{-12} g

14 Oscillations

1 C

2 D

3 D

4 D

6 a) C b) D

8 a) Amplitude = 4.0×10^{-2} m; frequency = 100 Hz so $\omega = 628$ rad s^{-1}
b) i) 1.6×10^4 m s^{-2}
ii) 25 m s^{-1}

9 a) 60 mm
c) i) 7.3 m s^{-2} ii) 0.66 m s^{-1}
f) $k = 49$ N m^{-1}
g) 0.087 J

10 c) $k = 4.0 \times 10^{-4}$ J m^{-2} (or N m^{-1})

15 Universal gravitation

1 D

2 a) 130 kJ
b) 290 J kg^{-1}

3 a) 82 m
b) i) No ii) No

4 a) 1.7 m s^{-2}
b) 890 N

5 8.7 N kg^{-1}

6 7.3×10^{-11} N m^2 kg^{-2}

7 $v = 1700$ m s^{-1}
$T = 6300$ s

8 b) 1.8×10^{27} kg

12 $g_r/g_{r\,+\,100\,\mathrm{km}} = 1.03$

16 Astrophysics

1 C

2 D

3 C

4 D

5 A

6 2.8×10^{-6} degrees; 4.8×10^{-8} rad

7 a) $\tan p = r/d$
b) i) 1.9×10^{17} m
ii) 3.7×10^{-6} degrees = 0.013 arcsec

8 kg s^{-3}

9 7.5×10^{18} m

10 a) Between 50 and 60×10^{27} W
b) About 1.0×10^{-9} W m^{-2}

11 About 1×10^{9} kg to 1 sig. fig.

13 $\lambda_m = 580$ nm when $T = 5.0$ kK

14 2.6×10^{6} m

15 a) Approximately 70–80 ly

17 Cosmology

1 A

2 a) D b) A

3 A

4 C

5 b) 0.004 32 u
c) 4.0 MeV to 2 sig. fig.

7 ~0.006 ly or 6×10^{13} m

8 c) 7.014 u

10 a) 3.5×10^{-12} m or 0.0035 nm (a red shift)

11 a) Approach velocity of Arcturus = 32 000 m s^{-1}
Recession velocity of Arcturus = 18 000 m s^{-1}

13 a) $(9.5 \pm 0.5) \times 10^{-27}$ kg m^{-3}
b) About 5 or 6 hydrogen atoms per cubic metre

18 A guide to practical work

1 b) iii) $n = -0.50$

Index

A

absolute thermodynamic temperature scale 97–8, 110
absolute zero 97, 110–12
absorption spectra 204, 206–7
acceleration
  centripetal 24
  of free fall 225
  in simple harmonic motion 151, 156, 157–8
activity (rate of decay) 139–40
addition of vectors 2, 18
air-track experiments 7, 16
alpha-particle scattering 68–9, 81
alpha radiation 131, 132, 134–5
  emission 136
aluminium 99
  measuring the specific heat capacity of 100–2
americium-241 136
amplitude 150, 156
amplitude resonance 168
analysis 216–20
angular displacement 21
angular velocity 21–3, 160
antimatter 82–3
  matter–antimatter annihilation 84–5
antiproton 85
apparent weightlessness 28, 179–80
arcsecs 186
area, estimating 48–9
astrophysics *see* stars
atmospheric pressure 114–15
atomic number (proton number) 68, 135–6
Avogadro constant 124

B

background radiation 128–9, 214
bar magnets 30, 53
  resultant field of two bar magnets 53
Barton's pendulums 167
baryons 86, 87
  conservation of total number of 88
base units 2
Becquerel, Henri 128
becquerels 139
Bell Burnell, Jocelyn 191
beta radiation 131, 133, 134–5
  β^+-emission 137
  β^--emission 136–7
Big Bang 208
Big Crunch 209
binding energy 199–200
black body radiators 193–4
black holes 191
Boltzmann, Ludwig 121
Boltzmann constant 121, 225
Boomerang Nebula 111
bottom quarks 86
brain imaging 84–5
bridges 171
Brown, Robert 124
Brownian motion 124
bubble chamber 75–6
buildings in earthquake zones 172
Burj Dubai skyscraper 172

C

camera flash unit 43–4
cancer 130
capacitance 39
  measuring for a capacitor 40
capacitors 39–52
  charging through a resistor 50
  discharge through a resistor 46–50
  energy storage of 40–3
  in the real world 43–4
  measuring capacitance of 40
capacitor key 44
car engines 172
carbon 199
carbon-14 dating 145
Celsius scale 97–8
centripetal acceleration 24
centripetal forces 23–7
Cepheid stars 188–9
chain reaction 202–3
change of state 109–10
charge 30, 31
  charging a capacitor through a resistor 50
  charging objects 32
  coulombmeter 44
  discharging a capacitor through a resistor 46–50
  force between two charges 36–7, 203
  like charges repel, unlike forces attract 31
charmed quarks 86
circular motion 21–9
  apparent weightlessness 28
  centripetal forces 23–7
  electron beams 59, 70–2
  terminology 21–3
closed universe 209–10
cloud chamber 75, 82
coaxial cable 44
Cockcroft, John 202
coils 54
collisions 5–7
  between molecules in kinetic theory 124
  elastic 15–17, 19, 81, 106–7
  electrons and protons 81
  inelastic 15–17, 81
  nuclear 78–9
  in two dimensions 18–19
conduction (heat) 96–7
continuous spectra 204
convection 97
cooling 101
copper 99
cosmic rays 129
cosmology 197–211
  Doppler shift 205–7
  expanding universe 204–9
  fate of the universe 209–10
  formation of stars 197–8
  Hubble constant 207–9
  nuclear binding energy 199–200
  stellar fusion 198, 203–4
  uranium fission 200–3
Coulomb force 36–7, 203
coulombmeter 40, 44
Coulomb's law 36–7
Curie, Marie 128
Curie, Pierre 128
current, magnetic field around wire carrying 53–4
cyclotron 72–3

D

damped oscillations 164–6
  investigating 165–6
  resonance in damped systems 168–72
dark energy 197, 210
dark matter 197, 210
data logger 62
data sheet 224
dating, radioactive 144–6
de Broglie, Prince Louis 88
decay constant 139–40, 142–3
Dees (cyclotron) 72
derived units 2
diagnosis 146
diffraction 89
digital stopwatch 213
dipoles 34–5
direct current (d.c.) electric motors 56–7
discharging a capacitor 46–50
disintegration processes 135–9

displacement
angular 21
in simple harmonic motion 151, 156, 157, 162
Doppler shift 205–7
down quarks 81, 86
drift chamber 75, 76
ductile materials 172

E

Earth
age from radioactive dating 145–6
electric field near surface 33
gravitational field 175, 176, 180, 181–2
magnetic field 54–5
nuclear fusion 204
radius 225
earthquake zones 172
efficiency of energy storage 42–3
Einstein, Albert 88
mass–energy equation 76–8
elastic collisions 15–17, 19, 81
ideal gas 106–7
elastic potential energy 13
electric fields 30–8
comparison with gravitational fields 182–3
radial 35
Coulomb's law 36–7
uniform 31–5
electric field strength 31, 32, 35
electric forces 30, 31
force between two charges 36–7, 203
electric motors 56–7
electric potential energy 41
electrical circuit, resonance in 170, 171
electrical work 108
electromagnetic induction 61–4
maglev train 110, 111
electrons 68, 86, 133
charge on 225
emission in beta decay 136–7
force on in a magnetic field 57–8
mass of 68, 225
scattering of protons 81
thermionic emission 70–1
see also beta radiation
electron antineutrino 137
electron beams 70
circular path 59, 70–2
in a magnetic field 58–60
electron diffraction 89
electron guns 58, 70
electron linacs 73–4, 81
electron neutrino 137
electron-volt (eV) 14, 225
emission spectra 204
energy 13–20
collisions in two dimensions 18–19
conservation 2, 13
Einstein's mass–energy equation 76–8
elastic and inelastic collisions 15–17
of ionising particles 134, 135
nuclear binding energy 199–200
in simple harmonic motion 162–4
storage by capacitors 40–3
transfer in resonance 167, 168
work and 13–15
energy resonance 168
energy spectrum for beta-decay electrons 136–7
equation of state for an ideal gas 121–2
equipotential lines 33, 35
equipotential surfaces 177, 180
errors, analysis of 216–18
eta mesons 88
evaporation 110
excited states 138
exponential change 45–6
log/linear graphs to test 46, 49–50, 217–18
exponential decay 39, 45–6
capacitor discharge 46–50
radioactive decay 142–3
exponential function 49–50
exponential growth 45
charging a capacitor 50

F

Faraday, Michael 30, 39, 53, 61
Faraday's law of electromagnetic induction 61
Fermi, Enrico 137
Fermilab proton linac 74
fiducial markers 153, 213
fields 30
see also electric fields; gravitational fields; magnetic fields
fission, nuclear 200–3
Fleming's left-hand rule 55, 59, 71
fluid pressure 114–15
see also pressure
force
centripetal 23–7
on a current-carrying wire in a magnetic field 54–6
electric 30, 31
between two charges 36–7, 203
force vs time graphs 9–10
impulsive 8–10
simple harmonic motion 151
forced oscillations 165, 167
free fall 28, 179
free oscillations 164
frequency
Doppler shift 205–6
natural 167, 168
simple harmonic motion 150–1
friction-compensated runway 5
fusion, nuclear 109, 198, 200, 203–4

G

galaxies 185
gamma radiation 131, 133–5
absorption experiment 214–16, 217–18
gamma emission 138–9
mass absorption coefficient 219–20
gas laws 115–21
Boyle's law 115–18
pressure and temperature 118–21
Geiger–Müller tube (GM tube) 133
geosynchronous satellites 179
glider experiments 7, 16
gravitation 175–84
Newton's law of 178–80
gravitational constant 178, 225
gravitational fields 30
comparison with electric fields 182–3
radial 180–2
uniform 175–7
gravitational field strength 175–6, 180, 225
variation 176
with distance from the centre of the Earth 181–2
gravitational potential energy (GPE) 13, 175–6
conversion of energy stored in a capacitor to 42–3
and kinetic energy in simple harmonic motion 162–4

H

half life 45
capacitor discharge 46, 48
radioactive decay 68, 140–4
experimental determination 143–4
Hall potential difference 58
health and safety 130–1, 214
heat 96–7
transfer and internal energy 107
transfer to surroundings 100, 101–2
helium 198, 204
nuclei 132, 199
see also alpha radiation
Hertzsprung–Russell diagram (H–R diagram) 189–91
Hooke's law 152–3
Hubble constant 146, 207–9, 225
Hubble's law 207
hydrogen 198, 199, 203–4
hydrogen bombs 203, 204
hysteresis 120, 121

I

ideal gas
equation of state for 121–2
internal energy of 106–7
ideal transformer 64–5
impulse 7–10
impulsive forces 8–10
momentum and 7–8
impulse–momentum equation 7, 8–9
Industrial Revolution 125
inelastic collisions 15–17, 81
internal energy 13, 15, 97, 106–13
absolute zero 110–12
change of state 109–10
heating and working 107–8
of an ideal gas 106–7
inverse-square laws
Coulomb's law 36–7
Newton's law of gravitation 178–80
radiation flux 188
iodine 146
iodine-131 143, 146
ionisation 75–6, 131
alpha radiation 132, 134
beta radiation 132, 134
gamma radiation 133
ionisation chamber 132

iron 125, 199
  stability of iron-56 200
isochronous motion 156
isothermals 116–17
isotopes 68, 135–6

J

Joule, James 13

K

kaons 88
Kelvin, Lord 97
kelvins 97–8
keyboard, PC 44
kinetic energy 13
  elastic and inelastic collisions 15–17
  of electrons accelerated across a potential difference 58
  of gas molecules and temperature 123–4
  and gravitational potential energy in simple harmonic motion 162–4
  internal energy 97, 106–7
  and mass of accelerated particles 77–8
  and momentum 14–15
  zero-point energy 111
kinetic theory 114–27
  equation of state for an ideal gas 121–2
  evidence for 124
  fluid pressure 114–15
  gas laws 115–21
  historical context 125
  kinetic model of temperature 123–4

L

Large Hadron Collider (LHC) 75, 110, 111
latent heat 109–10
lathes 172
lead 214–16
  mass absorption coefficient of 219–20
Leavitt, Henrietta 188–9
left-hand rule 55, 59, 71
Lenz's law 61
leptons 86
life cycle of a star 190–1
light, speed of 225
light from the stars 192–4
light year 185
linear accelerators (linacs) 73–4, 78, 81
linear momentum *see* momentum
lines of force 30
  gravitational 180
  magnetic 30, 53–4
liquid crystal displays 34–5
log/linear graphs 46, 49–50, 217–18
log/log graphs 217
luminosity 187–8
  H–R diagram 189–90
  Stefan–Boltzmann law 192–4

M

M51 galaxy (Whirlpool galaxy) 185
magnetic fields 30, 53–67
  d.c. electric motors 56–7
  electromagnetic induction 61–4
  electron beams 58–60
  lines of force 30, 53–4
  strength 54–6
  transformers 64–5
magnetic flux 60–1, 63–4
magnetic flux density 54–5, 60
magnetic levitating (maglev) trains 110, 111
magnetic resonance imaging (MRI) scanners 63, 172–3
main sequence stars 190–1, 192–3
  radius and distance from Earth 194
mass
  Einstein's mass–energy equation 76–8
  relativistic increase with speed 83
  rest mass 76
  units used in particle physics 83
mass absorption coefficient 219–20
mass deficit 199
mass–energy, principle of conservation of 76–8
mass number (nucleon number) 68, 135, 136
matter
  and antimatter 82–3
  creation and annihilation of 84–5
  structure of ordinary matter 86, 87
mean square speed 123
measurements 216
  choice of measuring instrument 212, 213, 214, 215, 216
mechanical work 107
medicine, nuclear 146–7
melting 109–10
mesons 81, 86, 87–8
microwave ovens 34, 172
Milky Way 185, 189, 194
Millennium Bridge, London 171
molybdenum-99 146–7
momentum 2, 4–12
  collisions 5–7
    elastic and inelastic 15–17
    in two dimensions 18–19
  conservation of 5–7
  electrons in circular motion 71
  and energy 13–20
  and impulse 7–10
  rockets 11
motion sensor 158
motors, d.c. electric 56–7
muons 86

N

natural frequency 167, 168
neptunium-237 136
neutral points 53
neutrinos 137, 198
neutrons 87
  mass 68
  nuclear fission 200, 201, 202–3
  quarks in 81, 82
neutron number 68
neutron stars 191
Newton's law of gravitation 178–80
Newton's laws of motion 2
  first law 24
  second law 7, 10, 24, 26
  third law 4, 7, 26
non-elastic collisions 15–17, 81
non-relativistic particles 14
nuclear binding energy 199–200
nuclear bomb 203
nuclear decay *see* radioactivity
nuclear fission 200–3
nuclear fuel, spent 142, 203
nuclear fusion 109, 200
  stellar 197, 203–4
nuclear medicine 146–7
nuclear model of the atom 69
nuclear reactors 142, 203
nuclear synthesis (nucleosynthesis) 191, 198
nucleon number 68, 135, 136
nuclides 68, 135

O

omega minus particle 87
open universe 209–10
oscillations 150–74
  damped 164–6, 169
  forced 165, 167
  free 164
  measurement of Young modulus 212–14
  resonance 167–73
  simple harmonic motion *see* simple harmonic motion

P

pair production 84
parallax, trigonometric 186–7, 189
parallelogram of vectors 18
particle accelerators 72–4
  cyclotron 72–3
  linear accelerators 73–4, 78, 81
particle detectors 75–6
particle interactions 78–9
particle physics 81–91
  baryons 86, 87, 88
  creation and annihilation of matter 84–5
  discovery of quarks 81–2
  matter and antimatter 82–3
  mesons 81, 86, 87–8
  standard model 86–7
  units 83
  wave–particle duality 88–9
pascals 114
Pauli, Wolfgang 136
pendulum, simple 154–5, 156–7, 162–4
percentage uncertainty 216–17
period 150–1
PET (positron emission tomography) scans 84–5
photographic emulsions 75
photons 84, 109, 198
  gamma radiation 133
physical quantities 2
pions 87
piston 161
Planck constant 88–9, 225

planetary nebula 191
planning practical work 212–16
plasma 109
Polaris 189
polonium 128
positrons 82–3
  annihilation 84
  emission in beta decay 137
potassium 139–40
  potassium-38 137
  potassium-40 145
potential difference
  Hall potential difference 58
  variation across a capacitor as it charges 40–1
potential energy
  elastic 13
  electric 41
  gravitational *see* gravitational potential energy
  internal energy 97, 106–7
potential gradient 33
practical work 212–20
  analysis 216–20
  implementation and measurements 216
  planning 212–16
prefixes for units 2
pressure 114
  atmospheric 114–15
  fluid pressure 114–15
  gas laws 115–21
    Boyle's law 115–18
    pressure and temperature law 118–21
protons 87
  cyclotron 72–3
  mass of 68, 225
  proton–antiproton annihilation 85
  quarks in 81, 82
  scattering by electrons 81
proton linac 74
proton number 68, 135–6
pulsars 191

Q

quantum mechanics 125
quarks
  baryons and mesons 87–8
  discovery of 81
  standard model 86–7

R

radial electric fields 35
  Coulomb's law 36–7
radial gravitational fields 180–2
radians 21
radiation (heat) 96
radiation badges 131
radiation flux 187–8
radioactive dating 144–6
radioactivity 128–49
  alpha radiation 131, 132, 134–5, 136
  background radiation 128–9, 214
  beta radiation 131, 133, 134–5, 136–7
  dangers of radiation 130–1
  discovery of 128
  disintegration processes 135–9
  gamma radiation 131, 133–5, 138
    absorption experiment 214–16, 217–18
  half life 68, 140–4
  modelling radioactive decay with dice 140–1
  nuclear medicine 146–7
  randomness of decay 139–40, 140–1
  spontaneous nature of decay 139–40
radium 128
radon 129
  radon-222 138–9
randomness of radioactive decay 139–40, 140–1
rate-meter 133
recoil 8
red giants 190, 191
red shift 189, 206–7, 207–8
research spending 195
resistors
  charging capacitors through 50
  discharging capacitors through 46–50
resolving vectors 2, 18
resonance 167–73
  of a damped system 169–71
  making use of 172–3
rest mass 76
resultant centripetal force 26–7
rockets 11
Röntgen, William 128
rotary motion sensor 158
rubber, hysteresis in 120, 121
rubber bung experiment for centripetal force 25
Rutherford, Ernest 69, 131, 132, 133, 202

S

safety precautions 130–1, 214
Sankey diagrams 15, 44
satellites 179
scaler 133
scintillators 85
semiconductors 125
sigma zero particle 87
simple harmonic motion 150–64
  energy in 162–4
  equations of 156–62
  simple pendulum 154–5, 156–7, 162–4
  spring 151–4
sketch graphs 160
solenoids 54
  electromagnetic induction 63–4
solids, atomic model of 106
spark chamber 76
spark counter 132
special relativity theory 78, 83
specific heat capacity 96–105
  measuring 99–103
spent nuclear fuel 142, 203
spontaneity of radioactive decay 139–40
springs 42
  simple harmonic motion of 152–4
spring constant 152–4
  finding from Hooke's law 152–3
  finding from simple harmonic motion 153–4
standard candles 188–9
standard model of fundamental particles 86–7
Stanford linear accelerator 74, 78, 81
stars 185–96, 204–5
  distance to 185–7
  formation 197–8
  Hertzsprung–Russell diagram 189–91
  life cycle of a star 190–1
  light from 192–4
  luminosity and flux 187–8
  nuclear fusion 198, 203–4
  research spending 195
  standard candles 188–9
steam engine 125
steel 125
Stefan–Boltzmann law 192–4
stellar parallax 186–7, 189
stopwatch, digital 213
straight line graphs 217–18
strange quarks 86
strontium-90 136, 137
Sun 187, 190, 191
super-capacitors 44
superconductivity 110–11, 173
supergiants 190, 191
supernova explosions 189, 191
Sutton Hoo burial ship 145

T

Tacoma Narrows Bridge 171
taus 86
technetium-99m 146–7
temperature
  heat and 96–7
  kinetic model of 123–4
  and pressure of a gas at constant volume 118–21
  units of 97–8
teslas 54
therapy, cancer 146
thermal equilibrium 110
thermionic emission 70–1
thermography 96–7
thunderstorms 33
time constant 46, 48
tokamak 109
top quarks 86
transformers 64–5
transistors 125
trigonometric parallax 186–7, 189
trolley collision experiments 5
tuning circuits 172
two-dimensional collisions 18–19

U

unified atomic mass unit (u) 83, 225
uniform electric fields 31–5
  dipoles in 34–5
uniform gravitational fields 175–7
units 2
universal molar gas constant 121
universe
  expanding 204–9
  fate of 209–10
up quarks 81, 86

uranium 128, 129, 199
fission 200–3

V

Van de Graaff accelerators 73
variable capacitor 44
vector quantities 2
addition of 2, 18
resolving 2, 18
velocity
angular velocity 21–3, 160
in simple harmonic motion 156–7
vernier callipers 213, 214, 215
Villard, Paul 131
volume of a gas 115–18

W

Walton, Ernest 202
water
specific heat capacity of 99, 225
measuring 102–3
molecules in a microwave oven 172
water-skiers 26
Watt, James 125
wave–particle duality 88–9
weak interaction 137
weightlessness, apparent 28, 179–80
Whirlpool galaxy (M51 galaxy) 185
white dwarfs 190, 191
Wien's law 192, 193
work
and energy 13–15
and internal energy 107–8

X

X-rays 128, 138
diffraction 89
safety precautions 130–1
xi particles 87

Y

Young modulus experiment 212–14, 216–17
yttrium-90 136, 137

Z

zero-point energy 111